Lecture Notes in Statistics 153

Edited by P. Bickel, P. Diggle, S. Fienberg, I
I. Olkin, N. Wermuth, S. Zeger

Springer
New York
Berlin
Heidelberg
Barcelona
Hong Kong
London
Milan
Paris
Singapore
Tokyo

Parimal Mukhopadhyay

Topics in Survey Sampling

 Springer

Parimal Mukhopadhyay
Applied Statistics Unit
Indian Statistical Institute
203 B.T. Road
Calcutta, 700 035
India

Library of Congress Cataloging-in-Publication Data

Mukhopadhyay, Parimal.
  Topics in survey sampling / Parimal Mukhopadhyay.
  p. cm. -- (Lecture notes in statistics ; 153)
  Includes bibliographical references and index.
  ISBN-13:978-0-387-95108-9   e-ISBN-13:978-1-4612-2088-6
  DOI: 10.1007/978-1-4612-2088-6

  1. Sampling (Statistics) I. Title. II. Lecture notes in statistics (Springer-Verlag) ; v.
153.

  QA276.6 .M77 2000
  519.5'2--dc21
                                                                    00-056271

Printed on acid-free paper.

Camera ready copy provided by the author.

9 8 7 6 5 4 3 2 1

ISBN-13:978-0-387-95108-9   Springer-Verlag New York Berlin Heidelberg   SPIN 10774724
A member of BertelsmannSpringer Science+Business Media GmbH

To My Family

Manju, Jayita, Pabak and Pralay

Preface

The aim of this book is to make a comprehensive review of some topics in Survey Sampling which have not been covered in details in books so far and to indicate new research areas wherever poosible. The book does not cover in details developments in fixed population model and super-population model-based prediction theoretic approaches. These have been discussed elaborately in Cassel et al (1977), Chaudhuri and Vos (1988), Sarndal et al (1992), Mukhopadhyay (1996, 1998 f), among others. The first two chapters make a quick tour in these areas to create the background for discussion in the subsequent chapters.

We concentrate in Chapters 3 and 4 on Bayes procedures and its modifications for predicting a finite population total. Apart from the book by Bolfarine and Zacks (1991), Ghosh and Meeden (1997), this aspect does not seem to have been covered in recent books. Chapter 3 considers Bayes and minimax prediction of population total under normal regression model using squared error loss function and the Linex loss function of Varian (1975) and Zellner (1980). The empirical Bayes (EB) prediction of total under normal models with and without use of covariates has been discussed. Applications of these procedures in small area estimation have been addressed. Important recent works in these areas have been discussed.

Chapter 4 considers different ramifications of Bayes procedures. Linear Bayes procedures due to La Motto (1978), Constrained Bayes procedures due to Louis (1984) and Ghosh (1992), Limited Translation Bayes estimation due to Effron and Morris (1971, 1972) have been discussed with reference to their applications in finite population sampling. Bayesian robustness under a class of alternative models as advocated by Bolfarine et al (1987) and robust Bayes estimation under a class of contaminated priors due to Ghosh and Kim (1993, 1997) have been addressed.

Chapter 5 considers design-based estimation, model-based prediction and design-model based robust strategies for predictiong a finite population variance along with Bayes and minimax prediction for the same. Prediction of a finite population regression coefficient under multiple regression model, asymptotic properties of a sample regression coefficient and estimation of a slope parameter in the linear regression model are also the subject matters of study of this chapter.

The next chapter considers the problems of prediction of finite population distribution function. The problems of prediction of finite population parameters (mean, variance, distribution function) under superpopulation models accommodating measurement errors have been reviewed in the following chpater. Both additive and multiplicative measurement errors have been dealt with.

The last chapter considers three special topics,- Calibration estimators due to Deville and Sarndal (1982), Post-stratification and Conditional unbiasedness under design-based approach.

As stated before, the idea of writing this book is to make a review of some of the ideas that have swept past in the field of survey sampling during the last few decades and to indicate new areas of research wherever possible. The topics have been chosen on the basis of their not being dealt with in details in many books, their importance and research potentials. In doing so we have tried to arrange the results systematically along with the relevant examples,- mostly theoretical, but some also empirical and based on live data. No novelty is claimed. This book can not be a stand-alone text book, but may serve as a supplementary reference for an advanced graduate course. We have assumed that the readers are acquainted with the developments 'n survey sampling at the level of Brewer and Hanif (1983), Cassel et al (1977) and Sarndal et al (1992). We have tried to be clear in the presentation to the best of our understanding and ability. We have also tried to cover many references which, however, is not claimed to be exhaustive.

The book was written at the Indian Statistical Institute, Calcutta, University of South Pacific, Suva, Fiji Islands and North Carolina State University, USA. I am indebted to the authorities of these organisations for their kind support for the work. My sincere thanks go to Prof. Thomas Gerig, Head, and Prof. Sastry Pantula, Assistant Head, Department of Statistics, NC State University for generously providing me the facilities at NCSU during the preparation of the manuscript.

April, 2000   *Parimal Mukhopadhyay*
Indian Statistical Institute,
Calcutta, India

Contents

Lecture Notes Editorial Policies

Lecture Notes in Statistics provides a format for the informal and quick publication of monographs, case studies, and workshops of theoretical or applied importance. Thus, in some instances, proofs may be merely outlined and results presented which will later be published in a different form.

Publication of the Lecture Notes is intended as a service to the international statistical community, in that a commercial publisher, Springer-Verlag, can provide efficient distribution of documents that would otherwise have a restricted readership. Once published and copyrighted, they can be documented and discussed in the scientific literature.

Lecture Notes are reprinted photographically from the copy delivered in camera-ready form by the author or editor. Springer-Verlag provides technical instructions for the preparation of manuscripts. Volumes should be no less than 100 pages and preferably no more than 400 pages. A subject index is expected for authored but not edited volumes. Proposals for volumes should be sent to one of the series editors or addressed to "Statistics Editor" at Springer-Verlag in New York.

Authors of monographs receive 50 free copies of their book. Editors receive 50 free copies and are responsible for distributing them to contributors. Authors, editors, and contributors may purchase additional copies at the publisher's discount. No reprints of individual contributions will be supplied and no royalties are paid on Lecture Notes volumes. Springer-Verlag secures the copyright for each volume.

Series Editors:

Professor P. Bickel
Department of Statistics
University of California
Berkeley, California 94720
USA

Professor P. Diggle
Department of Mathematics
Lancaster University
Lancaster LA1 4YL
England

Professor S. Fienberg
Department of Statistics
Carnegie Mellon University
Pittsburgh, Pennsylvania 15213
USA

Professor K. Krickeberg
3 Rue de L'Estrapade
75005 Paris
France

Professor I. Olkin
Department of Statistics
Stanford University
Stanford, California 94305
USA

Professor N. Wermuth
Department of Psychology
Johannes Gutenberg University
Postfach 3980
D-6500 Mainz
Germany

Professor S. Zeger
Department of Biostatistics
The Johns Hopkins University
615 N. Wolfe Street
Baltimore, Maryland 21205-21C
USA

Chapter 1

The Basic Concepts

1.1 INTRODUCTION

Sample survey, finite population sampling or survey sampling is a method
of drawing inference about the characteristic of a finite population by observing only a part of the population. Different statistical techniques have
been developed to achieve this end during the last few decades.

In this chapter we review some basic results in problems of estimating
a finite population total (mean) through a sample survey. We assume
throughout most of this chapter that the finite population values are fixed
quantities and are not realisations of random variables. The concepts will
be clear subsequently.

1.2 THE FIXED POPULATION MODEL

DEFINITION 1.2.1 A finite (survey) population $\mathcal{P}$ is a collection of a known
number N of identifiable units labelled $1, \ldots, N; \mathcal{P} = \{1, \ldots, N\}$, where i
stands for the physical unit labelled i. The integer N is called the size of
the population.

The following types of populations are, therefore, excluded from the coverage of the above definition: batches of industrial products of the same
specification (e.g. nails, screws) coming out from a production process,
as the units are not distinguishable individually; population of tigers in a
forest, as the population size is unknown. Collections of households in a

county, factories in an industrial complex and agricultural fields in a village
are examples of survey populations.

Let 'y' be a study variable having value y_i on $i(= 1, \ldots, N)$. As an example, in an agricultural population, y_i may be the yield of a farm i. The
quantity y_i is assumed to be fixed and nonrandom. Associated with $\mathcal{P}$, we
have, therefore, a vector of real numbers $\mathbf{y} = (y_1, \ldots, y_N)'$. The vector $\mathbf{y}$
constitutes the parameter for the model of a survey population, $\mathbf{y} \in R^N$,
the parameter space. In a sample survey one is often interested in estimat-
ing a parameter function $\theta(\mathbf{y})$, eg. populaton total, $T(\mathbf{y}) = T(= \sum_{i=1}^{N} y_i)$,

population mean $\bar{y}(= T/N)$,population variance $S^2 = \sum_{i=1}^{N}(y_i - \bar{y})^2/(N-1)$
by choosing a sample (a part of the population, defined below) from $\mathcal{P}$ and
observing the value of y only on the units in the sample.

DEFINITION 1.2.2 A sample is a part of the population.

A sample may be selected with replacement (wr) or without replacement
(wor) of the units already selected to the original population.
A sample when selected by a wr-sampling procedure may be written as a
sequence,
$$S = \{i_1, \ldots, i_n\}, \quad 1 \le i_t \le N \qquad (1.2.1)$$
where i_t denotes the label of the unit selected at the tth draw and is not
necessarily unequal to $i_{t'}$ for $t \ne t'(= 1, \ldots, N)$,. For a without replacement
sampling procedure, a sample when written as a sequence, is

$$S = \{i_1, \ldots, i_n\}, 1 \le i_t \le N, i_t \ne i_{t'} \text{ for } t \ne t'(= 1, \ldots, N) \qquad (1.2.2)$$

since repetition of units in S is not possible. Arranging the units in the
sample S in an increasing (decreasing) order of magnitudes of labels and
considering only the distinct units, a sample may also be written as a set
s. For a wr-sampling of n draws, a sample written as a set is, therefore,

$$s = (j_1, \ldots, j_{\nu(S)}), \quad 1 \le j_1 < \ldots < j_{\nu(S)} \le N \qquad (1.2.3)$$

where $\nu(S)$ is the number of distinct units in S. In a wor-sampling procedure, a sample of n-draws, written as a set is

$$s = (j_1, \ldots, j_n), 1 \le j_1 < \ldots < j_n \le N \qquad (1.2.4)$$

Thus, if in a wr-sampling $S = (7, 2, 7, 4)$, the corresponding s is $s = (2, 4, 7)$
with $\nu(S) = 3$. Similarly, if for a wor sampling procedure, $S = (4, 9, 1)$, the

corresponding s is $s = (1, 4, 9)$ with $\nu(S) = 3$. Clearly, information on the order of selection and repetition of units in the sample S is not available in s.

DEFINITION 1.2.3 Number of distinct units in a sample is its effective sample size. Number of draws in a sample is its nominal sample size. In (1.2.3), $\nu(S)$ is the effective sample size, $1 \leq \nu(S) \leq n$. For a *wor* -sample of n-draws, $\nu(S) = \nu(s) = n$.
Note that a sample is a sequence or set of some units from the population and does not include their y-values.

DEFINITION 1.2.4 The sample space is the collection of all possible samples and is often denoted as $\mathcal{S}$. Thus $\mathcal{S} = \{S\}$ or $\{s\}$ according as we are interested in S or s.

In a simple random sample with replacement (*srswr*) of n draws $\mathcal{S}$ consists of N^n samples S. In a simple random sample without replacement (*srswor*) of n draws $\mathcal{S}$ consists of $(N)_n$ samples S and $\binom{N}{n}$ samples s where $(a)_b = a(a-1)\ldots(a-b+1), a > b$. If the samples s of all possible sizes are considered in a *wor*- sampling procedure, there are 2^N samples in $\mathcal{S}$.

DEFINITION 1.2.5 Let $\mathcal{A}$ be the minimal σ-field over $\mathcal{S}$ and p be a probability measure defined over $\mathcal{A}$ such that $p(s)$ [or $p(S)$] denotes the probability of selecting s [or S], satisfying

$$p(s)[p(S)] \geq 0$$

$$\sum_S p(s)[\sum_S p(S)] = 1 \tag{1.2.5}$$

One of the main tasks of the survey statistician is to find a suitable $p(s)$ or $p(S)$. The collection $(\mathcal{S}, p)$ is called a sampling design (*s.d.*), often denoted as $D(\mathcal{S}, p)$ or simply p. The triplet $(\mathcal{S}, \mathcal{A}, p)$ is the probability space for the model of the finite population.

The expected effective sample size of a *s.d.* p is

$$E\{\nu(S)\} = \sum_{S \in \mathcal{S}} \nu(S)p(S) = \sum_{\mu=1}^{N} \mu P[\nu(S) = \mu] = \nu \tag{1.2.6}$$

We shall denote by ρ_ν the class of all fixed effective size $[FS(\nu)]$-designs i.e.

$$\rho_\nu = \{p : p(s) > 0 \Rightarrow \nu(S) = \nu\} \tag{1.2.7}$$

A *s.d.* p is said to be non-informative if $p(s)[p(S)]$ does not depend on the y-values. In this treatise, unless stated otherwise, we shall consider non-informative designs only. Informative designs have been considered by Basu (1969), Zacks (1969), Liao and Sedransk (1975), Stenger (1977), Bethlehem and Schuerhoff (1984), among others.

Basu (1958), Basu and Ghosh (1967) proved that all the information relevant to making inference about the poplation characteristic is contained in the set sample s and the corresponding y values. As such, unless otherwise stated, we shall consider samples as sets s only.

The quantities

$$\pi_i = \sum_{s \ni i} p(s), \; \pi_{ij} = \sum_{s \ni ij} p(s) \qquad (1.2.8)$$

$$\pi_{i_1,\ldots,i_k} = \sum_{s \ni i_1,\ldots,i_k} p(s)$$

are, respectively, the first order, second order,..,kth order inclusion-probabilities of units in the sample in a *s.d.* p. The following lemma states some relations among inclusion probabilities and expected effective sample size of a *s.d.*

LEMMA 1.2.1: For any *s.d.* p,

(i)

$$\pi_i + \pi_j - 1 \leq \pi_{ij} \leq \; \min \, (\pi_i, \pi_j)$$

(ii)

$$\sum_i \pi_i = \sum_{s \in \mathcal{S}} \nu(s) p(s) = \nu$$

(iii)

$$\sum_{i \neq j = 1}^{N} \pi_{ij} = \nu(\nu - 1) + V(\nu(s))$$

If $p \in \rho_\nu$

(iv)

$$\sum_{j(\neq i)=1}^{N} \pi_{ij} = (\nu - 1)\pi_i$$

(v)

$$\sum_{i \neq j = 1}^{N} \pi_{ij} = \nu(\nu - 1)$$

Result (i) is obvious. Results (ii), (iii) and (iv), (v) are, respectively, due to Godambe (1955), Hanurav (1962 a), Yates and Grundy (1953).

Further, for any *s.d.* p,

$$\theta(1 - \theta) \leq V\{\nu(S)\} \leq (N - \nu)(\nu - 1) \tag{1.2.9}$$

where $\nu = [\nu] + \theta, 0 \leq \theta < 1, \theta$ being the fractional part of ν. The lower bound in (1.2.9) is attained by a *s.d.* for which

$$P[\nu(S) = [\nu]] = 1 - \theta \text{ and } P[\nu(S) = \nu + 1] = \theta.$$

Mukhopadhyay (1975) gave a *s.d.* with fixed nominal sample size $n(> \nu)[p(S) > 0 \Rightarrow n(S) = n \; \forall \; S]$ such that $V\{\nu(S)\} = \theta(1 - \theta/(n - [\nu])$, which is very close to the lower bound in (1.2.9).

We shall denote by

$p_r(i_r)$=probability of selecting i_r at the rth draw

$p_r(i_r \mid i_1, \ldots, i_{r-1})$= conditional probability of selecting i_r at the rth draw given that $i_1, \ldots, i_{r-1}$ were drawn at the first,...,$(r-1)$th draw respectively;

$p_r(i_1, \ldots, i_r)$ = the joint probability that $(i_1, \ldots, i_r)$ are selected at the first,...,r-th draw respectively.

DEFINITION 1.2.6 A sampling scheme gives the conditional probability $p(i_r \mid i_1, \ldots, i_{r-1})$ of drawing a unit at any particular draw given the results of the earlier draws.

The following theorem shows that any sampling design can be attained through a draw-by-draw mechanism.

THEOREM 1.2.1 (Hanurav, 1962 b; Mukhopadhyay, 1972) For any given *s.d.*, there exists at least one sampling scheme which realises this design.

Suppose the values $x_1, \ldots, x_N$ of a closely related (to y) auxiliary variable x on units $1, 2, \ldots, N$, respectively, are available. As an example, in an agricultural survey, x may be the area of a plot under a specified crop and y the yield of the crop on that plot. The quantities $p_i = x_i/X, X = \sum_{i=1}^{N} x_i$ is called the size-measure of unit $i(= 1, \ldots, N)$ and is often used in selection of samples.

1.3   DIFFERENT TYPES OF SAMPLING DESIGNS

The sampling designs proposed in the literature can be generally grouped in the following categories.

- (a) Simple random sampling with replacement ($srswr$)

- (b) Simple random sampling without replacement($srswor$)

- (c) Probability proportional to size with replacement ($ppswr$) sampling: a unit i is selected with probability p_i at the rth draw and a unit once selected, is returned to the population before the next draw (r=1,2,..).

- (d) Unequal probability without replacement ($upwor$) sampling: A unit i is selected at the rth draw with probability proportional to $p_i^{(r)}$ and a unit once selected is removed from the population. Here,

$$p_1(i_1) = p_i^{(1)}$$

$$p_r(i_r \mid i_1, \ldots, i_{r-1}) = \frac{p_{i_r}^{(r)}}{1 - p_{i_1}^{(r)} - p_{i_2}^{(r)} - \ldots - p_{i_{r-1}}^{(r)}}, \quad r = 2, \ldots, n$$

(1.3.1)

The quantities $\{p_i^{(r)}\}$ are generally functions of p_i and the p_i-values of the units already selected. In particular, if $p_i^{(r)} = p_i \; \forall \; i = 1, \ldots, N$, the procedure may be called *probability proportional to size without replacement (ppswor)* sampling procedure. For $n = 2$, for this scheme,

$$\pi_i = p_i[1 + A - \frac{p_i}{1 - p_i}]$$

$$\pi_{ij} = p_i p_j (\frac{1}{1 - p_i} + \frac{1}{1 - p_j}), \quad \text{where } A = \sum_k p_k/(1 - p_k).$$

The sampling design may also be attained by an inverse sampling procedure where units are drawn wr, with probability $p_i^{(r)}$ at the rth draw, until for the first time n distinct units occur. The n distinct units each taken only once constitute the sample.

- (e) Rejective sampling: Draws are made wr and with probability $\{p_i^{(r)}\}$ at the rth draw. If all the units turn out distinct, the solution is taken as a sample; otherwise, the whole sample is rejected and fresh draws are made. In some situations $p_i^{(r)} = p_i \; \forall \; i$.

- (f) Systematic sampling with varying probability (including equal probability)

- (g) Sampling from groups: The population is divided into L groups either at random or following some suitable procedures and a sample of size n_h is drawn from the hth group by using any of the above-mentioned sampling designs such that the desired sample size $n = \sum_{h=1}^{L} n_h$ is attained. An example is the Rao-Hartley-Cochran (1962) sampling procedure.

Based on the above methods, there are many uni-stage or multi-stage stratified sampling procedures.

A FS(n)-*s.d.* with π_i proportional to p_i is often used for estimating a population total. This is because, an important estimator, the Horvitz Thompson estimator (HTE)has very small variance if y_i is proportional to p_i. (This fact will be clarified subsequntly). Such a design is called a πps design or *IPPS* (inclusion-probability proportional to size) design. Since $\pi_i \leq 1$, it is required that $x_i \leq X/n \ \forall \ i$ for such a design.

Many (exceeding sixty) unequal probability without replacement sampling designs have been suggested in the literature, mostly for use along with the *HTE*. Many of these designs attain the πps property exactly, some approximately. For some of these designs sample size is a variable. Again, some of these sampling designs are sequential in nature (eg., Chao (1982), Sunter (1977)). Mukhopadhyay (1972), Sinha (1973), Herzel (1986) considered the problem of realising a sampling design with pre-assigned sets of inclusion probabilities of first two orders.

Again, in a sample survey, all the possible samples are not generally equally preferable from the point of view of practical advantages. In agricultural surveys, for example, the investigators tend to avoid grids which are located further away from the cell camps, are located in marshy land, inaccessible places, etc. In such cases, the sampler would like to use only a fraction of the totality of all posible samples, allotting only a very small probability to the non-preferred units. Such designs are called *Controlled Sampling Designs* and have been considered by several authors (e.g. Chakravorty (1963), Srivastava and Saleh (1985), Rao and Nigam (1989, 1990), Mukhopadhyay and Vijayan (1996)).

For a review of different unequal probability sampling designs the reader may refer to Brewer and Hanif (1983), Chaudhuri and Vos (1988), Mukhopadhyay (1991, 1996), among others.

1.4   THE ESTIMATORS

After the sample is selected, data are collected from the field. Here again, data may be collected with respect to a 'sequence' sample or a 'set' sample.

DEFINITION 1.4.1 Data collected from the field through a sequence sample S are

$$d' = \{(k, y_k), k \in S\} \tag{1.4.1}$$

Data collected with respect to a set sample s are

$$d = \{(k, y_k), k \in s\} \tag{1.4.2}$$

Data are said to be unlabelled if after the collection of the data its label part is ignored. Unlabelled data may be represented by a sequence of the observed values (in the order the units appeared in S) or a set of the observed values (in the order the units appeared in s) without any reference to the labels of the units. Note that in this set of the observed values, the entries may not be distinct and are generally not ordered.

However, it is almost never possible to collect the data from the sampled units correctly and completely. For surveys involving human population the respondents may not be available during the time of survey. Also, some of the respondents may give incorrect information due to memory lapse or other factors. Familiar examples are statement of age, income, expenditure, data on family consumption. Again if the character of enquiry is of socially stigmatic nature (eg. abortion undergone by unmarried women, use of drugs, accumulated savings) the respondent may give intentionally false information. The investigators in the field may also fail to register correct information due to their own lapses, even where the data to be collected are of objective nature, eg. determinimg the percentage of area under specified crops in a field in an agricultural survey. Different techniques have been developed to minimise the occurance of non-response and adjust for the effects of nonresponse and different errors of measurement. Warner (1965) and his followers developed different randomized response designs to elicit information on the socially stigmatic characters. We shall not deal with the practical aspects of handling such measurement errors in surveys. However, in Chapter 6 we shall study the effects of measurement errors on usual estimates of population parameters under different superpopulation models.

Unless otherwise stated, we shall, therefore, assume throughout that the data are free from such types of errors due to nonresponse and errors of measurement and it has been possible to collect the information correctly and completely.

(Note that the data colected from the field may also be subject to errors at the subsequent stage of complilation for finding estimates, its variances, etc. Also, errors may arise due to inaccuracy in the frame from which a sample is selected. The reader may refer to Mukhopadhyay (1998 f) for a review of such non-sampling errors).

DEFINITION 1.4.2 An estimator $e = e(s, \mathbf{y})$ (or $e(S, \mathbf{y})$) is a function on $\mathcal{S} \times R^N$ such that for a given $(s, \mathbf{y})$ (or $(S, \mathbf{y})$) its value depends on $\mathbf{y}$ only through those i for which $i \in s$ (or S).

An estimator e is unbiased for T with respect to a sampling design p if

$$E_p(e(s, \mathbf{y})) = T \ \forall \ \mathbf{y} \in R^N \tag{1.4.3}$$

i.e.

$$\sum_{s \in \mathcal{S}} e(s, \mathbf{y}) p(s) = T \ \forall \ \mathbf{y} \in R^N$$

where E_p, V_p denote, respectively, expectation and variance with respect to the *s.d. p*. We shall often omit the suffix p when it is clear otherwise. This unbiasedness will sometimes be referred to as p-unbiasedness.

The mean square error (MSE) of e around T with respect to a *s.d. p* is

$$\begin{aligned} M(e) &= E(e - T)^2 \\ &= V(e) + (B(e))^2 \end{aligned} \tag{1.4.4}$$

where $B(e)$ denotes the design-bias, $E(e) - T$. If e is unbiased for $T, B(e)$ vanishes and (1.4.4) gives the variance of $e, V(e)$.

DEFINITION 1.4.3 A combination (p, e) is called a sampling strategy, often denoted as $H(p, e)$. This is unbiased for T if (1.4.4) holds and then its variance is $V\{H(p, e)\} = E(e - T)^2$.

A unbiased sampling strategy $H(p, e)$ is said to be better than another unbiased sampling strategy $H'(p', e')$ in the sense of having smaller variance, written as $H \succeq H'$, if

$$V\{H(p, e)\} \leq V\{H'(p', e')\} \ \forall \ \mathbf{y} \in R^N \tag{1.4.5}$$

with strict inequality holding for at least ony $\mathbf{y}$.

If the *s.d. p* is kept fixed, an unbiased estimator e is said to be better than another unbiased estimator e' in the sense of having smaller variance, written as $e \succeq e'$ if

$$V_p(e) \leq V_p(e') \ \forall \ \mathbf{y} \in R^N \tag{1.4.6}$$

with strict inequality holding for at least one $\mathbf{y}$.

A strategy H^* is said to be uniformly munimum variance unbiased $(UMVU)$ strategy in a class of unbiased strategies $\mathcal{H} = \{H\}$ if it is better than any other strategy $H \in \mathcal{H}$, i.e., if

$$V(H^*) \leq V(H) \ \forall \ \mathbf{y} \in R^N \tag{1.4.7}$$

with strict inequality holding for at least one $\mathbf{y}$, holds for all $H \neq H^* \in \mathcal{H}$.

For a fixed p, an estimator e^* is said to be UMVU-estimator in a class of unbiased estimators $\eta = \{e\}$ if it is better than any other estimator e in η, i.e., if

$$V_p(e^*) \leq V_p(e) \ \forall \ \mathbf{y} \in R^N \tag{1.4.8}$$

with strict inequality holding for at least one $\mathbf{y}$, holds for all $e \in \eta$.

We now consider different types of estimators for $\bar{y}$, first considering the case when the $s.d.$ is either $srswr$ or $srswor$, based on n draws.

- *Mean per unit estimator*

    (a)$srswr$:

    (i) sample mean, $\bar{y}_S = \sum_{i \in S} y_i/n$,

    (ii) mean of the distinct units, $\bar{y}'_S = \sum_{i \in s} y_i/\nu(s)$ where s is the set corresponding to the squence S. It will be shown in Section 1.5 that $\bar{y}'_S \succeq \bar{y}_S$.

    (b) $srswor$:

    (i) sample mean, $\bar{y}_s = \sum_{i \in s} y_i/n$

- *Ratio estimator*

    (a)$srswr$:

    (i)$\hat{\bar{y}}_{R(S)} = (\bar{y}_S/\bar{x}_S)\bar{X}, \bar{X} = X/N$

    (ii)$\hat{\bar{y}}'_{R(S)} = (\bar{y}'_S/\bar{x}'_S)\bar{X}$

    (b)$srswor$:

    (i) $\hat{\bar{y}}_R = (\bar{y}_s/\bar{x}_s)\bar{X}$

- *Product estimator*

    (a) $srswr$:

    (i)$\hat{\bar{y}}_{P(S)} = \bar{y}_S \bar{x}_S/\bar{X}$

(ii) $\hat{\bar{y}}'_{P(S)} = \bar{y}'_S \bar{x}'_S / \bar{X}$

(b)*srswor*:

(i) $\hat{\bar{y}}_P = \bar{y}_s \bar{x}_s / \bar{X}$

- *Difference estimator*

  (a)*srswr*:

  $\hat{\bar{y}}_{D(S)} = \bar{y}_S + d(\bar{X} - \bar{x}_S)$

  $\hat{\bar{y}}'_{D(S)} = \bar{y}'_S + d(\bar{X} - \bar{x}'_S)$

  (b)*srswor*:

  (i) $\hat{\bar{y}}_D = \bar{y}_s + d(\bar{X} - \bar{x}_s)$, where d is a constant.

- *Regression estimator*

  (a)*srswr*:

  (i)$\hat{\bar{y}}_{lr(S)} = \bar{y}_S + b'(\bar{X} - \bar{x}_S)$

  (ii)$\hat{\bar{y}}'_{lr(S)} = \bar{y}'_S + b''(\bar{X} - \bar{x}'_S)$

  (b)*srswor*:

  (i) $\hat{\bar{y}}_{lr} = \bar{y}_s + b(\bar{X} - \bar{x}_s)$

  where

  $$b' = \sum_{i \in S}(x_i - \bar{x}_S)(y_i - \bar{y}_S) / \sum_{1}^{n}(x_i - \bar{x}_S)^2$$

  $$b'' = \sum_{i \in s}(x_i - \bar{x}'_S)(y_i - \bar{y}'_S) / \sum_{i \in s}(x_i - \bar{x}'_S)^2,$$

  s being the set corresponding to S.

  $$b = \sum_{i \in s}(y_i - \bar{y}_s)(x_i - \bar{x}_s) / \sum_{i \in s}(x_i - \bar{x}_s)^2$$

- *Mean of the ratio estimator*

  (a)*srswr*:

  (i)$\hat{\bar{y}}_{MR(S)} = \bar{X}\bar{r}_S$ where $\bar{r}_S = \sum_{i \in S} r_i / n, r_i = y_i / x_i$

  (ii)$\hat{\bar{y}}'_{MR(S)} = \bar{X}\bar{r}'_S$ where $\bar{r}'_S = \sum_{i \in s} r_i / \nu(S)$, s being the set corresponding to S.

  (b) *srswor*:

  (i) $\hat{\bar{y}}_{MR} = \bar{X}\bar{r}$ where $\bar{r} = \sum_{i \in s} r_i / n$

Except for the mean per unit estimator and the difference estimator none of the above estimators is unbiased for $\bar{y}$. The estimators $\hat{\bar{y}}'_{R(S)}, \hat{\bar{y}}'_{P(S)}, \hat{\bar{y}}'_{D(S)},$ $\hat{\bar{y}}'_{lr(S)}, \hat{\bar{y}}'_{MR(S)}$ have not been considered in the literature. However, all these estimators are unbiased in large samples. Different modifications of ratio estimator, regression estimator, product estimator and estimators obtained by taking convex combinations of these estimators have been proposed in the literature. For $srswor$ an unbiased ratio estimator of $\bar{y}$ obtained by correcting $\hat{\bar{y}}_R$ for bias is

$$\hat{\bar{y}}_{HR} = \bar{r}\bar{X} + \frac{n(N-1)}{N(n-1)}(\bar{y}_s - \bar{r}\bar{x}_s)$$

Some other modifications of ratio estimator are due to Quenouille (1956), who used the $jacknife$ technique, Mickey(1959), Tin (1965) and Pascual (1961). Unbiased regression estimators have been developed following Mickey (1959) (see Rao (1969) for a review of ratio and regression estimators).

In $ppswr$-sampling an unbiased estimator of population total is the Hansen-Hurwitz estimator,

$$\hat{T}_{pps} = \sum_{i=1}^{n} \frac{y_i}{np_i} \qquad (1.4.9)$$

$$\begin{aligned} V(\hat{T}_{pps}) &= \frac{1}{n}\sum_{i=1}^{N}(\frac{y_i}{p_i} - T)^2 p_i \\ &= \frac{1}{2n}\sum_{i\neq j=1}^{N}(\frac{y_i}{p_i} - \frac{y_j}{p_j})^2 p_i p_j = V_{pps} \end{aligned} \qquad (1.4.10)$$

An unbiased estimator of V_{pps} is

$$v(\hat{T}_{pps}) = \frac{1}{n(n-1)}\sum_{1}^{n}(\frac{y_i}{p_i} - \hat{T}_{pps})^2 = v_{pps}$$

We shall call the combination $(ppswr, \hat{T}_{pps})$, a $ppswr$ strategy.

We now consider classes of linear estimators which are unbiased with respect to any $s.d.$. For any $s.d.$ p, consider a non-homogeneous linear estimator of T,

$$e'_L(s, \mathbf{y}) = b_{0s} + \sum_{i\in s} b_{si} y_i \qquad (1.4.11)$$

where the constant b_{0s} may depend only on s and b_{si} on (s, i) $(b_{si} = 0, i \notin s)$. The estimator e'_L is unbiased iff

$$\sum_{s\in S} b_{0s} p(s) = 0 \qquad (1.4.12.1)$$

$$\sum y_i \sum_{s \ni i} b_{si} p(s) = T \; \forall \; \mathbf{y} \; \in R^N \qquad (1.4.12.2)$$

Condition (1.4.12.1) implies for all practical purposes

$$b_{0s} = 0 \; \forall \; s : p(s) > 0 \qquad (1.4.13.1)$$

Condition (1.4.12.2) imples

$$\sum_{s \ni i} b_{si} p(s) = 1 \; \forall \; i = 1, \ldots, N \qquad (1.4.13.2)$$

Note that only the designs with $\pi_i > 0 \; \forall \; i$ admit an unbiased estimator of T.

It is evident that the Horvitz-Thompson estimator (HTE) e_{HT} is the only unbiased estimator of T in the class of estimators $\{\sum_{i \in s} b_i y_i\}$.

$$e_{HT} = \sum_{i \in s} y_i / \pi_i \qquad (1.4.14)$$

Its variance is

$$\begin{aligned} V(e_{HT}) &= \sum_{i=1}^{N} y_i^2 (1 - \pi_i)/\pi_i + \sum_{i \neq j=1}^{N} y_i y_j (\pi_{ij} - \pi_i \pi_j)/\pi_i \pi_j \\ &= V_{HT} \; (\text{say}) \end{aligned} \qquad (1.4.15)$$

If $p \in \rho_n$, (1.4.15) can be wrritten as

$$\sum_{i<j=1}^{N} (y_i/\pi_i - y_j/\pi_j)^2 (\pi_i \pi_j - \pi_{ij}) \qquad (1.4.16)$$

$$= V_{YG} \; (\text{say})$$

The expression V_{HT} is due to Horvitz and Thompson (1952) and V_{YG} is due to Yates and Grundy (1953). An unbiased estimator of V_{HT} is

$$\sum_{i \in s} \frac{y_i^2 (1 - \pi_i)}{\pi_i^2} + \sum_{i \neq j \in s} \frac{y_i y_j (\pi_{ij} - \pi_i \pi_j)}{\pi_i \pi_j \pi_{ij}} \qquad (1.4.17)$$

$$= v_{HT}$$

An unbiased estimator of V_{YG} is

$$\sum_{i<j \in s} (y_i/\pi_i - y_j/\pi_j)^2 \frac{\pi_i \pi_j - \pi_{ij}}{\pi_{ij}} \qquad (1.4.18)$$

$$= v_{YG}$$

The estimators v_{HT}, v_{YG} are valid provided $\pi_{ij} > 0 \ \forall \ i \neq j = 1 \ldots, N$. Both v_{HT}, v_{YG} can take negative values for some samples and this leads to the difficulty in interpreting the reliability of these estimators.

It is clear from (1.4.18) that sufficient conditions for v_{YG} to be nonnegative are

$$\pi_{ij} \leq \pi_i \pi_j, \ \ i \neq j = 1, \ldots, N \tag{1.4.19}$$

Sen (1953) suggested a biased non-negative variance estimator $v'(s)$ of e_{HT} such that $v'(s) = v_{HTs}[0]$ if $v_{HT} > 0[$ otherwise $]$. Sampling designs for which the conditions (1.4.19) hold have been discussed in the literature.

It may be noted that if $p \in \rho_n$ and $\pi_i \propto y_i \ \forall \ i, V_{HT} = 0$ (this is called the 'ratio estimator' property of HTE). In practice, y-values will be unknown and instead, values of some auxiliary variable x closely related to the main variable y will be known. In such cases if the assumption that x is proportional to y hold reasonably well, one may take $\pi \propto p_i$ in order to reduce the variance of e_{HT}. One may, therefore, reasonably expect that V_{HT} based on a πps design will be small provided the auxiliary variable x has been suitably chosen.

If we assume $b_{si} = b_s \ \forall \ i \in s$ and $b_{0s} = 0 \ \forall \ s$ in the class of estimators (1.4.11), then the condition

$$b_s = \frac{1}{M_1 p(s)} \ \forall \ s : p(s) > 0 \tag{1.4.20}$$

is required for e_L to be unbiased, where $M_i = \binom{N-i}{n-i}, i = 1, 2, \ldots.$ In (1.4.20), it has been assumed that all the M_1 samples have positive probability of selection. Therefore, the estimator

$$e_R' = \frac{\sum_s y_i}{M_1 p(s)} \tag{1.4.21}$$

is unbiased for T. Its variance is

$$V(e_R') = \sum_{i=1}^N y_i^2 \{ \frac{1}{M_1^2} \sum_{s \ni i} \frac{1}{p(s)} - 1 \} + \sum_{i \neq i'=1}^N y_i y_{i'} \{ \frac{1}{M_1^2} \sum_{s \ni (i,j)} \frac{1}{p(s)} - 1 \} \tag{1.4.22}$$

An unbiased estimator of $V(e_R')$ is

$$v(e_R') = \frac{\sum_s y_i^2}{M_1 p(s)} (\frac{1}{M_1} - 1) + \frac{\sum \sum_{i \neq i' \in s} y_i y_{i'}}{p(s)} (\frac{1}{M_1^2} - M_2) \tag{1.4.23}$$

The estimator (1.4.21) is a ratio estimator of T under the *s.d. p*. The case of special interest is when $p(s) \propto \sum_s p_i$. This *s.d* is known in the literature as Midzuno (1950)-Lahiri(1951)-Sen(1952) sampling design. In this case, (1.4.21) reduces to the ordinary ratio estimator $e_R = (\sum_s y_i)/(\sum_s p_i)$.

For *ppswor*-sampling an ordered estimator (i.e. an estimator depending on the order of selection of units in S) of T, suggested by Des Raj (1956) is

$$e_D(S) = \frac{1}{n} \sum_{r=1}^{n} t_r \qquad (1.4.24)$$

where

$$t_1 = y_{i_1}/p_{i_1}$$

$$\ldots$$

$$t_r = y_{i_1} + \ldots + y_{i_{r-1}} + \frac{y_{i_r}}{p_{i_r}}(1 - p_{i_1} - \ldots - p_{i_r}), \; r = 2, \ldots, n \qquad (1.4.25)$$

and the sample is $S = \{i_1, \ldots, i_n\}$. The estimator has the property that it has a non-negative variance estimator

$$v(e_D) = \frac{1}{n(n-1)} \sum_{r=1}^{n}(t_r - e_D)^2 = v_D \qquad (1.4.26)$$

and has a smaller variance than the *ppswr* strategy based on the same set of $\{n, p_i, i = 1, \ldots, N\}$-values. It will be shown in Section 1.5 that an estimator which depends on the sequence S can always be improved upon by confining to the corresponding set s only. By unordering $e_D(S)$, Murthy (1957) obtained an unordered estimator

$$e_M(s) = \sum{}^{\prime} \frac{e_D(S)p(S)}{p(s)} \qquad (1.4.27)$$

where $\sum{}^{\prime}$ dnotes summation over all S which correspond to their common s. It is shown that

$$e_M = \frac{1}{p(s)} \sum_{i \in s} y_i P(s \mid i) \qquad (1.4.28)$$

where $P(s \mid i)$ is the conditional probability of getting the unordered sample s, given that i was selected at the first draw.

An unbiased variance estimator of e_M is

$$v(e_M) = \frac{1}{2\{p(s)\}^2} \sum_{i \neq j \in s}\sum \{p(s)P(s \mid ij) - P(s \mid i)P(s \mid j)\}$$

$$p_i p_j (\frac{y_i}{p_i} - \frac{y_j}{p_j})^2 \qquad (1.4.29)$$

where $P(s \mid i,j)$ is the conditional probability of getting s given that (i,j) were selected at the first two draws.

For $n = 2, s = (1,2)$,

$$e_M = \frac{1}{2 - p_1 - p_2}\{(1 - p_2)\frac{y_1}{p_1} + (1 - p_1)\frac{y_2}{p_2}\} \qquad (1.4.30)$$

Its variance is

$$V(e_M) = \frac{1}{2}\sum_{i \ne j=1}^{N} p_i p_j \frac{(1 - p_i - p_j)}{(2 - p_i - p_j)}(\frac{y_i}{p_i} - \frac{y_j}{p_j})^2 \qquad (1.4.31)$$

Also,

$$v(e_M) = \frac{(1 - p_1)(1 - p_2)(1 - p_1 - p_2)}{(2 - p_1 - p_2)^2}(\frac{y_1}{p_1} - \frac{y_2}{p_2})^2 \qquad (1.4.32)$$

For *ppswor* samples of size 2, Murthy (1963) also proposed another estimator of T,

$$e'_M = \frac{1}{2 - p_i - p_j}[\frac{y_i}{\pi_i - p_i}(1 - p_i) + \frac{y_j}{\pi_j - p_j}(1 - p_j)]$$

for $s = (i,j), i < j$, and $\pi_i = p_i \sum_{j(\ne i)} p_j / (1 - p_j)$.

Mukhopadhyay (1977) considered generalisation of Des Raj estimators and Murthy estimator for application to any unequal probability without replacement sampling design.

We consider some further estimators applicable to any sampling design :

(a) *Generalised Difference Estimator* Basu (1971) considered an unbiased estimator of T,

$$e_{GD}(a) = \sum_s \frac{y_i - a_i}{\pi_i} + A, \ \ A = \sum a_i \qquad (1.4.33)$$

where $a = (a_1, \ldots, a_N)'$ is a set of known quantities. The estimator is unbiased and has less variance than e_{HT} in the neighbourhood of the point a.

(b) *Generalised Regression Estimator*

$$e_{GR} = \sum_s \frac{y_i}{\pi_i} + b(X - \sum_s \frac{x_i}{\pi_i}) \qquad (1.4.34)$$

where b is the sample regression coefficient of y on x. The estimator was first considered by Cassel, Sarndal and Wretman (1976) and is a generalisation of the linear regression estimator $\hat{\bar{y}}_{lr}$ to any $s.d\ p$.

(c) *Generalised Ratio Estimator*

$$e_{Ha} = X \frac{\displaystyle\sum_s y_i/\pi_i}{\displaystyle\sum_s x_i/\pi_i} \tag{1.4.35}$$

The estimator was first considered by Ha'jek (1959) and is a generalisation of $\hat{\bar{y}}_R$ to any $s.d\ p$.

The estimators e_{GR}, e_{Ha} are not unbiased for T. Besides these, specific estimaors have been suggested for specific procedures. An example is Rao-Hartley-Cochran (1962) estimator.

If we denote the strategy $(\pi ps, e_{HT})$ as H_1 and Hansen-Hurwitz strategy as H_2, both being based on the same set of $\{n, p_i, i = 1, \ldots, N\}$-values, then it follows that the conditions

$$\pi_{ij} \leq \frac{2(n-1)\pi_i\pi_j}{n} \tag{1.4.36}$$

$$\pi_{ij} > \frac{(n-1)\pi_i\pi_j}{n} \tag{1.4.37}$$

are, respectively, necessary conditions and sufficient conditions for $H_1 \succeq H_2$. Hence, for $n = 2$, nonnegativity of v_{YG} is a necessary condition for $H_1 \succeq H_2$. Gabler (1984) gave a sufficient condition for $H_1 \succeq H_2$.

1.5 SOME INFERENTIAL PROBLEMS UNDER FIXED-POPULATION SET-UP

Problems of inference about a fixed population mean through survey sampling under a fixed population set-up has been considered by various authors. Godambe (1955), Hanurav (1966), Lanke (1975), among others, proved that given any sampling design, there does not exist any uniformly minimum variance linear unbiased estimator for population mean, in the class of all linear unbiased estimators, except for the unicluster sampling design. Attention was, therefore, drawn to the search for admissible estimators for a fixed sampling design in a class of estimators [Godambe and Joshi (1965), Joshi (1965 a,b, 1969), among others] and admissible sampling strategies in a class of strategies [Joshi (1966), Scott (1975), Sekkappan and Thompson (1975), among others]. Godambe (1966) also observed that for a non-informative design, the likelihood function of the population parameter **y** is, in general, non-informative. The scale-load approaches of Hartley and Rao (1968, 1969), however, provide some important exceptions. Basu (1958), Basu and Ghosh (1967) were first to consider the concept of sufficiency and Rao-Blackwellisation in survey sampling.

In case the survey population is looked upon as a sample from a super-population, optimum sampling strategies are, however, available in certain classes and there exist lower bounds to the average variances of all unbiased strategies under some models [Godambe and Joshi (1965), Cassel et al (1976), Tam (1984), among others]. We will make a overview of these results in this section.

1.5.1   THE PDF OF DATA AND LIKELIHOOD FUNCTION

Let $D'(D)$ be a random variable whose one particular realisation is data $d'(d)$. Let also $\Psi'(\Psi)$ be a random variable having values $S(s)$ on $\mathcal{S}$.

A data point $d'(d)$ is said to be consistent with a chosen parameter vector **y** if $d'(d)$ can be obtained from **y**. For example, $d' = \{(3,6),(2,8),(3,6),(4,10)\}$ is consistent with $\mathbf{y} = \{11,8,6,10\}$ but not with $\mathbf{y}' = \{3,6,5,8\}$.

Let for a given $d'(d), \Omega_{d'}(\Omega_d)$ be the set of **y** for which $d'(d)$ is consistent.

The probability density functon (pdf) of D' is

$$
\begin{aligned}
f_{D'}(d';\mathbf{y}) &= P[\{(k,y_k), k \in S\} = d';\mathbf{y}] \\
&= P[\Psi' = S]P[D' = d' \mid \Psi' = S;\mathbf{y}] \\
&= P(S)\delta(d';\mathbf{y})
\end{aligned} \tag{1.5.1}
$$

where $\delta(d';\mathbf{y}) = 1(0)$ if d' is consistent (inconsistent) with **y**. Thus

$$
f_{D'}(d';\mathbf{y}) = p(S)(0) \text{ for } \mathbf{y} \in \Omega_{d'} \text{ (otherwise)}
$$

Similarly, pdf of D is

$$
f_D(d;\mathbf{y}) = p(s)(0) \text{ for } \mathbf{y} \in \Omega_d \text{ (otherwise)} \tag{1.5.2}
$$

1.5.2   LIKELIHOOD FUNCTION OF **y** :

Given the data $D' = d'$, the likelihood function of the parameter vector **y** is

$$
L(\mathbf{y} \mid d') = f_{D'}(d';\mathbf{y}) = p(S)(0) \text{ if } \mathbf{y} \in \Omega_{d'} \text{ (otherwise)} \tag{1.5.3}
$$

Similarly, the likelihood function of $\mathbf{y}$ given $D = d$ is

$$L(\mathbf{y} \mid d) = f_D(d; \mathbf{y}) = p(s)(0) \quad \text{if } \mathbf{y} \in \Omega_d \text{ (otherwise)} \tag{1.5.4}$$

The likelihood functions are, therefore, 'flat', taking values $p(s)(p(S))$ for $\mathbf{y} \in \Omega_d(\Omega_{d'})$ and zero for other values of $\mathbf{y}$. There does not exist any maximum likelihood solution of the parameter vector $\mathbf{y}$. The likelihood function only tells that given the data $d'(d)$, any $\mathbf{y} \in \Omega_{d'}(\Omega_d)$ is equiprobable. The likelihood functions (1.5.3) and (1.5.4), first considered by Godambe (1966), are, therefore, non-informative. However, if a superpopulation model ξ is postulated for the population vector, the likelihood function may give more information (see Ericson (1969 a), Royall (1976), Brecking et al (1990)). The use of superpopulation model in making inference from a finite population has been introduced towards the end of this section. Also, the scale-laod approach of Royall (1968), Hartley and Rao (1968, 1969) makes the likelihood function informative.

1.5.3 SUFFICIENCY, RAO-BLACKWELLISATION

The concept of sufficiency and Rao-Blackwellisation in survey sampling was first considered by Basu (1958). As in the traditional statistical theory, if a sufficient statistic is available, any estimator can be improved upon by Rao-Blackwellisation. In survey sampling D' forms the primary body of data. Any summarisation of data would have to be made necessarily over D'.

DEFINITION 1.5.1 A statistic $u(D')$ is a sufficient statistic for $\mathbf{y}$ if the conditional distribution of D' given $u(D') = u_0$ is independent of $\mathbf{y}$, provided the conditional distribution is well-defined.

Let $z(D')$ be a statistic defined over the range space of D' such that $z(D') = d$, i.e. z reduces d' obtained through a sequence sample S to the data d for the corresponding set s. As an example, if $d' = \{(3, 6), (4, 9), (3, 6)\}$, $z(d') = \{(3, 6), (4, 9)\}$.

THEOREM 1.5.1 (Basu and Ghosh, 1967) For any ordered design p, the statistic $z(D')$ is sufficient.

Proof Consider two data points d' of D' and d of D. Assume that the parameter vector $\mathbf{y} \in \Omega_{d'}$ and Ω_d. Otherwise, the conditional probability will not be defined.

$$P_{\mathbf{y}}[D' = d' \mid z(D') = d] = \frac{P_{\mathbf{y}}[(D' = d') \bigcap (z(D') = d)]}{P_{\mathbf{y}}[z(D') = d]} \tag{1.5.5}$$

provided $P_{\mathbf{y}}[z(D') = d] > 0$.

Case (a); $z(d') \neq d$. Here numerator in (1.5.5) is zero and hence (1.5.5) has value zero.

Case (b): $z(d') = d$. The numerator in (1.5.5) is

$$P_{\mathbf{y}}[D' = d'] = p(S) \quad \text{by (1.5.1)}$$

The denominator is

$$\sum{}' p(S) = p(s)$$

where $\sum'$ is over all S which correspond to the set s (corresponding to d). Hence, the value of (1.5.5) is $p(S)/p(s)$.

In either case, the conditional probability is independent of $\mathbf{y}$. Hence the proof.

Rao-Blackwellisation

For any estimator $e(D')$ for θ define $e(d) = E\{e(D') \mid z(D') = d\}$. Since $z(D')$ is sufficient for $\mathbf{y}$, $e(d)$ is independent of any unknown parameter, and depends on D' only through $z(D')$ and as such can be taken as an estimator of θ.

THEOREM 1.5.2 Let $e(D')$ be an estimator of θ. The estimator $e_1(d) = E\{e(D') \mid z(D') = d\}$ has the properties:

(i)
$$E(e) = E(e_1)$$

(ii)

$$MSE(e_1) \leq MSE(e) \text{ with strict inequality } \forall \, \mathbf{y} \in R^N \, iff \, P\{e \neq e_1; \mathbf{y}\} > 0$$

Proof:
$$E(e_1) = E[E\{e(D') \mid z(D') = d\}] = E\{e(D')\}$$
$$E(e - \theta)^2 = E(e_1 - \theta)^2 + E(e - e_1)^2 + 2E(e_1 - \theta)(e - e_1)$$
$$= MSE(e_1) + E(e - e_1)^2,$$

the last term on the right hand side vanishes.

Rao-Blackwellisation in survey sampling was first considered by Basu (1958). Some earlier works are due to Des Raj and Khamis (1958), Murthy (1957).

EXAMPLE 1.5.1

For a S drawn by *srswr* with n draws, let $d = \{(i_1, y_{i_1}), \ldots, (i_\nu, y_{i_\nu})\}, i_1 < \ldots, < i_\nu$, corresponding to data d'. The customary sample mean $\bar{y}_S$ depends on the multiplicity of units in S.

$$E\{\bar{y}_S \mid d\} = E\{\frac{1}{n} \sum_{i=1}^{n} y_i' \mid d\} = E(y_1' \mid d)$$

$$= \frac{1}{\nu} \sum_{j=1}^{\nu} y_{i_j} = \bar{y}_\nu,$$

where y_i' denotes the value of y on the unit selected at the ith draw $(i = 1, \ldots, n)$. Hence, $E(\bar{y}_\nu) = E(\bar{y}_S) = \bar{y}, V(\bar{y}_\nu) \leq V(\bar{y}_S)$.

The statistic D is not, however, a complete sufficient statistic. Hence, there may exist more than one estimator $e(D)$ for T. An example is the *HTE* which depends only on D and is unbiased for T. For further results, the reader may refer to Cassel et al (1977).

1.5.4 UNIFORMLY MINIMUM VARIANCE UNBIASED ESTIMATION

Godambe (1955) first observed that in survey sampling no UMVV estimator exists in the class of all linear unbiased estimators of population total, for any given p in general. The proof was subsequently improved upon by Hanurav (1966), Lanke (1975), among others. We shall recall here two important results without proof.

DEFINITION 1.5.2 A *s.d.* p is said to be a unicluster design if for any two samples s_1, s_2

$$\{p(s_1) > 0, p(s_2) > 0, s_1 \neq s_2|] \Rightarrow s_1 s_2 = \phi,$$

i.e. either two samples are identical or are disjoint.

A linear systematic sample is an example of a Unicluster design.

THEOREM 1.5.3 A *s.d.* p admits a UMVU estimator of T in the class of all linear unbiased estimators iff p is a unicluster design with $\pi_i > 0 \; \forall \; i$. For a unicluster *s.d*, the HTE is the UMVUE.

THEOREM 1.5.4 (Basu, 1971) For any non-census design p (with $\pi_i > 0 \; \forall \; i$), there does not exist any UMVUE of T in the class of all unbiased estimators.

Thus, in general, there does not exist any UMVU-estimator for any given *s.d.* p. Hence, there does not exist any UMVU sampling strategy in general.

1.5.5 ADMISSIBILITY OF ESTIMATORS

DEFINITION 1.5.3 For a fixed *s.d.p*, an estimator e is said to be an admissible estimator of T within a class C of estimators *iff* there does not exist any estimator in C which is uniformly better than e.

Clearly, within the same class C there may exist more than one admissible estimator. Admissibility ensures that an estimator is uniquely best in C at least at some point **y** in the parameter space. In the absence of a UMVU-estimator, one should choose an estimator within the class of all admissible estimators. However, a slightly inadmissible estimator may sometimes possess some practical advantages over an admissible estimator and may be used in preference to the later.

An important theorem is stated below without proof.

THEOREM 1.5.5 For any sampling design p, with $\pi_i > 0(\ \forall\ i)$, the generalised difference estimator $e_{GD}(a) = \sum_s \frac{y_i - a_i}{\pi_i} + A$ ($A = \sum_i a_i$) is admissible in the class of all unbiased estimators of T.

A corollary to this theorem is that Horvitz Thompson estimator e_{HT} is admissible in the class of all unbiased estimators of T.

1.5.6 AVERAGE VARIANCE OF A STRATEGY UNDER A SUPERPOPULATION MODEL

We now introduce the concept of superpopulation models in survey sampling. Assume that the value of y on i is a particular realisation of a random variable $Y_i(i = 1, \ldots, N)$. Hence the value $\mathbf{y} = (y_1, \ldots, y_N)$ of a survey population $\mathcal{P}$ may be looked upon as a particular realisation of a random vector $\mathbf{Y} = (Y_1, \ldots, Y_N)$ having a superpopulation model ξ_θ, indexed by a parameter vector $\theta, \theta \in \Theta$ (the parameter space). The class of priors $\{\xi_\theta, \theta \in \Theta\}$ is called a superpopulation model. The model ξ for $\mathbf{Y}$ is obtained through one's prior belief about $\mathbf{Y}$. As an example, in agricultural survey of yield of crops in N plots, if acreages $x_1, \ldots, x_N$ under the crop on these plots are assumed to be fixed over years, one may assume that the yield $\mathbf{y}$ in a particular year is a random sample from a prior distribution ξ of $\mathbf{Y}$, which may depend, among others, on $x_1, \ldots, x_N$. In particular, one may assume that ξ, the joint *pdf* of $\mathbf{Y}$ is such that $Y_1, \ldots, Y_N$ are independent with

$$\mathcal{E}(Y_i \mid x_i) = \beta x_i$$

$$\mathcal{V}(Y_i \mid x_i) = \sigma_i^2, \ i = 1, \ldots, N \qquad (1.5.6)$$

where $\beta, \sigma_i^2 (> 0)$ are constants. Here and subsequently, $\mathcal{E}, \mathcal{V}, \mathcal{C}$ will denote expectation, variance, covariance with respect to ξ.

The use of an appropriate superpopulation model distribution in survey sampling is justified by the fact that in surveys of usual interest (agricultural surveys, industrial surveys, cost of living surveys, employment surveys, traffic surveys, etc.), a **y** can not take any value in R^N, but takes values in a particular domain in R^N, some with higher probability. One may, therefore, postulate some reasonable superpopulation model ξ for **Y** and exploit ξ to produce suitable sampling strategies.

A good deal of inference in survey sampling emerges from the postulation of a suitable prior distribution ξ for **Y** and methodologies have been developed to produce optimal sampling strategies. We shall review some of these results based on frequentist approach in Chapter 2 of this treatise. The early uses of ξ are due to Cochran (1946), Yates (1949), Godambe (1955), Rao, Hartley and Cochran (1962), among others.

Average Variance under ξ

In most cases the expression for the variance of different strategies are complicated in nature and are not amenable to comparison; one may, therefore, take the average value of the variance under an assumed superpopulation model ξ and compare their average variances.

The average variance (AV) of an unbiased strategy (p, e) under ξ is given by $\mathcal{E}V_p(e)$. A strategy H_1 will be better than an unbiased strategy $H_2 (H_1 \succeq H_2)$ in the smaller average variance sense, if $AV(H_1) < AV(H_2)$.

We recall an important path-breaking result due to Godambe and Joshi (1965). The theorem shows that there exists a lower bound to the average variance of p−unbiased strategy under a very general superpopulation model ξ.

THEOREM 1.5.6 Consider model $\xi : Y_1, \ldots, Y_N$ are independent with $\mathcal{E}(Y_i) = \mu_i, \mathcal{V}(Y_i) = \sigma_i^2 (i = 1, \ldots, N)$. For any unbiased sampling strategy (p, e), with the value of first order inclusion probability π_i,

$$\mathcal{E}V(p, e) \geq \sum_{i=1}^{N} \sigma_i^2 (\frac{1}{\pi_i} - 1) \qquad (1.5.7)$$

COROLLARY 1.5.1 The lower bound (1.5.7) is attained by e_{HT} applied to a $FS(n)$-design with $\pi_i \propto \mathcal{E}(Y_i), i = 1, \ldots, N$. In particular, if $\mu_i = \beta x_i (\beta$

a constant), any $FS - \pi ps$ design applied to e_{HT} attains the lower bound in (1.5.7).

Thus, while, there does not exist any UMVUE of T under fixed population model, if the finite population is looked upon as a sample from a superpopulation, there exists a attainable lower bound to the avaerage variance of all p-unbiased strategies. For a given set of values of $\{\pi_i\}$, the lower bound is given by (1.5.7). This is the lower bound to the avarage variance of all p-unbiased strategies having the same set of π-values. The above lower bound is comparable to the Rao-Cramer lower bound of variance of regular estimators under the classical procedure of inference from an infinite population. Note that the superpopulation model ξ considered in Theorem 1.5. is very general. It only requires that the observations Y_i's are uncorrelated.

A strategy whose average variance attains the lower bound (1.5.7), at least approximately, under a class of superpopulation models is considered robust (Wright, 1983; Sarndal,1980 a,b; Brewer et al, 1988). Many of the model-based sampling strategies considered in the next chapter are robust, at least asymptotically. For further details on design-based procedures in survey sampling the reader may refer to Cochran (1977), Murthy (1977), Mukhopadhyay (1998 f), among others.

1.6 PLAN OF THE BOOK

The aim of this book is to make a comprehensive review of some topics in survey sampling which have not been covered in details in books so far and to indicate new research areas wherever possible. We do not discuss in details developments in fixed population model and superpopulation model-based prediction theoretic approaches which have given rise to a genesis of robust sampling strategies for predicting mean, total etc. These have been discussed elaborately in Cassel et al (1977), Chaudhuri and Vos (1988), Sarndal et al (1992), Mukhopadhyay (1996, 1998 f), among others. We make a brief tour in these areas in Chapters 1 and 2 respectively. These areas are not our main centres of attention. The first two chapters are like appetizers to the readers and have been introduced to make a review of the environment in which the theory of sample surveys have been developed in the modern times.

We concentrate on Bayes and related procedures in finite population sampling in Chapters 3 and 4. Apart from the books by Bolfarine and Zacks (1991), Ghosh and Meeden (1997), this aspect does not seem to have been covered in the recent books. After developing theory for Bayes and minimax prediction of finite population parameters under a general set up, we consider in section 3.3 Bayes and minimax prediction of T for the normal regression models under squared error loss function. Bayes prediction under the Linex loss function of Varian (1975) and Zellner (1980) has been considered subsequently. James-Stein estimator and associated estimators have been addressed. We have made detailed review of empirical Bayes (EB) prediction of T under simple location model. Subsequently, the normality assumption has been replaced by that of posterior linearity. EB-prediction of population total under normal model using covariates has been considered. The work of Ghosh et al (1998) using generalised linear models for simultaneous estimation of strata means opens a new vista of research. Application of Bayes procedures and Fay-Herriot modified procedures in small area estimation have been addressed. Live data from a survey carried in the urban area of Hugli district, West Bengal, India have been used to produce synthetic estimates, regression estimates and EB-estimates of population counts for 1997-98 for different municipal areas. Bayes prediction under random error variance model considered by Butar and Lahiri (1999, 2000) invokes further investigation in the area.

Chapter 4 considers different ramifications of Bayes procedures. Section 4.2 discusses Linear Bayes prediction due to Hartigan (1969) and Brunk (1980) and its application in finite population sampling. Restricted Linear Bayes procedure due to La Motte (1978) and its applications in finite population sampling as developed by Rodrigues (1989) have been considered subsequently. Constrained Bayes procedures due to Louis (1984) and Ghosh (1992), Limited Translation Bayes estimators due to Effron and Morris (1971, 1972) have been discussed with reference to finite population sampling. Bayesian robustness under a class of alternative models as developed by Bolfarine et al (1987) has been reviewed. Robust Bayes estimation under a class of contaminated priors due to Berger (1984), Berger and Berliner (1986) and their applications in finite population sampling due to Ghosh and Kim (1993, 1997) have been discussed.

The problem of estimation of a finite population variance does not seem to have received considerable attention in books on sample surveys. Chapter 5 considers design-based estimation, model-based prediction under Royall's (1970) approach and design-model based robust strategies for predicting a finite population variance. This chapter also considers Bayes and minimax procedures for the population variance. Prediction of a finite population regression coefficient under multiple regression model, asymptotic properties of a sample regression coefficient and estimation of slope parametrs in the linear regression model have also been addressed in this chapter.

The problems of prediction of finite population distribution function have not been covered in details in books. Chaper 6 considers design-based, model-based and design-model based prediction of a finite population distribution function in details. The next chapter considers prediction of finite population parameters (mean, variance, distribution function) under superpopulation models accommodating measurement errors. Both additive measurement errors and multiplicative measurement errors due to Hwang (1986) have been looked into.

The last chapter considers three special topics, Calibration estimators due to Deville and Sarndal (1982), Post-stratification and Conditional unbiasedness under design-based approach.

As stated before, the idea of writing the book is to make a review of some of the ideas that have swept past in the field of survey sampling during the last few decades and to indicate to the researchers new areas of study. The topics have been chosen on the basis of their not being dealt with hitherto in details in books, their importance and research potentials. In their elegant book, Bolfarine and Zacks (1992) have also discussed, among many topics, Bayes and minimax prediction of finite population total, variance and regression coefficient, EB-estimation of T, Restricted Linear Bayes prediction and Bayesian robustness as developed by Rodrigues (1987). Prediction of population total and variance under additive measurement models have also been considered. The marvellous book by Ghosh and Meeden (1991) develops noninformative Bayesian justification for commonly-used frequentist procedures in finite population sampling. The present treatise aims at reviewing some further researches in these areas. In doing so we have tried to be clear in the presentation to the best of our understanding and ability. We have also tried to cover many references, which however, is not claimed to be exhaustive.

Chapter 2

Inference under Frequentist Theory Approach

2.1 INTRODUCTION

In subsection 1.5.6 we considered the concept of superpopulation models in finite population sampling and introduced its uses in finding the average variance of a strategy for comparison among several strategies. The approach there was necessarily based on a fixed population model, as we had to confine to strategies having some desirable properties, like unbiasedness, admissibility, sufficiency, suggested by a fixed population set up.

Brewer (1963), Royall (1970, 1976), Royall and Herson (1973) and their co-workers considered the survey population as a random sample from a superpopulation and attempted to draw inference about the population parameter from a prediction-theorist's view-point. These model-dependent predictors are very sensitive to model mis-specification. Cassel et al (1976, 1977), Wright (1983), among others, therefore, considered model-based predictors, which are also based on sampling designs, to provide robust prediction under model failures. We make a brief review of some of these results in this section.

2.2   PRINCIPLES OF INFERENCE BASED ON THEORY OF PREDICTION

As in subsection 1.5.6 we assume that the value y_i on i is a realisation of a random variable $Y_i (i = 1, \ldots, N)$. For simplicity, we shall use the same symbol y_i to denote the population value as well as the random variable of which it is a particular realisation, the actual meaning being clear from the context. We assume throughout that there is a superpopulation distribution ξ_θ of $\mathbf{Y} = (Y_1, \ldots, Y_N)'$ indexed by a parameter vector $\theta \in \Theta$, the parameter space. Let $\hat{T}_s$ denote a predictor of T or $\bar{y}$ based on s (the specific one being clear from the context).

DEFINITION 2.2.1: $\hat{T}_s$ is model-unbiased or ξ-unbiased or m-unbiased predictor of $\bar{y}$ if

$$\mathcal{E}(\bar{T}_s) = \mathcal{E}(\bar{y}) = \bar{\mu} \text{ (say)} \ \forall \ \theta \in \Theta \text{ and } \forall s : p(s) > 0 \qquad (2.2.1)$$

DEFINITION 2.2.2: $\hat{T}_s$ is design-model unbiased (or $p\xi$-unbiased or pm-unbiased) predictor of $\bar{y}$ if

$$E\mathcal{E}(\bar{T}_s) = \bar{\mu} \ \forall \ \theta \in \Theta \qquad (2.2.2)$$

Clearly, a m-unbised predictor is necessarilly pm-unbiased.

For a non-informative design where $p(s)$ does not depend on the y-values order of operation $E, \mathcal{E}$ can always be interchanged.

Two types of mean square errors (mse's) of a sampling strategy $(p, \hat{T}_s)$ for predicting T has been proposed in the literature:

(a)
$$\mathcal{E}MSE(p, \hat{T}) = \mathcal{E}E(\hat{T} - T)^2 = M(p, \hat{T}) \text{ (say)}$$

(b)
$$\begin{aligned} EMS\mathcal{E}(p, \hat{T}) &= E\mathcal{E}(\hat{T} - \mu)^2 \text{ where } \mu = \sum \mu_k = \mathcal{E}(T) \\ &= M_1(p, \hat{T}) \text{ (say)} \end{aligned}$$

If $\hat{T}$ is p-unbiased for T $(E(\hat{T}) = T \ \forall \ \mathbf{y} \in R^N)$, M is model-expected p-variance of $\hat{T}$. If $\hat{T}$ is m-unbiased for T, M_1 is p-expected model-variance of $\hat{T}$.

It has been recommended that if one's main interest is in predicting the total of the current population from which the sample has been drawn, one should use M as the measure of uncertainty of $(p, \hat{T})$. If one's interest is in predicting the population total for some future population, which is of the same type as the present survey population (having the same μ), one is really concerned with μ, and here M_1 should be used (Sarndal, 1980 a). In finding an optimal predictor one minimises M or M_1 in the class of predictors of interest.

The following relations hold:

$$M(p, \hat{T}) = EV(\hat{T}) + E\{\beta(\hat{T})\}^2 + V(T) - 2\mathcal{E}\{(T - \mu)E(\hat{T} - \mu)\} \quad (2.2.3)$$

where $\beta(\hat{T}) = \mathcal{E}(\hat{T} - T)$, model-bias in $\hat{T}$.

It $\hat{T}$ is p-unbiased,

$$M(p, \hat{T}) = EV(\hat{T}) + E\{\beta(\hat{T})^2\} - V(T) \quad (2.2.4)$$

If $\hat{T}$ is p-unbiased as well as m-unbiased,

$$M(p, \hat{T}) = M_1(p, \hat{T}) - V(T) \quad (2.2.5)$$

Now, for the given data $d = \{(k, y_k), k \in s\}$, we have

$$T = \sum_s y_i + \sum_{\bar{s}} Y_i = \sum_s y_i + U_s \text{ (say)} \quad (2.2.6)$$

where $\bar{s} = \mathcal{P} - s$. Therefore, in predicting T one needs to only predict U_s, the part $\sum_s y_i$, being completely known.

A predictor

$$\hat{T}_s = \sum_s y_i + \hat{U}_s$$

will be m-unbiased for T if

$$\mathcal{E}(\hat{U}_s) = \mathcal{E}(\sum_{\bar{s}} Y_i) = \sum_{\bar{s}} \mu_i = \mu_{\bar{s}} \text{ (say)} \quad \forall \; \theta \in \Theta, \; \forall \; s : p(s) > 0 \quad (2.2.7)$$

In finding an optimal $\hat{T}$ for a given p, one has to minimise $M(p, \hat{T})$ (or $M_1(p, \hat{T})$) in a certain class of predictors. Now, for a m-unbiased $\hat{T}$,

$$\begin{aligned} M(p, \hat{T}) &= E\mathcal{E}(\hat{U}_s - \sum_{\bar{s}} Y_k)^2 \\ &= E\mathcal{E}\{(\hat{U}_s - \mu_{\bar{s}}) - (\sum_s Y_k - \mu_{\bar{s}})\}^2 \\ &= E[V(\hat{U}_s) + V(\sum_{\bar{s}} Y_k) - 2\mathcal{C}(\hat{U}_s, \sum_{\bar{s}} Y_k)] \end{aligned} \quad (2.2.8)$$

If Y_i are independent, $\mathcal{C}(\hat{U}_s, \sum_{\bar{s}} Y_k) = 0$ ($\hat{U}_s$ being a function of $Y_k, k \in s$ only). Hence, in this case, for a given s, the optimal m-unbiased predictor of T (in the minimum $\mathcal{E}(\hat{T} - T)^2$-sense) is (Royall, 1970),

$$\hat{T}_s^+ = \sum_s y_k + \hat{U}_s^+ \quad (2.2.9)$$

where

$$\mathcal{E}(\hat{U}_s^+) = \mathcal{E}(\sum_{\bar{s}} y_k) = \mu_{\bar{s}} \qquad (2.2.10.1)$$

$$\mathcal{V}(\bar{U}_s^+) \leq \mathcal{V}(\hat{U}_s') \qquad (2.2.10.2)$$

for any $\hat{U}_s'$ satisfying (2.2.10.1). It is clear that $\hat{T}_s^+$, when it exists , does not depend on the sampling design (unlike, the design-based estimator, eg. e_{HT}).

An optimal design-predictor pair $(p, \hat{T}_s^+)$ in the class $(\rho, \hat{\tau})$ is the one for which

$$M(p^+, \hat{T}_s^+) \leq M(p, \hat{T}_s') \qquad (2.2.11)$$

for any $p \in \rho$, a class of sampling designs and $\hat{T}'$, any other m-unbiased predictor $\in \hat{\tau}$.

After $\hat{T}_s$ has been derived via (2.2.9) - (2.2.10.2), an optimum sampling design is obtained through (2.2.11). This approach is, therefore, completely model-dependent, the emphasis being on the correct postulation of a superpopulation model that will efficiently describe the physical situation at hand and thereby, generating $\hat{T}_s$. After $\hat{T}_s$ has been specified, one makes a pre-sampling judgement of eficiency of $\hat{T}_s$ with respect to different sampling designs and obtain p^+ (if it exists). The choice of a suitable sampling design is, therefore, relegated to a secondary importance in this prediction-theoretic approach.

EXAMPLE 2.2.1

Consider the polynomial regression model:

$$\mathcal{E}(y_k \mid x_k) = \sum_{j=0}^{J} \delta_j \beta_j x_k^j$$

$$\mathcal{V}(y_k \mid x_k) = \sigma^2 v(x_k), \quad k = 1, \ldots, N \qquad (2.2.12)$$

$$\mathcal{C}(y_k, y_{k'} \mid x_k, x_{k'}) = 0, \quad k \neq k' = 1, \ldots, N \qquad (2.2.13)$$

where x_k's are assumed fixed (non-stochastic) quantities, $\beta_j(j = 1, \ldots, J), \sigma^2$ are unknown quantities, $v(x_k)$ is a known function of $x_k, \delta_j = 1(0)$ if the ternm x_k^j is present (absent) in $\mathcal{E}(y_k) = \mu_k$. The model (2.2.12),(2.2.13) has been denoted as $\xi(\delta_0, \delta_1, \ldots, \delta_J; v(x))$ by Royall and Herson (1973). The best linear unbiased predictor (BLUP) of T under this model is, therefore,

$$\hat{T}_s^*(\delta_0, \ldots, \delta_J) = \sum_s y_k + \sum_{j=0}^{J} \delta_j \hat{\beta}_j^* \sum_{\bar{s}} x_k^j \qquad (2.2.14)$$

where $\hat{\beta}_j^*$ is the BLUP of β_j under $\xi(\delta_0, \ldots, \delta_J; v(x))$ as obtainable from Gauss-Markoff theorem.

Under model $\xi(0, 1; v(x))$,

$$\hat{T}_s^*(0, 1; v(x)) = \sum_s y_k + \{(\sum_s x_k y_k/v(x_k))(\sum_s x_k^2/v(x_k))^{-1}\}\sum_{\bar{s}} x_k \tag{2.2.15}$$

$$\mathcal{E}(\hat{T}_s^* - Y)^2 = \sigma^2(\sum_{\bar{s}} x_k)^2 / \sum_s \frac{x_k^2}{v(x_k)} + \sigma^2\sum_{\bar{s}} v(x_k) \tag{2.2.16}$$

It follws, therefore, that if

- $v(x_k)$ is a monotonically non-decreasing function of x

- $v(x)/x^2$ is a monotonically non-increasing function of x

the strategy $(p, \hat{T}^*)$ will have mnimum average variance in the class of all strategies $(p, \hat{T}), p \in \rho_n, \hat{T} \in \mathcal{L}_m$, the class of all linear m-unbiased predictors under ξ, where the sampling design p^* is such that

$$p^*(s) = 1(0) \text{ for } s = s^* \text{ (otherwise) }, \tag{2.2.17}$$

s^* having the property

$$\sum_{s^*} x_k = \max_{s \in \mathcal{S}_n} \sum_s x_k \tag{2.2.18}$$

$$\mathcal{S}_n = \{s : \nu(s) = n\} \tag{2.2.19}$$

Consider the particular case, $v(x) = x_k^g$. Writing $\hat{T}^*(0, 1; x^g)$ as $\hat{T}_g^*$, we have,

$$\begin{aligned}
\hat{T}_0^* &= \sum_s y_k + \{(\sum_s x_k y_k)(\sum_{\bar{s}} x_k)\}/\sum_s x_k^2 \\
\hat{T}_1^* &= \sum_s y_k + \{(\sum_s y_k)(\sum_{\bar{s}} x_k)\}/\sum_s x_k = \frac{\bar{y}_s}{\bar{x}_s} X \\
\hat{T}_2^* &= \sum_s y_k + \{\sum_s (y_k/x_k) \sum_{\bar{s}} x_k\}/\nu(s)
\end{aligned} \tag{2.2.20}$$

For each of these predictors p^* is the optimal sampling design in ρ_n. $\hat{T}_1^*$ is the ordinary ratio predictor $\hat{T}_R = (\bar{y}_s/\bar{x}_s)X$ and is optimal m-unbiased under $\xi(0, 1; x)$. This result is in agreement with the design-based result in Cochran (1977, p. 158 -160), where the ratio estimator is shown to be the BLU-estimator if the population is such that the relation between y and x is a straight line through the origin and the variance of y about this line is proportional to the (fixed) values of x. However, the new result is that p^* is the optimal sampling design to use $\hat{T}_R$ where as Cochran considered *srswor* only. It will be seen that if the assumed superpopulation model $\xi(0, 1; x)$ does not hold, $\hat{T}_R$ will be model-biased for any sample (including s^*) in general and a *srswor* design may be used to salvage the unbiasedness of the ratio predictor under a class of alternative models.

EXAMPLE 2.2.2

Consider now prediction under multiple regression models. Assume that apart from main variable y we have $(r+1)$ closely related auxiliary variables $x_j(j = 0, 1, \ldots, r)$ with known values $x_{kj} \; \forall \; k = 1 \ldots, N$. The variables $y_1, \ldots, y_N$ are assumed to have a joint distribution ξ such that

$$\mathcal{E}(y_k \mid x_k) = \beta_0 x_{k0} + \beta_1 x_{k1} + \ldots + \beta_r x_{kr}$$

$$\mathcal{V}(y_k \mid \mathbf{x}_k) = \sigma^2 v_k \tag{2.2.21}$$

$$\mathcal{C}(y_k, y_{k'} \mid x_k, x_{k'}) = 0 \quad (k \neq k')$$

where $x_k = (x_{k0}, x_{k1}, \ldots, x_{kr})', \beta_0, \beta_1, \ldots, \beta_r, \sigma^2(> 0)$ are unknown parameters, v_k a known function of x_k. If $x_{k0} = 1 \; \forall \; k$, the model has an intercept term β_0. Assuming without loss of generaliy that $s = (1, \ldots, n)$, we shall write

$$\mathbf{y} = (y_s \; y_{\bar{s}})', \; \beta = (\beta_0, \ldots, \beta_r)'$$

$$X = \begin{bmatrix} x_{10} & x_{11} & \cdots & x_{1r} \\ x_{20} & x_{21} & \cdots & x_{2r} \\ \cdot & \cdot & \cdots & \cdot \\ x_{N0} & x_{N1} & \cdots & x_{Nr} \end{bmatrix} = \begin{bmatrix} X_s \\ X_{\bar{s}} \end{bmatrix} \tag{2.2.22}$$

$$V = \begin{bmatrix} V_s & 0 \\ 0 & V_{\bar{s}} \end{bmatrix}$$

X_s being a $n \times (r + 1)$ submatrix of X corresponding to $k \in s, (X_{\bar{s}}$ defined similarly) $V_s(V_{\bar{s}})$ being a $n \times n((N - n) \times (N - n))$ submatrix of V corresponding to $k \in s(k \in \bar{s})$. The multiple regression model (2.2.22) is, therefore,

$$\mathcal{E}(\mathbf{y}) = X\beta, \quad \mathcal{D}(\mathbf{y}) = \sigma^2 V \tag{2.2.23}$$

$\mathcal{D}(.)$ denoting model-dispersion matrix of $(.)$. We shall denote

$$X_j = \sum_k x_{kj}, \; x_{js} = \sum_{k \in s} x_{kj}, \; \bar{x}_{js} = x_{js}/n,$$

$$x_s = (x_{0s}, \ldots, x_{rs})', \quad \bar{x}_s = (\bar{x}_{0s}, \ldots, \bar{x}_{rs})'$$

and $x_{j\bar{s}}, \bar{x}_{j\bar{s}}, x_{\bar{s}}, \bar{x}_{\bar{s}}$ similarly. The model (2.2.23) will be denoted as $\xi(X, v)$. For a given s, the BLUP of T is

$$\hat{T}_s^*(X, v) = \sum_s y_k + x_{\bar{s}}'\hat{\beta}_s^* \qquad (2.2.24)$$

where $\hat{\beta}_s^*$ is the generalised least square predictor of β,

$$\hat{\beta}_s^* = (X_s'V_s^{-1}X_s)^{-1}(X_s'V_s^{-1}y_s) \qquad (2.2.25)$$

Hence (Royall, 1971),

$$\hat{T}_s^*(X, v) = [1_n' + (1_{N-n}'X_{\bar{s}})(X_s'V_s^{-1}X_s)^{-1}X_s'V_s^{-1}]y_s \qquad (2.2.25)$$

where $1_q' = (1, \ldots, 1)_{q\times 1}'$. Also,

$$\begin{aligned} M(p, \hat{T}^*) &= E[\mathcal{V}(x_{\bar{s}}'\hat{\beta}^*) + \mathcal{V}(\sum_{\bar{s}} y_k)] \\ &= \sigma^2 E[\{x_{\bar{s}}'(X_s'V_s^{-1}X_s)^{-1}x_{\bar{s}}\} + \sum_{\bar{s}} v_k] \end{aligned} \qquad (2.2.26)$$

2.2.1 PROJECTION PREDICTORS

We considered in (2.2.6) the problem of predicting $\sum_{k\in\bar{s}} Y_k$ only, because the part $\sum_{k\in s} y_k$ of T is completely known when the data are given and found optimal strategies that minimise $M(p, \hat{T})$. However, in predicting the total of a fnite population of the same type as the current survey population, one's primary interest is in estimating the superpopulation total $\mu = \sum_k \mu_k$.

For a given s, a m-unbiased predictor of T will, therefore, be

$$\hat{T} = \sum \hat{y}_k \qquad (2.2.27)$$

where

$$\mathcal{E}(\hat{y}_k) = \mu_k, k = 1, \ldots, N \qquad (2.2.28)$$

The predictors (2.2.27) are called '*projection predictors*' (Cochran 1977, p.159; Sarndal, 1980 a).

Under $\xi(\delta_0, \ldots, \delta_J; v(x))$, BLU-projection predictor of T is

$$\hat{T}^*(\delta_0, \ldots, \delta_J; v(x)) = \sum_{k=1}^N \sum_0^J \delta_j \hat{\beta}_j^* x_k^j \qquad (2.2.29)$$

where $\hat{\beta}_j^*$ is as defined in (2.2.14). Under $\xi(X,v)$,

$$\widehat{T}^*(X,v) = x_0'\hat{\beta}_s^* = 1'X\hat{\beta}_s^* \qquad (2.2.30)$$

where $x_0' = (x_{00}, x_{01}, \ldots, x_{0r}), x_{0j} = \sum_k x_{kj}, \hat{\beta}_s^*$ is given in (2.2.25).

2.3 ROBUSTNESS OF MODEL-DEPENDENT OPTIMAL STRATEGIES

The model-dependent optimal predictor $\hat{T}(\xi)$ will, in general, be biased and not optimal under an alternative model ξ'. Suppose from practical considerations we assume that the model is $\xi(\delta_0, \ldots, \delta_J; v(x))$ and use the predictor $\hat{T}(\xi)$ which is BLUP under ξ. The bias of this predictor under a different model ξ' for a particular sample s is

$$\mathcal{E}_{\xi'}\{\hat{T}_s^*(\xi) - T\} = B\{\hat{T}_s^*(\xi), \xi'\} = B_{\delta_0', \ldots, \delta_J'; v'(x)}(\hat{T}_s^*(\delta_0, \ldots, \delta_J; v(x)) \quad (2.3.1)$$

ξ'-bias of $\hat{T}^*(\xi)$ for a particular sampling design p is

$$\sum_{s \in \mathcal{S}} \mathcal{E}_{\xi'}\{\hat{T}_s^*(\xi) - T\}p(s) \qquad (2.3.2)$$

To preserve the property of unbiasedness of $\hat{T}^*(\xi)$ even when the true model is ξ', we may choose the sampling design in such a way that $\hat{T}^*(\xi)$ remains also unbiased under ξ'. With this end in view, Royall and Herson (1973) introduced the concept of balanced sampling design.

Another way to deal with the situation may be as follows. Of all the predictors belonging to a subclass $\bar{\tau}(\xi)$ say, we may choose one $\hat{T}^0(\xi)$, which is least subject to bias even when the model is ξ'. Thus, for a given s, we should use $\hat{T}^0(\xi)$ such that

$$\mid B(\hat{T}_s^0(\xi), \xi') \mid < \mid B(\hat{T}_s(\xi), \xi') \mid$$

$\forall \, \hat{T}(\xi) \in \bar{\tau}(\xi)$, the choice of subclass $\bar{\tau}(\xi)$ being made from other considerations, e.g. from the point of view of mse, etc. The predictor $\hat{T}_s^0(\xi)$ will be the most bias-robust *wrt* the sample s and the alternative model ξ' among a class of competing predictors $\hat{T}(\xi) \in \bar{\tau}(\xi)$.

2.3.1   BIAS OF $\hat{T}_g^*$

We have

$$B\{\hat{T}_g^*, \xi'(\delta_0', \ldots, \delta_J'; v(x))\}$$

$$= \sum_{j=0}^{J} \delta_j' H_g(j, s)\beta_j$$

where,

$$H_g(j, s) = [\sum_s x_k^{j+1-g} \sum_{\bar{s}} x_k - \sum_s x_k^{2-g} \sum_{\bar{s}} x_k^g] / \sum_s x_k^{2-g} \qquad (2.3.3)$$

$$= I_g(j, s) / \sum_s x_k^{2-g} \qquad (2.3.4)$$

which is independent of the form of the variance function in ξ' (Mukhopadhyay, 1977). Note that $H_g(1, s) = 0$.

DEFINITION 2.3.1 (Royall and Herson, 1973). A sampling design $\bar{p}(L)$ is a balanced s.d. of order L (if it exists) if $\bar{p}(s) = 1(0)$ for $s = s_b(L)$ (otherwise), where $s_b(L)$, called a *balanced sample* of order L is such that

$$\bar{x}_{s_b(L)}^{(j)} = \bar{x}_{\bar{s}_b(L)}^{(j)}, j = 1, \ldots, L \qquad (2.3.5)$$

and

$$\bar{x}_s^{(h)} = \sum_s x_k^h / n, \ \bar{x}_{\bar{s}}^{(h)} = \sum_{\bar{s}} x_k^h / (N - n) \qquad (2.3.6)$$

If there are K such samples (2.3.5), $\bar{p}$ chooses each such sample with probability K^{-1}.

It follows from (2.3.3) and (2.3.4) that the ratio predictor $\hat{T}_1^* = \hat{T}^*(0, 1; 1)$ which is optimal under the model $\xi(0, 1; x)$ remains (m-)unbiased under alternative class of models $\xi(\delta_0, \ldots, \delta_J; v(x))$, when used along with a balanced sampling design $\bar{p}(J)$.

In general, consider the bias of $\hat{T}^*(0, 1; v(x))$ under $\xi(\delta_0, \ldots, \delta_J; V(x))$. This is given by

$$\sum_0^J \delta_j \beta_j [(\sum_{\bar{s}} x_k)(\frac{\sum_s x_k^{j+1}/v(x_k)}{\sum_s x_k^2/v(x_k)}) - \sum_{\bar{s}} x_k^j] \qquad (2.3.7)$$

Hence, if a sample $s^*(J)$ satisfies

$$(\sum_s x_k^j)/(\sum_{\bar{s}} x_k) = (\sum_s x_k^{j+1}/v(x_k))/(\sum_s x_k^2/v(x_k)),$$

$$j = 0, 1, \ldots, J \qquad (2.3.8)$$

the predictor $\hat{T}^*(0, 1; v(x))$ remains unbiased under $\xi(\delta_0, \ldots, \delta_J; V(x))$ on $s^*(J)$ for any $V(x)$.

Samples $s^*(J)$ satisfying (2.3.8) may be called *generalised balanced samples* of order J (Mukhopadhyay and Bhattacharyya, 1994; Mukhopadhyay, 1996).

For $v(x) = x^2$, (2.3.8) reduces to

$$n^{-1} \sum_s x_k^{j-1} = (\sum_{\bar{s}} x_k^j)/(\sum_{\bar{s}} x_k), \; j = 0, \ldots, J \qquad (2.3.9)$$

Samples satisfying these conditions have been termed *over-balanced samples* $s_0(J)$ (Scott et al, 1978).

The following theorem shows that $\hat{T}^*(0, 1; v(x))$ remains BLUP under $\xi(\delta_0, \ldots, \delta_J; V(x))$ for $s = s^*(J)$ when $V(x)$ is of a particular form.

THEOREM 2.3.1 (Scott, et al, 1978) For $s = s^*(J), \hat{T}^*(0, 1; v(x))$ is BLUP under $\xi(\delta_0, \ldots, \delta_J; V(x))$ provided

$$V(x) = v(x) \sum_0^J \delta_j a_j x^{j-1} \qquad (2.3.9)$$

where a_j's are arbitrary non-negative constants.

It is obvious that all types of balanced samples are rarely available in practice. Royall and Herson (1973), Royall and Pfeffermann (1982) recommended simple random samples as approximately balanced samples $s_b(J)$. Mukhopadhyay (1985 a) showed that simple random sampling and $pps\sqrt{x}$ sampling are asymptotically equivalent to balanced sampling designs $\bar{p}(J)$ for using the ratio predictor. Mukhopadhyay (1985 b) suggested a post-sample predictor which remains almost unbiased under alternative polynomial regression models. For further details on the robustness of the model-dependent optimal predictors the reader may refer to Mukhopadhyay (1977, 1996), among others.

2.4   A CLASS OF PREDICTORS UNDER MODEL $\xi(X, v)$

Sarndal (1980 b) considered different choices of $\hat{\beta}$ under $\xi(X, v)$. Considering $\hat{\beta}^*$ in (2.2.25) (dropping the suffix s), we note that a predictor $\hat{\beta}$ of β

is of the form

$$\hat{\beta} = (Z'_s X_s)^{-1} Z'_s Y_s \tag{2.4.1}$$

where Z'_s is a $n \times (r+1)$ matrix of weight z_{kj} to be suitably chosen such that a predictor $\hat{\beta}$ of β has some desirable properties. The matrix $Z'_s X_s$ is of full rank. Different choies of Z_s are :

(a)$\pi^{-1}weighted$: Z_s and the corresponding $\hat{\beta}$ may be called π^{-1} weighted if

$$\Pi_s^{-1} 1_n \in \mathcal{C}(Z_s) \tag{2.4.2}$$

ie. if

$$\Pi_s^{-1} 1_n = Z_s \lambda, \tag{2.4.3}$$

$\mathcal{C}(Z_s)$ denoting the column space of $Z_s, \lambda = (\lambda_0, \lambda_1, \ldots, \lambda_r)'$, a vector of constants λ_j. Here $\Pi = \text{Diag}(\pi_k; k = 1, \ldots, N), \Pi_s = \text{Diag}(\pi_k; k \in s)$. If $Z_s = \Pi_s^{-1} X_s$,

$$\hat{\beta} = (X'_s \Pi_s^{-1} X_s)^{-1} X'_s \Pi_s^{-1} y_s = \hat{\beta}(\Pi^{-1}) = \hat{\beta}_\Pi (\text{say}) \tag{2.4.4}$$

(b)$BLU - weighted$. Here $Z_s = V_s^{-1} X_s$ when

$$\hat{\beta} = \hat{\beta}^* = \hat{\beta}(V^{-1}) \text{ (say)} \tag{2.4.5}$$

(c)$weighted\ by\ an\ arbitrary\ matrix\ Q$ Here $Z_s = Q_s X_s$, where Q is an arbitrary $N \times N$ diagonal matrix of weights and Q_s is a submatrix of Q corresponding to units $k \in s$. Therefore,

$$\hat{\beta} = (X'_s Q_s X_s)^{-1} (X'_s Q_s y_s) = \hat{\beta}(Q) \text{ (say)} \tag{2.4.6}$$

For $\xi(X, v)$, Cassel, Sarndal and Wretman (1976), Sarndal (1980 b) suggested a generalised regression (*greg*) predictor of T,

$$\sum_s \frac{y_k}{\pi_k} + \sum_{j=0}^r (X_j - \sum_s \frac{x_{kj}}{\pi_k})\hat{\beta}_j^*$$

$$\begin{aligned}
&= 1'_n \Pi_s^{-1} y_s + (1'X - 1'_s \Pi_s^{-1} X_s)\hat{\beta}^* \\
&= \hat{T}_{GR}(V^{-1}) = \hat{T}_{GR}^* \text{ (say)}
\end{aligned} \tag{2.4.7}$$

where $\hat{\beta}_j^*$ is the BLUP of β_j obtained from the jth element of $\hat{\beta}^*$ given in (2.2.25). If β is known, (2.4.7) is a generalised difference predictor, studied by Cassel et al (1976), Isaki and Fuller (1982), among others.

For arbitrary weights Q in $\hat{\beta}(Q)$, generalised regression predictor of T is

$$
\begin{aligned}
\hat{T}_{GR}(Q) &= \sum_s \frac{y_k}{\pi_k} + \sum_0^r \hat{\beta}_j(Q)(X_j - \sum_s \frac{x_{kj}}{\pi_k} \\
&= 1'_n \Pi_s^{-1} Y_s + (1'X - 1'_s \Pi_s^{-1} X_s)\hat{\beta}(Q)
\end{aligned} \tag{2.4.8}
$$

Some specific choices of Q are $V^{-1}, \Pi^{-1}, \Pi^{-1}V^{-1}$.

EXAMPLE 2.4.1

Consider the model $\xi : \mathcal{E}(y_k) = \beta, \mathcal{V}(y_k) = \sigma^2, \mathcal{C}(y_k, y_{k'}) = 0, k \neq k' = 1, \ldots, N$. For predicting $\bar{y}$,

$$
\hat{\bar{y}}_{GR}(\Pi^{-1}) = (\sum_{k \in s} \frac{1}{\pi_k})^{-1}(\sum_{k \in s} \frac{y_k}{\pi_k}),
$$

$$
\hat{\bar{y}}_{GR}^* = \sum_{k \in s} \frac{y_k}{N\pi_k} + \bar{y}_s\{1 - \sum_{k \in s} \frac{1}{N\pi_k}\}.
$$

For designs with $\pi_k = n/N \; \forall \; k$, both the predictors coincide.

EXAMPLE 2.4.2

The model is $\xi : \mathcal{E}(y_k) = \beta_0 + \beta x_k, \mathcal{V}(y_k) = \sigma^2 v_k, \mathcal{C}(y_k, y_{k'}) = 0, k \neq k' = 1, \ldots, N$. Here,

$$
\hat{\bar{y}}(\Pi^{-1}) = (\sum_s \frac{1}{\pi_k})^{-1}(\sum_s \frac{y_k}{\pi_k}) + \hat{\beta}(\Pi^{-1})(X - (\sum_s x_k/\pi_k)/(\sum_s 1/\pi_k))
$$

where

$$
\hat{\beta}(\Pi^{-1}) = \{\sum_s \pi_k^{-1}(\sum_s (x_k y_k)\pi_k^{-1}) - (\sum_s y_k \pi_k^{-1})(\sum_s x_k \pi_k^{-1})\}
$$

$$
\{\pi_k^{-1}(\sum_s x_k^2 \pi_k^{-1}) - (\sum_s x_k \pi_k^{-1})^2\}^{-1}
$$

Also, the *greg*-predictor of $\bar{y}$ is,

$$
\hat{\bar{y}}^* = \sum_s \frac{y_k}{N\pi_k} + \hat{\beta}^*(\bar{X} - \sum_s \frac{x_k}{N\pi_k}) + \hat{\beta}_0^*(1 - \sum_s \frac{1}{N\pi_k})
$$

where $\hat{\beta}^*$'s are the generalised least squares predictors of β's.

Wright (1983) considered a (p, Q, R) strategy for predicting T as a combination of sampling design p and a predictor

$$\hat{T}(Q, R) = \sum_{k \in s} r_k \{ y_k - \sum_0^r x_{kj} \hat{\beta}_j(Q) \} + \sum_0^r X_j \hat{\beta}_j(Q) \qquad (2.4.9)$$

$$= 1'\Delta Ry + (I - \Delta R) X \hat{\beta}(Q)$$

where $\Delta = \text{Diag}\,(\delta_k; k = 1, \ldots, N), \delta_k = 1(0)$ if $k \in (\notin)s, R = \text{Diag}\,(r_k; k = 1, \ldots, N), R_s = \text{Diag}\,(r_k; k \in s), r_k$ being a suitable weight and I is a unit matrix of order N. For different choices of Q and R one gets different predictors. Some choices of Q are, as before, $V^{-1}, \Pi^{-1}, (V\Pi)^{-1}$ and of R are $0, I$, and Π^{-1}. The choice $R = 0$ gives projection predictors of the type (2.2.29); $R = \Pi^{-1}$ gives the class of generalised regression predictors of the type (2.4.8) considered by Cassel, et al (1976, 1977), Sarndal (1980 b).

EXAMPLE 2.4.3

Consider the model $\xi : \mathcal{E}(y_k) = \beta x_k, \mathcal{V}(y_k) = v_k, \mathcal{C}(y_k, y_{k'}) = 0, k \neq k' = 1, \ldots, N$. Here

$$\hat{T}(\Pi^{-1}V^{-1}, \Pi^{-1}) = \sum_s \frac{y_k}{\pi_k} + (\sum_s \frac{x_k y_k}{v_k \pi_k})$$

$$(\sum_s \frac{x_k^2}{v_k \pi_k})^{-1} (X - \sum_s \frac{x_k}{\pi_k})$$

2.5 ASYMPTOTIC UNBIASED ESTIMATION OF DESIGN-VARIANCE OF $\hat{T}_{GR}$

We shall now address the problem of asymptotic unbiased estimation of design -variance of $\hat{T}_{GR}(\Pi^{-1}V^{-1})$ under $\xi(X, v)$. Consider a more general problem of estimation of A linear functions $\mathbf{F} = (F_1, F_2, \ldots, F_A)' = C'\mathbf{y}$ where $F_a = C_a'\mathbf{y}, C_a = (C_{1a}, \ldots, C_{Nq})', C$ a $N \times Q$ matrix $((C_{ka})), C_{ka}$ being known constants. Consider the following estimates of F_a :

$$\hat{T}_a = C_{as}' \Pi_s^{-1}(y_s - \hat{y}_s) + C_a'\hat{y} \qquad (2.5.1)$$

where C_{as}' is the row vector $(C_{ka}, k \in s)$ and $\hat{y}_k = x_k'\hat{\beta}_s$ with

$$\hat{\beta}_s = \hat{\beta}(V^{-1}\Pi^{-1}) = (X_s'V_s^{-1}\Pi_s^{-1}X_s)^{-1}X_s'V_s^{-1}\Pi_s^{-1}y_s \qquad (2.5.2)$$

The estimator (2.5.1) is an extension of generalised regression estimator $\hat{T}_{GR}(V^{-1}\Pi^{-1})$ of T. Let $\hat{T} = (\hat{T}_1, \ldots, \hat{T}_A)'$. Then

$$\hat{T} = C_s'\Pi_s^{-1}(y_s - \hat{y}_s) + C'\hat{y} \tag{2.5.3}$$

where C_s is the part of C corresponding to $k \in s$. Now

$$\hat{T} = G_s'\Pi_s^{-1}y_s \tag{2.5.4}$$

where

$$\begin{aligned}
G_s' &= C_s' - M_s'H_s^{-1}X_s'V_s^{-1} \\
M_s' &= C_s'\Pi_s^{-1}X_s - C'X \\
H_s &= X_s'V_s^{-1}\Pi_s^{-1}X_s
\end{aligned} \tag{2.5.5}$$

Thus $\hat{T}_a = \sum_{k \in s} g_{ska}y_k/\pi_k$, g_{ska} being the (k, a) th component of G_s. The following two methods have been suggested by Sarndal (1982) for estimating the design-dispersion matrix $D(\hat{T}) = (cov_p(\hat{T}_a, \hat{T}_b))$

(a)TAYLOR EXPANSION METHOD :

An estimate of $Cov_p(\hat{T}_a, \hat{T}_b)$ is approximately the Yates-Grundy estimator of covariance,

$$\sum_{k<l\in s}\sum(\pi_{kl}/\pi_k\pi_l - 1)(z_{ka}/\pi_k - z_{la}/\pi_l)(z_{kb}/\pi_k - z_{lb}/\pi_l) \tag{2.5.6}$$

$$= v_T(a, b) \text{ (say)}$$

where

$$z_{ka} = C_{ka}e_k, \quad e_k = Y_k - x_k'\hat{\beta}_s \tag{2.5.7}$$

Writing

$$YG_s(d_{ka}, d_{kb}) = \sum_{k<l\in s}(\pi_{kl}/\pi_k\pi_l - 1)(d_{ka}/\pi_k - d_{la}/\pi_l)(d_{kb}/\pi_k - d_{lb}/\pi_l) \tag{2.5.8}$$

as the YG-transformation of $(d_{ka}, d_{kb}), k \in s$, we have

$$v_T(a, b) = YG_s(z_{ka}, z_{kb}) \tag{2.5.9}$$

(b) MODEL METHOD :

Here an approximate estimate of $Cov_p(\hat{T}_a, \hat{T}_b)$ is

$$v_M(a, b) = YG_s(z_{ka}^*, z_{kb}^*) \tag{2.5.10}$$

where

$$z_{ka}^* = g_{ska}e_k \qquad (2.5.11)$$

Both $((v_T))$ and $((v_M))$ are approximately p-unbiased for $D(\hat{T})$ where $\hat{T} = (\hat{T}_a, \ldots, \hat{T}_A)'$, whether the assumed model $\xi(X, v)$ is true or false. The model is used here only to generate the form of the estimator $\hat{T}$ and hence $((v_T))$ and $((v_M))$.

EXAMPLE 2.5.1

Consider the model $\xi(0, 1; x)$. Here

$$\hat{\beta}_s = \tilde{y}_s/\tilde{x}_s, \tilde{y}_s = (\sum_s y_k/\pi_k)/(\sum_s 1/\pi_k); \hat{\tilde{y}} = \bar{X}\hat{\beta}_s$$

(Ha'jek's predictor). Two estimators of $V_p(\hat{\tilde{y}})$ are

(i)

$$v_T(\hat{\tilde{y}}) = YG_s(e_k^2/N^2)$$

(ii)

$$v_M(\hat{\tilde{y}}) = YG_s\{(\tilde{X}e_k)^2/(\tilde{x}_s\tilde{\pi}_s)^2\},$$

where

$$e_k = y_k - \hat{\beta}_s x_k, \ \tilde{\pi}_s = \sum_s 1/\pi_k.$$

Under srs, $v_M = \hat{\hat{v}}$ (say) $= (\bar{X}/\bar{x})^2 v_T$, where, $v_T = \hat{v}$ (say) $= \{(1 - f)/n(n-1)\} \sum_s(y_k - \hat{\beta}_s x_k)^2$, the usual estimator of mean square estimator of ratio estimator of population mean. Royall and Cumberland (1978) derived robust variance estimator $v_h = h_s\hat{v}$ where,

$$h_s = \{(\bar{x}_s - \bar{X})/\bar{x}_s\}^2/1 - \sum_s(x_k - \bar{x}_s)^2/n(n-1)\bar{x}_s^2\}$$

If n is large and $N >> n, h_s \simeq (\bar{x}/\bar{x}_s)^2$ and $v_h \simeq \hat{\hat{v}}$.

For further details on inferential problems in survey sampling the reader may refer to Sarndal et al (1992), Mukhopadhyay (1996), among others.

Chapter 3

Bayes and Empirical Bayes Prediction of a Finite Population Total

3.1 INTRODUCTION

We shall consider in this chapter Bayes and Empirical Bayes (EB) prediction of a finite population total. Following Bolfarine (1989) we develop in section 3.2 theory for Bayes and minimax prediction of finite population parameters under a general set up. Section 3.3 considers Bayes and minimax prediction of a finite population total for the normal regression models under squared error loss function. The next section considers Bayes prediction under the Linex loss function of Varian (1975) and Zellner (1986). Section 3.5 reviews James-Stein estimation and associted procedures. Subsequently, we consider empirical Bayes (EB) prediction under simple location model. In this model no covariate is used. Ghosh and Meeden (1986) considered EB-estimation of a finite population total under simple location model by using information from past surveys under a normal Bayesian set up. Mukhopadhyay (1998 c) developed an alternative EB-estimator in the balanced case. Ghosh and Lahiri (1987 a,b), Tiwari and Lahiri (1989) considered the simultaneous EB-estimation of strata population means and variances replacing the normality assumption by the assumption of posterior linearity. These results have been reviewed in Section 3.6. The next section considers EB-prediction of T under models using covariates. Lahiri and Peddada (1992) considered the normal theory Bayesian analysis under multiple linear regression models and obtained EB-Ridge estimators. Ghosh et al (1998) used generalised linear models for simultaneous estimation of population means for several strata. These works have been reviewed in this section. In the following section we review some Bayesian prediction procedures in small area problems. This section also considers modification of James-Stein estimators due to Fay and Herriot (1979) and its applications in small area problems. Results obtained from a survey conducted by the Indian Statistical Institute, Calcutta (1999) are discussed in this context. Finally, we consider Bayes prediction of several (infinite) population means and variances under random error variance model.

3.2 BAYES AND MINIMAX PREDICTION OF FINITE POPULATION PARAMETERS

As in Chapter 2, we shall assume that the value y_i on unit i is a realisation of a random variable $Y_i (i = 1, \ldots, N)$. Therefore, the value $\mathbf{y} = (y_1, \ldots, y_N)'$ of the survey population will be looked upon as a particular realisation of a random vector $\mathbf{Y} = (Y_1, \ldots, Y_N)'$ having a superpopulation distribution $\xi_\theta \in \Theta$ (the parameter space). A sample s of size n is selected from $\mathcal{P}$ according to some sampling design p. Let $\bar{s} = r = \mathcal{P} - s$. After the sample has been selected, we may re-order the elements of $\mathbf{y}$ so that we may write $\mathbf{y} = (y_s', y_r')'$ where y_s denotes the set of observations in the sample. Clearly, y_s is a realised value of the random vector Y_s. As before, we shall use the symbol y_i to denote the population value as well as the random variable Y_i of which it is a realisation, the actual meaning being clear from the context. Let $g(\mathbf{y}) = g(y)$ be the finite population quantity to be predicted. Examples of $g(y)$ are population total, $T = \sum_{i=1}^N y_i$, population variance, $S_y^2 = \sum_{i=1}^N (y_i - \bar{y})^2/(N-1)$. Let $\hat{g}(y_s)$ be a predictor of $g(y)$. The loss involed in predicting $g(y)$ by $\hat{g}(y_s)$ is $L(\hat{g}(y_s), g(y))$ where $L(.,.)$ is a suitable loss function. For the given data y_s and for a given value of the parameter vector θ average loss or risk associated with an estimator $\hat{g}(y_s)$ is

$$r(\hat{g}(y_s) \mid \theta) = \int L(\hat{g}(y_s), g(y)) dF(y_r \mid y_s, \theta) \qquad (3.2.1)$$

where $dF(y_r \mid y_s, \theta)$ denotes the conditional distribution of y_r given y_s, θ and the integral is taken over all possible values of y_r.

Assume that θ has prior distribution Λ over the space Θ. The average or expected risk involved with the estimator $\hat{g}(y_s)$ is

$$r_\Lambda(\hat{g}(y_s)) = \int_\Theta r(\hat{g}(y_s \mid \theta) \Lambda(\theta) d\theta$$

$$= \int E[L(\hat{g}(y_s), g(y)) \mid y_s] dP_{Y_s}(y_s) \qquad (3.2.2)$$

where $P_{Y_s}(.)$ denotes the predictive distribution of y_s.

When there is no reason for confusion, we shall, henceforth, use the symbols E, V to denote expectation, variance with respect to superpopulation model, prior distribution, posterior distribution, etc. When the sampling design is also involved, we shall use E_p, V_p or E, V to denote expectation, variance with respect to p and often E, V or $\mathcal{E}, \mathcal{V}$, to denote the same with respect to model distribution, the actual meaning being clear from the context.

A Bayes predictor of $g(y)$ in the class $\mathcal{G}$ of predictors with respect to prior Λ of θ is given by $\hat{g}_\Lambda(y_s)$ where

$$r_\Lambda(\hat{g}_\Lambda(y_s)) = \min{}_{\hat{g} \in \mathcal{G}} r_\Lambda(\hat{g}(y_s)) \qquad (3.2.3)$$

The Bayes predictor $\hat{g}_\Lambda(y_s)$ has the minimum average risk among a class of predictors based on the prior Λ.

In the absence of knowledge about the prior distribution of θ, a measure of performance of $\hat{g}$ is

$$\rho(\hat{g}) = \sup{}_{\theta \in \Theta} r(\hat{g}(y_s) \mid \theta) < \infty \qquad (3.2.4)$$

A predictor $\hat{g}_M(y_s)$ is said to be minimax in the class $\mathcal{G}'$ of predictors if

$$\rho(\hat{g}_M(y_s)) = \inf{}_{\hat{g} \in \mathcal{G}'} \rho(\hat{g}(y_s)) \qquad (3.2.5)$$

and if

$$\inf{}_{\hat{g} \in \mathcal{G}'} \sup{}_{\theta \in \Theta} r(\hat{g}(y_s), g(y)) = \sup{}_{\theta \in \Theta} \inf{}_{\hat{g} \in \mathcal{G}'} r(\hat{g}(y_s), g(y))$$

Considering squared error loss functions, $L[\hat{g}(y_s), g(y)] = [\hat{g}(y_s) - g(y)]^2$ and prior density $\Lambda(\theta)$ it follows that Bayes predictor of $g(y)$ is given by

$$\begin{aligned} \hat{g}_\Lambda(y_s) &= \int g(y)[\int_\Theta f(y_r \mid y_s, \theta) d\theta] dy_r \\ &= \int g(y) f^*(y_r \mid y_s) dy_r \\ &= E_\Lambda[g(y) \mid y_s)] \end{aligned} \qquad (3.2.6)$$

where $f^*(y_r \mid y_s)$ is the conditional marginal (predictive) density of y_r given y_s and $E_\Lambda(. \mid y_s)$ denotes conditional expectation of $(.)$ given y_s with respect to predictive distribution of $(.)$, when Λ is the prior density of θ. Throughout this section we shall consider the square error loss function only.

LEMMA 3.2.1 If $\hat{g}_\Lambda(y_s)$ is the Bayes predictor of $g(y)$ under squared error loss function with respect to prior Λ, then the Bayes risk of $\hat{g}_\Lambda(y_s)$ is given by

$$r_\Lambda = \int Var[g(y) \mid y_s]dP_{Y_s}(y_s).$$

In particular, if $Var[g(y) \mid y_s]$ is independent of y_s, then

$$r_\Lambda = Var[g(y) \mid y_s]$$

Proof. We have

$$\hat{g}_\Lambda(y_s) - g(y) = E[g(y) \mid y_s] - g(y)$$

and hence

$$E\{[g(y) - \hat{g}_\Lambda(y_s)]^2 \mid y_s\} = Var[g(y) \mid y_s]$$

when the result follows by (3.2.2).

A prior distribution Λ for θ is said to be least favourable if $r_\Lambda \geq r_{\Lambda'}$ for any other distribution Λ'.

The following theorems on minimax prediction in finite population follow from the corresponding theorems in minimax estimation theory (e.g., Zacks, 1971, Lehmann, 1983).

THEOREM 3.2.1 If $\hat{g}_\Lambda(y_s)$ is a Bayes predictor and if $r(\hat{g}_\Lambda(y_s) \mid \theta)$ is independent of θ, then $\hat{g}_\Lambda(y_s)$ is a minimax predictor and Λ is a least favourable distribution.

Theorem 3.2.1 is inapplicable when a least favorable distribution does not exist. In this case one may consider the limiting Bayes risk method for finding a minimax predictor.

Let $\{\Lambda_j\}$ be a sequence of prior distributions for $\theta; \hat{g}_{\Lambda_j}(y_s)$, the Bayes predictor of $g(y)$ corresponding to Λ_j and

$$\begin{aligned} r_j &= \int E_\theta[\hat{g}_{\Lambda_j}(y_s) - g(y)]^2 d\Lambda_j(\theta) \\ &= \int E\{[\hat{g}_{\Lambda_j}(y_s) - g(y)]^2 \mid y_s\}dP_{Y_s}(y_s) \end{aligned}$$

the corresponding Bayes risk.

THEOREM 3.2.2 Let $\{\Lambda_j, j = 1, 2, \ldots\}$ be a sequence of prior densities over Θ and let $\{\hat{g}_{\Lambda_j}(y_s), j = 1, 2, \ldots\}$ and $\{r_{\Lambda_j}(\hat{g}_{\Lambda_j}(y_s)), j = 1, 2, \ldots, \}$ be the corresponding Bayes predictor and Bayes risk. Suppose $r_j \to r$ as $j \to \infty$. If $\hat{g}(y_s)$ be a predictor such that

$$\sup {}_\theta E_\theta[\hat{g}(y_s) - g(y)]^2 \leq r$$

then $\hat{g}(y_s)$ is a minimax predictor and the sequence $\{\Lambda_j\}$ is least favorable.

THEOREM 3.2.3 Let **y** be jointly distributed according to a distribution F, belonging to a family of distributions $\mathcal{F}_1$. Suppose that $\hat{g}_M$ is a minimax predictor of g when $F \in \mathcal{F}_0$ ($\in \mathcal{F}_1$). If

$$\sup{}_{F \in \mathcal{F}_0} r(\hat{g}_M, F) = \sup{}_{F \in \mathcal{F}_1} r(\hat{g}_M, F)$$

where $r(\hat{g}_M, F) = E_F[L(\hat{g}_M, g(y))]$, expectation being taken with respect to joint predictive distribution of $y \in F$, then $\hat{g}_M$ is minimax for $\mathcal{F}_1$.

EXAMPLE 3.2.1

Suppose that y_k's are independent with $P_\theta(y_k = 1) = \theta = 1 - P_\theta(y_k = 0)$. We are interested in estimating the population total $T = \sum_{k=1}^{N} y_k$. Consider prior distribution of θ as Beta $B(a, b)$. The posterior distribution of θ given y_s is then $B(a + n\bar{y}_s, a + b + n)$ where $\bar{y}_s = \sum_{k \in s} y_k / n$. The Bayes predictor of T is

$$\hat{T}_B = E(T \mid y_s) = n\bar{y}_s + \sum_r E(y_i \mid y_s)$$

$$= n\bar{y}_s + \sum_r E[E(y_i \mid y_s, \theta) \mid y_s]$$

$$= n\bar{y}_s + \sum_r E(\theta \mid y_s)$$

$$= n\bar{y}_s + (N - n)\frac{a + n\bar{y}_s}{n + a + b}$$

To find a minimax predictor using Theorem 3.2.1, consider the risk function $r(\hat{T}_B \mid \theta)$ taking a squared error loss function.

$$r(\hat{T}_B \mid \theta) = E_\theta(\hat{T}_B - T)^2$$

$$= Var_\theta(\hat{T}_B - T) + \{E_\theta(\hat{T}_B - T)\}^2$$

$$= (N - n)^2 \{ \frac{a^2}{(n + a + b)^2} + \theta[\frac{n}{(n + a + b)^2} + \frac{1}{N - n} - \frac{2a(a + b)}{(n + a + b)^2}]$$

$$- \theta^2[\frac{n}{(a + b + n)^2} + \frac{1}{N - n} - \frac{(a + b)^2}{(n + a + b)^2}]\}$$

Hence, $\hat{T}_B$ is a Bayes predictor with constant risk iff

$$n(N - n) + (n + a + b)^2 - 2a(a + b)(N - n) = 0 \qquad (i)$$

and

$$-n(N - a) - (n + a + b)^2 + (a + b) + (a + b)(N - n) = 0 \qquad (ii)$$

These give

$$a = \frac{\sqrt{n}\{\sqrt{n} + \sqrt{n + N(N - n - 1)}\}}{2(N - n - 1)} = \frac{a + b}{2}$$

Hence, the minimax predictor of T is

$$\hat{T}_M = n\bar{y}_s + \frac{\{\sqrt{n}\bar{y}_s + \frac{\sqrt{n} + \sqrt{n + N(N - n - 1)}}{2(N - n - 1)}\}}{\frac{\sqrt{n}\{(N - n) + \sqrt{n + N(N - n - 1)}\}}{N - n - 1}}$$

with risk function

$$r(\hat{T}_M \mid \theta) = \frac{(N - n)^2\{\sqrt{n} + \sqrt{n + N(N - n - 1)}\}}{4\{\sqrt{n}(N - n) + \sqrt{n + N(N - n - 1)}\}}$$

Bolfarine (1987) compared the performance of the expansion estimator $\hat{T}_E = N\bar{y}_s$ with $\hat{T}_M$ and found out range of values of θ where $\hat{T}_M$ is better than $\hat{T}_E$.

EXAMPLE 3.2.2

Consider the model

$$y_i = x_i\beta + e_i, \quad e_i \sim N(0, \sigma^2 x_i), (\sigma^2 \text{ known}) \qquad (i)$$

and a prior $\beta \sim N(\nu, R)$. It follows that the posterior distribution of β is

$$N(\nu^*, \frac{\sigma^2 R}{n\bar{x}_s R + \sigma^2}) \qquad (ii)$$

where

$$\nu^* = \frac{nR\bar{y}_s + \nu\sigma^2}{n\bar{x}_s R + \sigma^2}$$

Hence, under squared error loss, Bayes predictor of T is

$$\hat{T}_\Lambda = n\bar{y}_s + (N - n)\bar{x}_{\bar{s}}\nu^*$$

with Bayes risk

$$r(\hat{T}_\Lambda) = \sigma^2 \sum_r x_i[\frac{RX + \sigma^2}{n\bar{x}_s R + \sigma^2}]$$

where $X = \sum_{i=1}^{N} x_i$. Note that as $R \to \infty, r(\hat{T}_\Lambda) \to \sigma^2(\sum_r x_i / \sum_s x_i)X = E(\hat{T}_R - T)^2$, where $\hat{T}_R$ is the ratio predictor, $(\bar{y}_s / \bar{x}_s)X$ and the expectation is taken with respect to model (i). Therefore, by theorem 3.2.2, $\hat{T}_R$ is a minimax predictor of T under (i). Also, an optimal sampling design to base $\hat{T}_R$ is a purposive sampling design p^* which selects sample s^* with probability one, where s^* is such that

$$\sum_{s^*} x_i = \max_{s \in \mathcal{S}_n} \sum_s x_i,$$

$$\mathcal{S}_n = \{s : n(s) = n\},$$

$n(s)$ denoting the effective size of the sample s.

It follows, therefore, that under simple location model ($x_i = 1 \ \forall \ i$), mean per unit estimator $N\bar{y}_s$ is minimax for T.

In this section we have developed Bayes and minimax prediction of finite population parameters under a general set up. In the next section we shall consider Bayes and minimax prediction of T under regression model with normality assumptions.

3.3 BAYES PREDICTION OF A FINITE POPULATION TOTAL UNDER NORMAL REGRESSION MODEL

Consider the model

$$\mathbf{y} = X\beta + e$$

$$e \sim N(0, V) \tag{3.3.1}$$

denoted as $\psi(\beta, V)$, where $X = ((x_{kj}, k = 1, \ldots, N; j = 1, \ldots, p))$, x_{kj} is the value of the auxiliary variable x_j on unit k, $e = (e_1, \ldots, e_N)'$, $\beta = (\beta_1, \ldots, \beta_p)'$, a $p \times 1$ vector of unknown regression coefficients and V is a $N \times N$ positive definite matrix of known elements. It is further assumed that

$$\beta \sim N(\nu, R) \tag{3.3.2}$$

The model $\psi(\beta, V)$ together with the prior (3.3.2) of β will be denoted as ψ_R. After the sample s is selected we have the partitions of $\mathbf{y}, X$ and V as follows:

$$\mathbf{y} = \begin{bmatrix} y_s \\ y_r \end{bmatrix}, \quad X = \begin{bmatrix} X_s \\ X_r \end{bmatrix}, \quad V = \begin{bmatrix} V_s & V_{sr} \\ V_{rs} & V_r \end{bmatrix} \tag{3.3.3}$$

We have the following theorems.

THEOREM 3.3.1 Under the Bayesian model ψ_R, the Bayes predictive distribution of y_r given y_s is multivariate normal with mean vector

$$\begin{aligned} E_{\psi_R}[y_r \mid y_s] &= \eta_r(y_s) \text{ (say)} \\ &= X_r\hat{\beta}_R + V_{rs}V_s^{-1}(y_s - X_s\hat{\beta}_R) \end{aligned} \tag{3.3.4}$$

and covariance matrix

$$Var_{\psi_R}[y_r \mid y_s] = \Sigma_r \text{ (say)}$$

$$= V_r - V_{rs}V_s^{-1}V_{sr} + (X_r - V_{rs}V_s^{-1}X_s)$$
$$(X_s'V_s^{-1}X_s + R^{-1})^{-1}(X_r - V_{rs}V_s^{-1}X_s)' \tag{3.3.5}$$

where

$$\hat{\beta}_R = (X_s'V_s^{-1}X_s + R^{-1})^{-1}(X_s'V_s^{-1}y_s + R^{-1}\nu) \tag{3.3.6}$$

Proof We have

$$\begin{bmatrix} y_s \\ y_r \end{bmatrix} \sim N\left[\begin{bmatrix} X_s \\ X_r \end{bmatrix}\beta, \begin{bmatrix} V_s & V_{sr} \\ V_{rs} & V_r \end{bmatrix}\right] \tag{3.3.7}$$

Hence, conditional distribution of y_r given y_s is normal with

$$\begin{aligned} E(y_r \mid y_s) &= E_{\beta|y_s}(y_r \mid y_s, \beta) \\ &= E_{\beta|y_s}[X_r\beta + V_{rs}V_s^{-1}(y_s - X_s\beta)] \\ &= E_{\beta|y_s}[(X_r - V_{rs}V_s^{-1}X_s)\beta + V_{rs}V_s^{-1}y_s] \end{aligned} \tag{3.3.8}$$

To find the conditional expectation $E(\beta \mid y_s)$ consider the joint distribution of β and y_s. It follows that

$$\begin{bmatrix} \beta \\ y_s \end{bmatrix} \sim N\left[\begin{bmatrix} \nu \\ X_s\beta \end{bmatrix}, \begin{bmatrix} R & RX_s' \\ X_sR & X_sRX_s' + V_s \end{bmatrix}\right] \tag{3.3.9}$$

Hence,

$$E(\beta \mid y_s) = \nu + RX_s'(X_sRX_s' + V_s)^{-1}(y_s - X_s\nu) \tag{3.3.10}$$

Substituting this in (3.3.8) and on simplification (3.3.4) follows. Again,

$$\begin{aligned} V(y_r \mid y_s) &= E_{\beta|y_s}[V(y_r \mid y_s, \beta)] + V_{\beta|y_s}[E(y_r \mid y_s, \beta)] \\ &= V_r - V_{rs}V_s^{-1}V_{sr} + (X_r - V_{rs}V_s^{-1}X_s) \end{aligned}$$

$$(R - RX_s'(X_sRX_s' + V_s)^{-1}X_sR)(X_r - V_{rs}V_s^{-1}X_s)' \tag{3.3.11}$$

The result (3.3.5) follows on simplification (using Binomial Inversion theorem).

The theorem 3.3.1 was considered by Bolfarine, Pereira and Rodrigues (1987), Lui and Cumberland (1989), Bolfarine and Zacks (1991), among others. Royall and Pfeffermann (1982) considered the special case of a non-informative prior which is obtained as the limit of $N(\nu, R)$ when $R \to \infty$.

THEOREM 3.3.2 Consider the model ψ_R with $V = \sigma^2 W$ where W is a known diagonal matrix with $W_{rs} = 0$, but σ^2 unknown. Consider non-informative prior distribution for (β, σ^2) according to which the prior density is

$$\xi(\beta, \sigma^2) \propto \frac{1}{\sigma^2} \qquad (3.3.12)$$

In this case, the posterior distribution of y_r given y_s is normal with

$$E_{\psi_R}[y_r \mid y_s] = X_r \hat{\beta}_s \qquad (3.3.13)$$

and

$$Var_{\psi_R}[y_r \mid y_s] = \frac{\phi}{\phi - 2} \hat{\sigma}_s^2 [W_r + X_r (X_s' W_s^{-1} X_s)^{-1} X_r'] \qquad (3.3.14)$$

where

$$\begin{aligned} \hat{\beta}_s &= (X_s' W_s^{-1} X_s)^{-1} W_s^{-1} y_s \\ \hat{\sigma}_s^2 &= (y_s - X_s \hat{\beta}_s)' W_s^{-1} (y_s - X_s \hat{\beta}_s)/\phi \end{aligned} \qquad (3.3.15)$$

with $\phi = n - p$.

Proof Replacing R^{-1} and V_{rs} by 0 in (3.3.4) we get (3.3.13), since $E[y_r \mid y_s]$ is independent of σ. Again,

$$Var_{\psi_R}[y_r \mid y_s, \sigma] = \sigma^2 [W_r + X_r (X_s' W_s^{-1} X_s)^{-1} X_r']$$

Hence,

$$Var_{\psi_R}[y_r \mid y_s] = E_{\psi_R}[\sigma^2 \mid y_s][W_r + X_r (X_s' W_s^{-1} X_s)^{-1} X_r'].$$

The result (3.3.14) follows observing that

$$E_{\psi_R}[\sigma^2 \mid y_s] = \frac{\phi}{\phi - 2} \hat{\sigma}_s^2$$

We now consider prediction of linear quantities $g_L(y) = l'y$ where $l = (l_s', l_r')'$ is a vector of constants.

THEOREM 3.3.3 For any linear quantity $g_L = l'y$, the Bayes predictor under the squared error loss and any ψ_R model for which $Var_{\psi_R}[y_r \mid y_s]$ exists, is

$$\hat{g}_{BL} = l_s' y_s + l_r' E_{\psi_R}[y_r \mid y_s] \qquad (3.3.16)$$

The Bayes risk of this predictor is

$$E_{\psi_R}[(\hat{g}_{BL} - g_L)^2 \mid y_s] = l_r' Var_{\psi_R}[y_r \mid y_s] l_r \qquad (3.3.17)$$

Proof Follows easily from the definition of Bayes predictor and lemma 3.2.1.

COROLLARY 3.3.1 The Bayes predictor of population total $T(\mathbf{y})$ under the normal regression model (3.3.1) - (3.3.2) is

$$\hat{T}_B(y_s) = 1_s' y_s + 1_r'[X_r \hat{\beta}_R + V_{rs} V_s^{-1}(y_s - X_s \hat{\beta}_R)] \qquad (3.3.18)$$

The Bayes prediction risk of $\hat{T}_B$ is

$$E_{\psi_R}[(\hat{T}_B(y_s) - T(y)]^2 = 1_r'(V_r - V_{rs} V_s^{-1} V_{sr}) 1_r$$
$$+ 1_r'(X_r - V_{rs} V_s^{-1} X_s)(X_s' V_s^{-1} X_s + R^{-1})^{-1} \qquad (3.3.19)$$
$$(X_r - V_{rs} V_s^{-1} X_s)' 1_r$$

THEOREM 3.3.4 Consider the normal superpopulation model ψ_R, with $V_{rs} = 0$. The minimax predictor of T with repect to the squared error loss is

$$\hat{T}_M = 1_s' y_s + 1_r' X_r \hat{\beta}_s \qquad (3.3.20)$$

with prediction risk

$$E_\psi[\hat{T}_M - T]^2 = 1_r' V_r 1_r + 1_r' X_r (X_s' V_s^{-1} X_s)^{-1} X_r' 1_r \qquad (3.3.21)$$

Proof Consider a sequence of prior distributions $N(\nu, R_k)$ for β such that $||R_k|| = k$, when the norm of the covariance matrix $||R|| = $ trace R. The corresponding Bayes predictor converges (vide (3.3.6) and (3.3.16)), as $k \to \infty$, to the best linear unbiased predictor (BLUP) of Royall (1976)

$$\hat{T}_{BLUP} = 1_s' y_s + 1_r' X_r \hat{\beta}_s$$

Moreover, the Bayes prediction risk $r(\hat{T}_{R_k}; \nu, R_k)$ converges, as $k \to \infty$, to the prediction risk of $\hat{T}_{BLUP}$, namely,

$$E_\psi[\hat{T}_{BLUP} - T]^2 = 1_r' V_r 1_r + 1_r' X_r (X_s' V_s^{-1} X_s)^{-1} X_r' 1_r \qquad (3.3.22)$$

Since this prediction risk is independent of β, $\hat{T}_{BLUP}$ is, by theorem 3.2.2 a minimax predictor of T.

Note that when the values y_i are considered as fixed quantities (those belonging to s as observed and belonging to r, unobserved but fixed), a statistic $\hat{\theta}(\mathbf{y})$ is considered as an estimator for $\theta(\mathbf{y})$ (a constant), while if the y_i's are considered as random variables, the same is considered as a predictor for the $\theta(\mathbf{y})$ which itself is now a random variable.

So far we have considered squared error loss function only. The next section consideres Bayes prediction of T under a asymmetric loss function.

3.4 BAYES PREDICTION UNDER AN ASYMMETRIC LOSS FUNCTION

In some estimation problems, use of symmetric loss functions may be inappropriate. For example, in dam construction an underestimate of the peak water level is usually much more serious than an overestimate, - see, for example, Aitchison and Dunsmore (1975), Varian (1975), Berger (1980). Let $\Delta = \hat{\phi} - \phi$ denote the scalar estimation error in using $\hat{\phi}$ to estimate ϕ. Varian (1975) introduced the following loss function

$$L(\Delta) = b[e^{a\Delta} - a\Delta - 1], \quad a \neq 0, b > 0 \tag{3.4.1}$$

where a and b are two parameters. For $a = 1$, the function is quite asymmetric with overestimation being more costly than underestimation. When $a < 0, L(\Delta)$ rises almost exponentially when $\Delta < 0$ and almost linearly when $\Delta > 0$. For small values of $\mid a \mid$, the function is almost symmetric and not far from a squared error loss function.

Let $p(\phi \mid D)$ be the posterior pdf for $\phi, \phi \in \Phi$, the parameter space, where D denotes the sample and prior information. Let E_ϕ denote the posterior expectation.

$$E_\phi L(\Delta) = b[e^{a\phi} E_\phi e^{-a\phi} - a(\hat{\phi} - E_\phi \phi) - 1] \tag{3.4.2}$$

The value of $\hat{\phi}$ that minimises (3.4.2), denoted by $\hat{\phi}_B$, is

$$\hat{\phi}_B = -(1/a) \log (E_\phi e^{-a\phi}), \tag{3.4.3}$$

provided, of course, $E_\phi e^{-a\phi}$ exists and is finite. The risk function of $\hat{\phi}_B$, $R(\hat{\phi}_B)$ is defined as the prior expectation of $L(\Delta_B)$ and the Bayes risk as the posterior expectation of $L(\Delta_B)$ where $\Delta_B = \hat{\phi}_B - \phi$. Clearly, $R(\hat{\phi}_B)$ will depend on the parameter ϕ involved in the model.

When ϕ has a normal posterior pdf with mean m and variance v,

$$\hat{\phi}_B = m - av/2 \tag{3.4.4}$$

is the Bayesian estimate relative to the Linex loss function (3.4.1). When $|a|v/2$ is small, $\hat{\phi}_B \simeq m$, the posterior mean which is optimal relative to the squared error loss.

LEMMA 3.4.1 Let ϕ have a normal posterior *pdf* with mean m and variance v. Under the Linex loss function (3.4.1) the Bayes estimate $\hat{\phi}_B$ has Bayes risk

$$BR(\hat{\phi}_B) = b[a^2 v/2]$$

The proof is straightforward.
For further details the reader may refer to Zellner (1986).

3.4.1   BAYES PREDICTOR OF T

From (3.4.3) the Bayes predictor of T, with repect to the Linex loss function (3.4.1) is given by

$$\hat{T}_{BL} = -\frac{1}{a} \log \{E[e^{-aT} \mid y_s]\} \tag{3.4.5}$$

After y_s has been observed, we may write $T = 1'_s y_s + 1'_r y_r$. Therefore, predictor $\hat{T}_{BL}$ in (3.4.5) may be written as

$$\hat{T}_{BL} = n\bar{y}_s - \frac{1}{a} \log \{E(e^{-a1'_r y_r} \mid y_s)\} \tag{3.4.6}$$

We now consider the models (3.3.1), (3.3.2). Now,

$$E\{e^{-a1'_r y_r} \mid y_s\} = E\{E[e^{-a1'_r y_r} \mid y_s, \beta] \mid y_s\}$$

$$= e^{-a1'_r[X_r\hat{\beta}+V_{rs}V_s^{-1}(y_s-X_s\hat{\beta})]+a^2 U/2} \tag{3.4.7}$$

where

$$U = 1'_r[V_r - V_{rs}V_s^{-1}V_{sr} + (X_r - V_{rs}V_s^{-1}X_s)$$
$$(X'_s V_s^{-1} X_s + R^{-1})^{-1}(X_r - V_{rs}V_s^{-1}X_s)]1_r$$

and $\hat{\beta}$ is the usual least square estimator of β. Therefore, the Bayes predictor of T under loss function (3.4.1) is (Bolfarine, 1989)

$$\hat{T}_{BL} = n\bar{y}_s + 1'_r[X_r\hat{\bar{\beta}} + V_{rs}V_s^{-1}(y_s - X_s\hat{\bar{\beta}})] - (a/2)U \tag{3.4.8}$$

where

$$\hat{\bar{\beta}} = E(\beta \mid y_s) = (X'_s V_s^{-1} X_s + R^{-1})^{-1}(X'_s V_s^{-1} y_s + R^{-1}\nu) \tag{3.4.9}$$

The Bayes risk of $\hat{T}_{BL}$ with respect to the Linex loss function (3.4.1) is

$$R_b(\hat{T}_{BL}) = b(a^2/2)U \tag{3.4.10}$$

In the particular case, when $V_{rs} = 0$ and R is so large that $R^{-1} \simeq 0$, the Bayes predictor (3.4.8) reduces to

$$\hat{T}'_{BL} = n\bar{y}_s + 1'_r X_r \hat{\beta} - (a/2)1'_r[V_r + X_r(X'_s V_s^{-1} X_s)^{-1} X'_r]1_r \qquad (3.4.11)$$

with Bayes risk (which is also the risk under squared error loss function),

$$R_L(\hat{T}'_{BL}) = (ba^2/2)1'_r[V_r + X_r(X'_s V_s^{-1} X_s)^{-1} X'_r]1_r = bA \qquad (3.4.12)$$

It follows that the risk (with respect to the Linex loss) of Royall's (1970) optimal predictor

$$\hat{T}_{BLUP} = n\bar{y}_s + 1'_r X_r \hat{\beta} \qquad (3.4.13)$$

which is also the optimal predictor of T with respect to the squared error loss and non-informative prior on β is given by

$$R_L(\hat{T}_{BLUP}) = b(e^A - 1) > R_L(\hat{T}'_{BL})$$

It follows, therefore. that $\hat{T}_{BLUP}$ is *inadmissible* with respect to the Linex loss function. It follows similarly that $\hat{T}'_{BL}$ of (3.4.11) is inadmissible with respect to the squared error loss function.

EXAMPLE 3.4.1

Consider a special case of the models (3.3.1), (3.3.2):

$$y_i = x_i\beta + e_i, i = 1, \ldots, N$$

$$V = \sigma^2 \operatorname{Diag}(x_1, \ldots, x_N), \ \beta \sim N(\nu, R) \qquad (i)$$

Here,

$$\hat{T}_{BL} = n\bar{y}_s + \hat{\hat{\beta}} \sum_r x_i - \frac{a}{2}(\sigma^2 \sum_r x_i + \frac{(\sum_r x_i)^2}{\sum_s x_i/\sigma^2 + 1/R}) \qquad (ii)$$

where

$$\hat{\hat{\beta}} = (\sum_s y_i/\sigma^2 + \nu/R)/(\sum_s x_i/\sigma^2 + 1/R).$$

The Bayes risk is

$$R_L(\hat{T}_{BL}) = (ba^2/2)(\sigma^2 \sum_r x_i + \frac{\sum_r x_i)^2}{\sum_r x_i/\sigma^2 + 1/R}) \qquad (iii)$$

Thus, as in the case of Royall's optimum strategy, a purposive sample s^* will be an optimal sample to base $\hat{T}_{BL}$. If R is so large that $R^{-1} \simeq 0$, predictor in (ii) reduces to

$$\hat{T}_{RL} = \hat{T}_R - \frac{a}{2}\frac{N(N-n)}{n}\frac{\bar{x}\bar{x}_r}{\bar{x}_s}\sigma^2 \qquad (v)$$

where $\hat{T}_R = \frac{\bar{y}_s}{\bar{x}_s}\bar{X}$ and $\bar{x}_r = \frac{1}{N-n}\sum_r x_i$. It follows that

$$R_L(\hat{T}_{RL}) = \frac{ba^2}{2}\frac{N(N-n)}{n}\frac{\bar{x}\bar{x}_r}{\bar{x}_s}\sigma^2$$

Note that as $R \to \infty, R_L(\hat{T}_{BL}) \to R_L(\hat{T}_{RL}) = \sup{}_\beta R_L(\hat{T}_{RL})$. Hence, it follows from theorem 3.2.2 that the predictor $\hat{T}_{RL}$ is a minimax predictor with respect to the Linex loss under model (i).

Bolfarine (1989) also considered Bayes predictor of T under Linex loss function (3.4.1) in two-stage sampling.

In the next section we consider James-Stein estimator and its competitors for population parametrs. Application of these estimators in estimating finite population parameters will be considered in sections 3.6 and 3.8.

3.5 JAMES-STEIN ESTIMATOR AND ASSOCIATED ESTIMATORS

Suppose we have m independent samples, $y_i \, ind \, N(\theta_i, B)$ where B is a known constant. We wish to estimate $\theta = (\theta_1, \ldots, \theta_k)'$ using the $SSEL$

$$L(\delta(\mathbf{y}), \theta) = \sum_{i=1}^{k}(\delta_i(\mathbf{y}) - \theta_i)^2 \qquad (3.5.1)$$

where $\delta(\mathbf{y}) = (\delta_1(\mathbf{y}), \ldots, \delta_m(\mathbf{y}))'$ and $\delta_i(\mathbf{y})$ is an estimate of θ_i. The maximum likelihood estimator (mle) of θ is

$$\mathbf{y} = (y_1, \ldots, y_m)' \qquad (3.5.2)$$

and its risk is

$$r(\mathbf{y} \mid \theta) = E_\theta \sum_{i=1}^{m}(y_i - \theta_i)^2 = mB \qquad (3.5.3)$$

The estimator $\mathbf{y}$ has minimum variance among all unbiased estimators or among all translation-invariant estimators (i.e. estimators $\psi(\mathbf{y})$ with the property $\psi(\mathbf{y} + a) = \psi(\mathbf{y}) + a \,\forall\, \mathbf{y}$ and all vectors of constants a).

Stein (1955) proposed the following non-linear and biased estimator of θ_i,

$$\delta_{iS} = (1 - \frac{CB}{S})y_i \qquad (3.5.4)$$

where C is a positve constant and

$$S = \sum_{i=1}^{k} y_i^2 \qquad (3.5.5)$$

The optimum choice of C was found to be

$$C = (m - 2), \text{ for } m > 2 \qquad (3.5.6)$$

and for this choice

$$r(\delta_S \mid \theta) < r(\mathbf{y} \mid \theta) = mB \; \forall \; \theta \qquad (3.5.7)$$

The estimator $\delta_S = (\delta_{1S}, \ldots, \delta_{mS})'$, therefore, dominates $\mathbf{y}$ with respect to the loss function in (3.5.1). The estimator δ_S , in effect shrinks, the *mle*, $\mathbf{y}$ towards 0 i.e. each component of $\mathbf{y}$ is reduced by the same factor. The amount of shrinkage depends on the relative closeness of $\mathbf{y}$ towards 0; for y_i near 0 the shrinkage is substantial, while for y_i far away from 0, the shrinkage is small and δ_{iS} is essentially equal to y_i.

The estimator δ_S can be interpreted in Bayesian perspective as follows. Suppose θ has a prior, $\theta \sim N(0, AI)$. Then Bayes estimate of θ is

$$\delta_B = (1 - \frac{B}{B + A})\mathbf{y} \qquad (3.5.8)$$

Now, under predictive distribution of $\mathbf{y}$ with respect to prior distribution of θ above,

$$E(\frac{(m - 2)B}{S}) = \frac{B}{B + A} \qquad (3.5.9)$$

Thus the James-Stein estimator

$$\delta_{JS} = (1 - \frac{(m - 2)B}{S})\mathbf{y} \qquad (3.5.10)$$

may be said to be an Empirical Bayes (EB) estimator corresponding to the Bayes estimator δ_B in (3.5.8).

Another extention of the problem arises when $y_i \underset{\sim}{ind} N(\theta_i, B_i)$ where B_i is known. Assume that $\theta_i \underset{\sim}{ind} N(0, AB_i)$. One may then calculate JS-estimator of θ_i by considering the transformed variables $y_i/\sqrt{B_i}$ in place of y_i. This JS-estimator

$$\delta^* = (\delta_1^*, \ldots, \delta_m^*)'$$

dominates the *mle* **y** with respect to the risk function

$$r(\delta^* \mid \theta) = \sum_{i=1}^{m} E(\delta_i^* - \theta_i)^2 / B_i \qquad (3.5.11)$$

$\forall\ \theta$. This estimator will be most suitable against a Bayes prior in which the variance of the prior distribution is proportional to the sampling variance.

Stein (1960) conjectured that a 'positive-part' version of the estimator δ_{iS} in (3.5.4) would be superior to δ_{iS}. Baranchik (1970) showed that the conjecture is in fact true.. The positive-part Stein-rule estimator is δ_S^+ where

$$\delta_{iS}^+ = \left\{ \begin{array}{ll} \delta_{iS} & \text{if } \delta_{iS} > 0 \\ 0 & \text{otherwise} \end{array} \right. \qquad (3.5.12)$$

This estimator will be denoted as

$$[1 - \frac{(m-2)B}{S}]^+ y_i \qquad (3.5.13)$$

Lindley (1962) proposed the modified Stein-rule estimator

$$\delta_{iL}^+ = \bar{y}_s + [1 - \frac{(m-3)B}{\sum(y_i - \bar{y}_s)^2}]^+ (y_i - \bar{y}_s), m > 3 \qquad (3.5.14)$$

where $\bar{y}_s = \sum_{i=1}^{k} y_i / m$. This estimator shrinks the *mle* y_i towards the mean $\bar{y}_s$, rather than towards zero.

Stein-rule estimator is scale-invariant (i.e. multiplying y_i by a constant c, changes δ_{iS} to $c\delta_{iS}$) but not translation-invariant (i.e. if $y_i^* = y_i + d$, then $\delta_{iS} \neq \delta_{iS} + d$). Lindley-Stein estimator δ_{iL}^+ is equivariant with respect to a change of origin and scale.

Assuming a multiple regression model, the Stein-rule estimator can be derived as follows. Consider the linear model

$$\mathbf{y} = X\beta + u$$

$$u \sim N(0, \sigma^2 I)$$

when **y** is a $n \times 1$ vector of random observations, X is a $n \times p$ matrix of known values $x_{jk}(j = 1, \ldots, p; k = 1, \ldots, n)$ of auxiliary variables x_j, β a vector of p unknown regression coefficients and u a $(n \times 1)$ vector of random error components. The Stein-rule estimator of β is

$$\hat{\beta}_S = [1 - \frac{g\sigma^2}{b'X'Xb}]b \qquad (3.5.15)$$

where $g(>0)$ is a suitable constant and

$$b = (X'X)^{-1}X'\mathbf{y}$$

is the least squares estimate of β. The James-Stein estimator of β is

$$\hat{\beta}_{JS} = [1 - \frac{g\hat{u}'\hat{u}}{(n-p)b'X'Xb}]b, \qquad (3.5.16)$$

where σ^2 has been estimated by

$$s^2 = \hat{u}'\hat{u}/(n-p) = (\mathbf{y} - Xb)'(\mathbf{y} - Xb)/(n-p)$$

An immediate generalisation of (3.5.10) follows in the following Hierarchical Bayes (HB) set up. Suppose that

$$y_i \, ind \, N(\theta_i, B), \quad \theta_i \mid \beta \, ind \, N(x_i'\beta, B) \qquad (3.5.17)$$

$$\beta \sim \text{uniform(improper) over} R^p$$

where B is a known constant, $x_i' = (x_{i1}, \ldots, x_{ip})$ is a set of p known real numbers, $\beta = (\beta_1, \ldots, \beta_p)'$ a set of p unknown regression coefficients. The UMVU-estimator of y_i in the frequentist set up is

$$y_i^* = x_i'(X'X)^{-1}X'\mathbf{y} \qquad (3.5.18)$$

where $X = (x_1, \ldots, x_n)'$ and $(X'X)$ is of full rank. The Bayes estimate of θ_i is

$$\begin{aligned} \delta_{iB}^* &= y_i^* + (1 - \frac{B}{B+F})(y_i - y_i^*) \\ &= \frac{B}{B+F}y_i^* + \frac{F}{B+F}y_i \end{aligned} \qquad (3.5.19)$$

In this case, the J-S estimator of θ_i is, for $p < m - 2$,

$$\begin{aligned} \delta_{iJS}^* &= y_i^* + [1 - \frac{(m-p-2)B}{S^*}](y_i - y_i^*) \\ &= \frac{(m-p-2)B}{S^*}y_i^* + [1 - \frac{(m-p-2)D}{S^*}]y_i \end{aligned} \qquad (3.5.20)$$

where

$$S^* = \sum_i (y_i - y_i^*)^2 \qquad (3.5.21)$$

Efron and Morris (1971,1972) pointed out that both Bayes estimators $\delta_B^* = (\delta_{1B}^*, \ldots, \delta_{mB}^*)'$ and EB estimators $\delta_{JS}^* = (\delta_{1JS}^*, \ldots, \delta_{mJS}^*)'$ might do poorly for estimating individual θ_i's with unusually large and small values (vide exercise 2 of chapter 4). To overcome this problem they suggested 'limlited translation' estimators discussed in section 4.4.

In practical situations, often the sampling variance of y_i about θ_i is not proportional to prior variance of θ_i about 0 (as has been assumed in deriving the estimator (3.5.8)). An approach is to develop an estimator that closely resembles the Bayes estimator for the prior distribution $\theta_i \, ind \, N(0, A)$. Efron and Morris (1973) first proposed an extention of (3.5.4) in this direction (see exercise 3).

Another suggestion is due to Carter and Ralph (1974). They showed that for the model

$$y_i \, ind \, N(\theta_i, D_i), \quad \theta_i \, ind \, N(\nu, A), \quad i = 1, \ldots, k \qquad (3.5.22)$$

with D_i known but θ_i, ν unknown, the weighted sample mean

$$\nu^* = \sum_i \frac{y_i}{A + D_i} \Big/ \sum_i \frac{1}{A + D_i} \qquad (3.5.23)$$

has the property, assuming A to be known,

$$E(\sum_i (y_i - \nu^*)^2 / (A + D_i)) = k - 1, \qquad (3.5.24)$$

where expectation is taken over joint distribution of y and θ. They suggested estimating A as the unique solution A^* to the equation

$$\sum_i \frac{(y_i - \nu^*)^2}{A^* + D_i} = k - 1 \qquad (3.5.25)$$

subject to the condition $A^* > 0$. They considered $A^* = 0$ if no positive solution of (3.5.23) and (3.5.24) exists. An estimate of θ_i is then given by

$$\delta_{iCR} = \frac{A^*}{A^* + D_i} y_i + \frac{D_i}{A^* + D_i} \nu^* \qquad (3.5.26)$$

We shall consider the applications of these procedures in estimating a finite population total in the next section and section 3.8.

3.6   EMPIRICAL BAYES PREDICTOR OF POPULATION TOTAL UNDER SIMPLE LOCATION MODEL

In this section we do not use any covariate in the model and attempt to find EB-estimator of population total. Consider the simple location model

$$y_i = \theta + \epsilon_i, \quad i = 1, \ldots, N \qquad (3.6.1)$$

where θ and ϵ_i are independently distributed with

$$\theta \sim N(\mu, \tau^2), \quad \epsilon_i \sim NID(0, \sigma^2) \tag{3.6.2}$$

and the phrase *NID* denotes independently and normally distributed. The joint prior (unconditional) distribution of $\mathbf{y} = (y_1, \ldots, y_N)$ is

$$N_N(\mu \mathbf{1}_N, \sigma^2 I_N + \tau^2 J_N) \tag{3.6.3}$$

where $\mathbf{1}_q$ is a $q \times 1$ vector of 1's , I_q is a unit vector of order q and $J_q = \mathbf{1}_q \mathbf{1}_q'$. The joint conditional distribution of y_r given y_s is , therefore, $(N - n)$ - variate normal with mean vector

$$\{(M\mu + n\bar{y}_s)/(M + n)\}\mathbf{1}_{N-n}$$

and dispersion matrix

$$\sigma^2\{\mathbf{1}_{N-n} + (M + n)^{-1}J_{N-n}\}$$

where $M = \sigma^2/\tau^2$. The Bayes predictor of population mean $\bar{y}$ under squared error loss function, is by formula (3.3.16),

$$\hat{\bar{y}}_B = E[\bar{y} \mid (s, y_s)] = N^{-1}[n\bar{y}_s + (N - n)(B\mu + (1 - B)\bar{y}_s)] \tag{3.6.4}$$

where $B = M/(M + n)$ (Ericson, 1969 a). The parameters μ and B (and hence M) are generally unknown and require to be estimated.

Ghosh and Meeden (1986) used the estimators from the analysis of variance to estimate the parameter B. They assumed that the data are available from (m-1)- previous surveys from the same (similar type of) population. It is assumed that in the j-th survey ($j = 1, \ldots, m$, including the present or mth survey) the survey character was $y^{(j)}$, taking values $y_i^{(j)}$ on the ith unit in the population $\mathcal{P}_j$ of size N_j, the sample being s_j of size n_j. The relevant superpopulation model is assumed to be

$$y_i^{(j)} = \theta^{(j)} + \epsilon_i^{(j)}, \quad i = 1, \ldots, N_j; j = 1, \ldots, m, \tag{3.6.5}$$

$\theta^{(j)}, \epsilon_i^{(j)}$ being independently distributed, $\theta^{(j)} \sim N(\mu, \tau^2), \epsilon_i^{(j)} \sim NID(0, \sigma^2)$. Let

$$y_{s_j} = (y_1^{(j)}, \ldots, y_{n_j}^{(j)})'$$

where, without loss of generality, we kake $s_j = (1, \ldots, n_j), (j = 1, \ldots, m)$. Then $y_{s_1}, \ldots, y_{s_m}$ are marginally independently distributed with

$$y_{s_j} \sim N_{n_j}(\mu \mathbf{1}_{n_j}, \sigma^2 I_{n_j} + \tau^2 J_{n_j}) \tag{3.6.6}$$

Let

$$n_T = \sum_{j=1}^{m} n_j, \ \bar{y}_s = \sum_j n_j \bar{y}^{(j)} / \sum_j n_j, \ \bar{y}^{(j)} = \sum_{i=1}^{n_j} y_i^{(j)} / n_j$$

Define

$$BMS(\text{ Between Mean Square }) = \sum_{j=1}^{m} n_j (\bar{y}^{(j)} - \bar{y}_s)^2 / (m - 1)$$

$$WMS(\text{ Within Mean Square }) = \sum_{j=1}^{m} \sum_{i=1}^{n_j} (y_i^{(j)} - \bar{y}^{(j)})^2 / (n_T - m) \quad (3.6.7)$$

Lemma 3.6.1

$$E(WMS) = \sigma^2, \ V(WMS) = 2\sigma^4 / (n_T - m) \quad (3.6.8)$$

$$E(BMS) = \sigma^2 + g\tau^2 / (m - 1)$$

where

$$g = (g(n_1, \ldots, n_m)) = n_T - (\sum_{j=1}^{m} n_j^2) / n_T$$

and

$$V(BMS) = \frac{2}{(m-1)^2} [\sum_{j=1}^{m} \tau_j^4 + (\sum_{j=1}^{m} n_j \tau_j^2)^2 / n_T^2 - 2 \sum_{j=1}^{m} n_j \tau_j^4 / n_T] \quad (3.6.9)$$

where

$$\tau_j^2 = \sigma^2 + \tau^2 n_j \ (j = 1, \ldots, m). \quad (3.6.10)$$

Proof The result (3.6.8) follows from the fact that $(n_T - m)WMS/\sigma^2 \sim \chi^2_{(n_T - m)}$. Observe that $\sqrt{n_j}(\bar{y}^{(j)} - \mu) \sim NID(0, \tau_j^2)$. Hence, $(m - 1)BMS \sim Z'AZ$ where $Z \sim N(0, I_m)$ and $A = D - uu'$ with $D = \text{Diag} (\tau_1^2, \ldots, \tau_m^2)$ and $u = (\sqrt{n_1}\tau_1, \ldots, \sqrt{n_m}\tau_m)'/\sqrt{n_T}$. The expressions for expectation and variance of BMS follows from the fact, $E(Z'AZ) = \text{tr} (A)$ and $V(Z'AZ) = 2 \text{ tr} (A^2)$, and on simplification, where $tr(H)$ denotes the trace of the matirx H.

Consider the folowing assumptions, called the assumptions A.

- (1)$n_j \geq 2$

- (2) $\sup_{j=1,\ldots,m} n_j = K < \infty$

it can be easily proved that under assumptions A, a consistent estimator of $M^{-1} = \tau^2/\sigma^2$ is

$$max\{0, (BMS/WMS - 1)(m-1)g^{-1}\} \qquad (3.6.11)$$

The authors modified the above estimator slightly so that the resulting estimator is identical with the James-Stein estimator in the balanced case (i.e. when $n_1 = \ldots = n_m = n$). Thus, they proposed the estimator

$$\hat{M}^{-1} = max\{0, (\frac{(m-1)BMS}{(m-3)WMS} - 1)(m-1)g^{-1}\}, \quad (m \geq 4) \qquad (3.6.12)$$

The estimator $\hat{M}^{-1}$ is consistent for M^{-1}. It follows that $\hat{B}_j = \frac{1}{1+n_j\hat{M}^{-1}}$ is consistent for $B_j = \frac{1}{1+n_jM^{-1}}$.

To estimate μ we note that the *mle* of μ for known M is obtained from the joint *pdf* of $(y_{s_1}, \ldots, y_{s_m})$ as

$$\tilde{\mu} = \sum_{j=1}^{m}(1 - B_j)\bar{y}^{(j)}/\sum_{j=1}^{m}(1 - B_j) \qquad (3.6.13)$$

Consequently, an EB-estimator of μ is

$$\hat{\mu} = \begin{cases} \sum_{j=1}^{m}(1 - \hat{B}_j)\bar{y}^{(j)}/\sum_{j=1}^{m}(1 - \hat{B}_j) & \text{if } \hat{M}^{-1} \neq 0 \\ m^{-1}\sum_{j=1}^{m}\bar{y}^{(j)} & \text{if } \hat{M}^{-1} = 0 \end{cases} \qquad (3.6.14)$$

An EB- predictor of population mean $\hat{\bar{y}}_{EB}$ is obtained by replacing μ and B in (3.6.4) by $\hat{\mu}$ and $\hat{B}$ respectively, where $\hat{B} = (1+n\hat{M}^{-1})^{-1}$. Therefore,

$$\hat{\bar{y}}_{EB} = N^{-1}[n\bar{y}_s + (N-n)(\hat{B}\hat{\mu} + (1-\hat{B})\bar{y}_s)] \qquad (3.6.15)$$

Alternative EB-estimators of μ and M are available in the balanced case. In this case, the *mle* of μ is $\hat{\hat{\mu}} = \bar{\bar{y}} = \sum_{j=1}^{m}\bar{y}^{(j)}/m$. Also, the *mle* of M^{-1} is

$$\hat{\hat{M}}^{-1} = max\{0, (\frac{(m-1)BMS}{mWMS} - 1)n^{-1}\} \qquad (3.6.16)$$

These can be taken up as EB-estimators and hence, an alternative EB-predictor of $\bar{y}$ can be obtained using (3.5.4). In the balanced case, $g = n(m-1)$ and $\hat{M}^{-1}$ differs from $\hat{\hat{M}}^{-1}$ only in the coefficient of WMS (m being replaced by $(m-3)$). Clearly, the asymptotic (as $m \to \infty$) performance of both the estimators are similar.

Mukhopadhyay (1998 c) suggested an alternative EB -predictor of $\bar{y}$ in the balanced case. Let

$$S = ((S_{\lambda\lambda'}))_{n \times n}$$

where

$$S_{\lambda\lambda'} = \sum_{j=1}^{m}(y_\lambda^{(j)} - \bar{y}_{(\lambda)})(y_{\lambda'}^{(j)} - \bar{y}_{(\lambda')}), \lambda, \lambda' = 1, \ldots, n, \quad (3.6.17)$$

$$\bar{y}_{(\lambda)} = \sum_{j=1}^{m} y_\lambda^{(j)}/m$$

At any stage j, the observations $\{y_\lambda^{(j)}, \lambda = 1, \ldots, n\}$ can be labelled $\lambda = 1, \ldots, n$, randomly. For every j, y_{s_j} has an exchangeable $(n-1)$- variate normal distribution with dispersion matrix $\Sigma = \sigma^2 I_n + \tau^2 J_n$ and any random permutation of observations at the jth stage can be considered while computing S. Therefore, S follows Wishart distribution $W(n, m-1, \Sigma)$. The *mle* of $(1 + M)^{-1}$ is then given by (vide Muirhead (1982), p. 114)

$$(1 + \dot{M})^{-1} = \frac{\sum \sum_{\lambda \neq \lambda'=1}^{n} S_{\lambda\lambda'}}{(m-2) \sum_{\lambda=1}^{n} S_{\lambda\lambda}} \quad (3.6.18)$$

An EB-estimator of M^{-1} is, therefore,

$$\dot{M}^{-1} = max\{\frac{\sum_{\lambda \neq \lambda'=1}^{n} S_{\lambda\lambda'}}{(m-2) \sum_{\lambda=1}^{n} S_{\lambda\lambda} - \sum_{\lambda \neq \lambda'=1}^{n} S_{\lambda\lambda'}}, 0\} \quad (3.6.19)$$

An EB-estimator of $\mu, \dot{\mu}$ is obtained by replacing $\dot{M}$ by $\dot{M}$ in (3.6.14). This gives another EB -predictor $\tilde{\bar{y}}_{EB}$ of $\bar{y}$ from (3.6.15).

EXAMPLE 3.6.1

Consider the data on number of municiplal employees in 1984 (ME84) in 48 municipalities in Sarndal et al (1992, p. 653-654). It is assumed that $m = 4$ and the municipalities labelled $1, \ldots, 12$ constitute the population at the first stage, the next 12 municipalities, population at the second stage, etc. Six random samples of size four each are selected from each of the populations at the first three stages. For the present population which consists of municipalities $37, \ldots, 48$, two samples are drawn, $s_4^{(1)} = (38, 40, 44, 48)$ and $s_4^{(2)} = (37, 41, 43, 45)$. Each of the samples from the earlier stages is combined with $s_4^{(i)}(i = 1, 2)$ thus giving two sets of 216 samples each and average bias (AB) and average mse (AM) of the estimators

$\hat{\bar{y}}_{EB}$ and $\tilde{\bar{y}}_{EB}$ are calculated for each of these sets of samples. It is observed that

$$AB(\hat{\bar{y}}_{EB}^{(1)}) = 62616.6 \qquad AB(\hat{\bar{y}}_{EB}^{(2)}) = 150360.8$$
$$AM(\hat{\bar{y}}_{EB}^{(1)}) = 2372605776.6 \quad AM(\hat{\bar{y}}_{EB}^{(2)}) = 70514314056.7$$
$$AB(\tilde{\bar{y}}_{EB}^{(1)}) = 22283.8 \qquad AB(\tilde{\bar{y}}_{EB}^{(2)}) = 26056.1$$
$$AM(\tilde{\bar{y}}_{EB}^{(1)}) = 1193933317.0 \quad AM(\tilde{\bar{y}}_{EB}^{(2)}) = 1806187631.7$$

where $\hat{\bar{y}}_{EB}^{(k)}(k = 1, 2)$ denotes the estimator based on the k-th set of samples and similarly for $\tilde{\bar{y}}_{EB}^{(k)}$. The estimator $\tilde{\bar{y}}_{EB}$ was found to be consistently better than $\hat{\bar{y}}_{EB}$ both in the sense of average bias and average mse.

To compare the performance of two predictors, say EB-predictor e_{EB} and an arbitrary predictor e of a statistic $\lambda(\mathbf{y})$ vis-a-vis the Bayes predictor e_B under a prior ξ, Effron and Morris (1973) (also, Morris, 1983) introduced the concept of *relative savings loss* (RSL). Let $r(\xi, e)$ denote the average risk of the predictor e under ξ with respect to squared error loss, as defined in (3.2.2),

$$r(\xi, e) = E(e - \lambda(\mathbf{y}))^2 \tag{3.6.20}$$

where the expectation is taken with respect to prior distribution as well as over all $\mathbf{y}$ containing y_s. The RSL of e_{EB} wrt an arbitrary predictor e under the prior ξ is given by

$$
\begin{aligned}
RSL(\xi; e_{EB}, e) &= \frac{r(\xi, e_{EB}) - r(\xi, e_B)}{r(\xi, e) - r(\xi, e_B)} \\
&= 1 - \frac{r(\xi, e) - r(\xi, e_{EB})}{r(\xi, e) - r(\xi, e_B)}
\end{aligned}
\tag{3.6.21}
$$

$RSL(\xi; e_{EB}, e)$ measures the increase in average risk in using EB estimator e_{EB} instead of Bayes estimator e_B with respect to the same in using an arbitrary estimator e. The ratio RSL < 1 means e_{EB} is better than e in the sense of smaller average risk . The ratio RSL$\to 0$ means EB estimator e_{EB} is asymptotically equivalent to Bayes estimator e_B i.e. e_{EB} is asymptotically equivalent in the sense of Robbins (1955). It follows that

$$r(\xi, e) = r(\xi, e_B) + E(e - e_B)^2 \tag{3.6.22}$$

Using $\lambda(\mathbf{y}) = \bar{y} = \sum_{i=1}^{N} y_i/N$ and $e = \hat{\bar{y}}_{EB}$ (given in (3.6.15)), $\hat{\bar{y}}_0$ and $\hat{\bar{y}}_1$ in succession, where $\hat{\bar{y}}_0 = \bar{\bar{y}} = \sum_{j=1}^{m} \bar{y}^{(j)}/m$ and $\hat{\bar{y}}_1 = \sum_j n_j \bar{y}^{(j)} / \sum_j n_j$ and denoting the superpopulation model (3.6.1), (3.6.2) as ς, we have the followings theorems.

THEOREM 3.6.1 Under the prior ς,

$$RSL(\varsigma; \hat{\tilde{y}}_{EB}, \hat{\tilde{y}}_0) = E[(\hat{B}_m - B_m)(\bar{y}^{(m)} - \mu)$$

$$-\hat{B}_m(\hat{\mu} - \mu)]^2[B_m^2 E(\bar{y}^{(m)} - \mu)^2]^{-1} \tag{3.6.23}$$

$$RSL(\varsigma; \hat{\tilde{y}}_{EB}, \hat{\tilde{y}}_1) = (1 - f_m)^2 E[(\hat{B}_m - B_m)(\bar{y}^{(m)} - \mu) -$$

$$\hat{B}_m(\hat{\mu} - \mu)]^2[E(\bar{\bar{y}} - \hat{\tilde{y}}_B)^2]^{-1} \tag{3.6.24}$$

where $f_m = n_m/N_m$, $\hat{\mu}$ is given in (3.6.14) and $\hat{\tilde{y}}_B$ by (3.6.4).

THEOREM 3.6.2 Under assumption A,

$$RSL(\varsigma; \hat{\tilde{y}}_{EB}, \hat{\tilde{y}}_k) \to 0 \text{ as } m \to \infty, \text{ for } k = 0, 1 \tag{3.6.25}$$

It follows, therefore, that $r(\varsigma; \hat{\tilde{y}}_{EB}) \to r(\varsigma, \hat{\tilde{y}}_B)$ as $m \to \infty$, so that the proposed estimator $\hat{\tilde{y}}_{EB}$ is asymptotically optmum in the sense of Robbins (1955). The property (3.6.25) readily extends to the estimator $\tilde{\tilde{y}}_{EB}$.

In a similar set up (as in (3.6.1) and (3.6.2)) Ghosh and Lahiri (1987 a) considered the simultaneous EB -prediction of strata means in stratified random sampling. Let y_{hi} be the value of y on the ith unit in the hth stratum $\mathcal{P}_h$ of size $N_h(h = 1, \ldots, L; \sum N_h = N)$. The problem is to find a EB-predictor of $\gamma = (\gamma_1, \ldots, \gamma_L)'$, where $\gamma_h = \sum_{i=1}^{N_h} y_{hi}/N_h$, the popula-

tion mean for the hth stratum, on the basis of a stratified random sample $s = \bigcup_{i=1}^{L} s_h, s_h$, the sample from the stratum h being taken as $(1, \ldots, n_h)$ without loss of generality $(n = \sum_h n_h)$, with the sum of squared error loss (SSEL) function

$$L(c, \gamma) = \frac{1}{L} \sum_{h=1}^{L} (c_h - \gamma_h)^2 \tag{3.6.26}$$

where $c = (c_1, \ldots, c_L)'$ is a predictor of γ. Let, as before, $y_{s_h} = (y_{h1}, \ldots, y_{hn_h})'$, $\bar{y}_h = \sum_{1}^{n_h} y_{hi}/n_h$. However, the normality assumption (3.6.2) is replaced by the following general assumptions:

(a) Conditional on $\theta = (\theta_1, \ldots, \theta_L)', y_{h1}, \ldots, y_{hn_h}$ are iid with distribution depending only on θ_h and $E(y_{hi} \mid \theta_h) = \theta_h, V(y_{hi} \mid \theta_h) = \mu_2(\theta_h)$ $(h = 1, \ldots, L)$

(b) θ_h's are iid with $E(\theta_h) = \mu, V(\theta_h) = \tau^2$

(c) $0 < E[\mu_2(\theta_1)] = \sigma^2 < \infty$

We call this set of assumptions as C. The following assumptions of *posterior linearity* is also made.

$$E(\theta_h \mid y_{s_h}) = \sum_{i=1}^{n_h} a_{hi} y_{hi} + b_h \quad (h = 1, \ldots, L) \tag{3.6.27}$$

where the a_{hi} and b_h are constants not depending on y_{sh}. Since y_{hi}'s are conditionally independently and identically distributed (*iid*) given θ_h, it follows (Goldstein, 1975) that

$$E(\theta_h \mid y_{s_h}) = a_h \bar{y}_h + b_h \quad (h = 1, \ldots, L) \tag{3.6.28}$$

where a_h's are some constants. It follows from Ericson (1969 b) or Hartigan (1969) that

$$\begin{aligned} a_h &= \frac{\tau^2}{\tau^2 + \sigma^2/n_h} = \frac{n_h}{M + n_h} = 1 - B_h, \\ b_h &= B_h \mu \end{aligned} \tag{3.6.29}$$

where

$$M = \sigma^2/\tau^2 \quad \text{and} \quad B_h = M/(M + n_h).$$

Hence, following (3.6.4), Bayes estimator of γ is $\hat{\gamma}_B = (\hat{\gamma}_B^{(1)}, \ldots, \hat{\gamma}_B^{(L)})'$ where

$$\hat{\gamma}_B^{(h)} = E(\gamma_h \mid y_{s_h}) = \bar{y}_h - f_h B_h(\bar{y}_h - \mu) \tag{3.6.30}$$

and $f_h = 1 - n_h/N_h$.

Considering EB estimation of γ, first assume that τ and σ^2 and hence B_h are known, but μ is unknown. Since $E(y_{s_h}) = \mu 1_{n_h}$, the best linear unbiased estimator (BLUE) of μ is obtained by minimising

$$\sum_{h=1}^{L} (y_{s_h} - \mu 1_{n_h})'(\sigma^2 I_{n_h} + \tau^2 J_{n_h})^{-1}(y_{s_h} - \mu 1_{n_h}) \tag{3.6.31}$$

with respect to μ. If the underlyimg distributions are normal, the BLUE of μ is identical with its *mle*, and is given by

$$\mu^* = \sum_{h=1}^{L}(1 - B_h)\bar{y}_h / \sum_{h=1}^{L}(1 - B_h). \tag{3.6.32}$$

The EB estimator of $\gamma_h, \hat{\gamma}_B^{(h)}$ is obtained by replacing μ by μ^* in (3.6.30). In the balanced case ($n_h = n \; \forall \; h$), μ^* reduces to $\sum_{1}^{L} \sum_{1}^{m} y_{hi}/mL = \bar{y}_s$. In this case EB estimator of γ_h is

$$\hat{\gamma}_{EB}^{(h)} = \bar{y}_h - f_h B(\bar{y}_h - \bar{y}_s), h = 1, \ldots, L \tag{3.6.33}$$

which is the finite population analogue of an estimator of Lindley and Smith (1972) obtained under normality assumtions and a hierarchical Bayes approach.

In case both M and μ are unknown, estimators similar to those in (3.6.11) -(3.6.15), (3.6.18), etc. can be obtained under the relevant assumptions.

Ghosh and Lahiri (1987 b) extended the study to simultaneous prediction of variances of several strata $\mathbf{S}^2 = (S_1^2, \ldots, S_L^2)'$ where $S_h^2 = \sum_{i=1}^{N_h}(y_{hi} - \gamma_h)^2/(N_h - 1), h = 1, \ldots, L$ under the SSEL (3.6.26). Apart from the assumptions in C it is further assumed that $\mu_2(\theta_j)$ is polynomial of (at most) order 2,

$$\mu_2(\theta_j) = d_0 + d_1\theta_j + d_2\theta_j^2 \tag{3.6.34}$$

where at least one of $d_i(i = 0, 1, 2)$ is non-zero. The Bayes posterior linearity (3.6.27) is also assumed. The Bayes estimator of $\mathbf{S}^2$ is $\hat{\mathbf{S}}_B^2 = (\hat{S}_{1(B)}^2, \ldots, \hat{S}_{L(B)}^2)'$ where

$$
\begin{aligned}
\hat{S}_{h(B)}^2 &= E(S_h^2 \mid y_{s_h}) \\
&= (N_h(N_h - 1))^{-1} E\{\sum \sum_{i \neq i'=1}^{n_h} (y_{hi} - y_{hi'})^2/2 \\
&\quad + \sum_{i=1}^{n_h} \sum_{i'=n_h+1}^{N_h} (y_{hi} - y_{hi'})^2/2 \\
&\quad + \sum \sum_{i \neq i'=n_h+1}^{N_h} (y_{hi} - y_{hi'})^2/2 \mid y_{s_h}\}
\end{aligned}
$$

In the special case where $d_1 = d_2 = 0$ and $d_0 = \sigma^2$,

$$\hat{S}_{h(B)}^2 = (N_h - 1)^{-1}[(n_h - 1)s_h^2 + f_h\{\sigma^2(N_h - B_h) + n_h(\bar{y}_h - \mu)^2\}] \tag{3.6.35}$$

where $s_h^2 = \sum_{i=1}^{n_h}(y_{hi} - \bar{y}_h)^2/(n_h - 1)$.

The EB-estimator of S_h^2, $\hat{S}_{h;EB}^2$ is obtained by replacing μ and B_h by their estimators. We shall consider prediction of finite population variance S_y^2 in details in chapter 5.

NOTE 3.6.1

The case $d_1 = d_2 = 0, d_0 = \sigma^2(> 0)$ holds in the normal superpopulation model set up (Ericson, 1969 a; Ghosh and Meeden, 1986). If $y_{hi} \mid \theta_h \underset{\sim}{iid} Poisson(\theta_h)(i = 1, \dots, n_h), d_0 = 0, d_1 = 1, d_2 = 0$. Morris (1983) characterised all possible conditional distributions $y_{hi} \mid \theta_h$ belonging to natural exponential family with quadratic variance functions (QVF).

Tiwari and Lahiri (1989) extended the results of Ghosh and Lahiri (1987 a,b). They retained the posterior linearity property (3.5.20) of the posterior expected values of the stratum superpopulation mean in each stratum but allowed the prior mean and variance of each stratum superpopulation mean to vary from stratum to stratum. Thus, in their formulation, assumptions (a) of C holds but (b) and (c) are replaced respectively by

- (b)' θ_h's are independent with $E(\theta_h) = e_h\mu$ and $V(\theta_h) = f_h\tau^2(h = 1, \dots, L)$

- (c)' $E[\mu_2(\theta_h)] = g_h\sigma^2 < \infty, (h = 1, \dots, L)$

where the constants e_h, f_h, g_h are assumed to be positive.

In this section we have not considered use of any covariate in the regression equation. The next section considers EB-prediction of T under normal models using covariates.

3.7 EB-PREDICTION UNDER NORMAL MODEL USING COVARIATES

Lahiri and Peddada (1992) considered Bayesian analysis in finite population sampling under multiple regression model using assumptions of linearity. Consider the following assumptions:

(i)
$$y_i \mid \beta \sim N(x_i'\beta, \sigma^2), i = 1, \dots, N$$

(ii)
$$\beta \sim N_p(\nu, \tau^2 H) \tag{3.7.1}$$

where $x_i = (x_{i1}, \dots, x_{ip})', H$ is a $p \times p$ positive-definite matrix and σ^2, τ^2 are constants. A sample of size n, say, $(1, \dots, n)$ (without loss of generality) is selected from the population following some non-informative sampling design p. Under squared error loss function, $L(a, \theta) = (a - \theta)^2$, Bayes predictor for $\bar{y}$ is

$$\bar{\hat{y}}_B^* = E(\bar{y} \mid s, \mathbf{y}_s)$$

$$= N^{-1} E[\sum_{i=1}^{n} y_i + \sum_{i=n+1}^{N} E(y_i \mid s, \mathbf{y}_s)] \qquad (3.7.2)$$

$$= N^{-1} [n\bar{y}_s + \sum_{i=n+1}^{N} x_i' E(\beta \mid s, y_s)]$$

Under model (3.7.1),

$$E(\beta \mid s, y_s) = [(\sigma^2)^{-1} I + (\tau^2)^{-1} (X_s' X_s)^{-1} H^{-1}]^{-1}$$

$$[(\sigma^2)^{-1} b + (\tau^2)^{-1} (X_s' X_s)^{-1} H^{-1} \nu] \qquad (3.7.3)$$

where

$$b = (X_s' X_s)^{-1} X_s' y_s$$

Using (3.7.2) and (3.7.3), Bayes predictor is

$$\bar{\hat{y}}_B^* = \bar{y}_s f + (1 - f) \bar{x}_{\bar{s}}' [(\sigma^2)^{-1} I + (\tau^2)^{-1} (X_s' X_s)^{-1} H^{-1}]^{-1}$$

$$[(\sigma^2)^{-1} b + (\tau^2)^{-1} (X_s' X_s)^{-1} H^{-1} \nu] \qquad (3.7.4)$$

where

$$\bar{x}_{\bar{s}} = (\bar{x}_{1\bar{s}}, \ldots, \bar{x}_{p\bar{s}})', \quad \bar{x}_{j\bar{s}} = \sum_{i=n+1}^{N} x_{ij} / (N - n), \quad f = n/N$$

Following Arnold (1981), Zellner (1986), Ghosh and Lahiri (1987), the authors considered the natural conjugate prior of β when

$$H = (X_s' X_s)^{-1} \qquad (3.7.5)$$

In this case,

$$\bar{\hat{y}}_B^* = f\bar{y}_s + (1 - f)\bar{x}_{\bar{s}}' [\nu + (1 - C)(b - \nu)] \qquad (3.7.6)$$

where

$$C = \frac{\sigma^2}{\tau^2 + \sigma^2}$$

When τ^2 is very large compared to σ^2, $\bar{\hat{y}}_B^*$ tends to the classical regression estimator

$$\bar{\hat{y}}_{reg} = f\bar{y}_s + (1 - f)b'\bar{x}_{\bar{s}} \qquad (3.7.7)$$

When τ^2 is very small compared to $\sigma^2, \hat{\bar{y}}_B^*$ tends to

$$\hat{\bar{y}}_B' = f\bar{y}_s + (1-f)\bar{x}_{\bar{s}}'\nu \tag{3.7.8}$$

We now assume that ν is known but σ^2, τ^2 are unknown. Assuming without loss of generality $\nu = 0$, we have

$$\hat{\bar{y}}_B^* = f\bar{y}_s + (1-f)(1-C)\bar{x}_{\bar{s}}'b \tag{3.7.9}$$

A *ridge estimator* of C is

$$\hat{C}_K = \frac{K\hat{\sigma}^2}{b'X_s'X_sb} \tag{3.7.10}$$

where

$$\hat{\sigma}^2 = y_s'(I_n - X_s(X_s'X_s)^{-1}X_s')y_s/(n-p) \tag{3.7.11}$$

and $\tau^2 + \sigma^2$ is estimated by

$$\frac{1}{p-2}b'X_s'X_sb \tag{3.7.12}$$

and K is a suitable constant. A EB-ridge type estimator of $\bar{y}$ is

$$\hat{\bar{y}}_{EB(K)} = f\bar{y}_s + (1-f)(1-\hat{C}_K)\bar{x}_{\bar{s}}'b \tag{3.7.13}$$

For K=0, one gets $\hat{\bar{y}}_{reg}$. For $p \geq 3$,

$$\hat{\sigma}^2 = \frac{\hat{\sigma}^2(n-p)}{n-p+2} \tag{3.7.14}$$

is the best scale-invariant estimator of σ^2. Also, in this case, (3.7.12) is the uniformly minimum variance unbiased estimator. Therefore, for $p \geq 3$ the authors proposed

$$K^* = \frac{(n-p)(p-2)}{n-p+2} \tag{3.7.15}$$

Therefore, a EB-ridge estimator of $\bar{y}$, using K^* is

$$\hat{\bar{y}}_{EB}^* = f\bar{y}_s + (1-f)(1-\hat{C})\bar{x}_{\bar{s}}'b \tag{3.7.16}$$

where

$$\hat{C} = \hat{C}_{K^*}.$$

Again, since $C \leq 1$, the positive-part ridge estimator of C is

$$\hat{C}^+ = \min(1, \hat{C}) \tag{3.7.17}$$

which gives the positive-part EB-ridge estimator $\widetilde{\hat{y}}_{EB}^{+*}$ by using $\widehat{C}^+$ in place of $\widehat{C}$ in (3.7.16). For another reason for choice of optimum value K^* of K the equation (3.7.20) may be seen.

The authors compare the EB -estimators $\widetilde{\hat{y}}_{EB}^*$ with the classical regression estimator $\hat{\hat{y}}_{reg}$ in terms of the RSL introduced by Efron and Morris (1973) under the model (3.7.1)(both (i) and (ii)) denoted as ξ. The RSL of $\widehat{\hat{y}}_{EB}^*$ with respect to $\hat{\hat{y}}_{reg}$ is

$$RSL(\xi; \widehat{\hat{y}}_{EB}^*, \hat{\hat{y}}_{reg}) = \frac{r(\xi, \widehat{\hat{y}}_{EB}^*) - r(\xi, \hat{\hat{y}}_B^*)}{r(\xi, \hat{\hat{y}}_{reg}) - r(\xi, \hat{\hat{y}}_B^*)} \qquad (3.7.18)$$

where $r(\xi, e)$ is defined in (3.6.20).

The authors obtained expressions for $r(\xi, \hat{\hat{y}}_B^*)$ and $r(\xi, \hat{\hat{y}}_{EB(K)})$ for $p \geq 3$. It is shown that

$$RSL(\xi, \hat{\hat{y}}_{EB(K)}, \hat{\hat{y}}_{reg}) = \frac{n-p+2}{(n-p)p(p-2)}K^2 - 2K/p + 1 \qquad (3.7.19)$$

which is minimised for fixed n, p for

$$K = \frac{(n-p)(p-2)}{n-p+2} \qquad (3.7.20)$$

the minimum value being $\frac{2n}{p(n-p+2)}$. The estimator $\widehat{\hat{y}}_{EB}^*$ is the best in the class of all estimators of the form $\hat{\hat{y}}_{EB(K)}$ for $p \geq 3$ for fixed n and fixed $p(\geq 3)$. Also

$$RSL(\xi, \widehat{\hat{y}}_{EB}^*, \hat{\hat{y}}_{reg}) = 1 - \frac{(p-2)(n-p)}{p(n-p+2)} \qquad (3.7.21)$$

$$< 1 \; \forall \; n > p \geq 3$$

For fixed p, $RSL(\xi, \widehat{\hat{y}}_{EB}^*, \hat{\hat{y}}_{reg})$ is a decreasing function of n with

$$lim_{n \to \infty} RSL(\xi; \widehat{\hat{y}}_{EB}^*, \hat{\hat{y}}_{reg}) = 2/p \qquad (3.7.22)$$

For fixed $n(\geq 4)$, $RSL(\xi; \widehat{\hat{y}}_{EB}^*, \hat{\hat{y}}_{reg})$ is a decresing function of p provided $3 \leq p \leq 1 + [n/2]$ and is increasing in p if $p \geq 2 + [n/2]$, where $[x]$ is the integer part of x.

Therefore, $\hat{\hat{y}}_{EB}^*$ is always better than $\hat{\hat{y}}_{reg}$ so long as $n > p(\geq 3)$. For fixed p, $\widehat{\hat{y}}_{EB}^*$ has increasingly higher precision compared to $\hat{\hat{y}}_{reg}$ as n increases. For fixed n, $\widehat{\hat{y}}_{EB}^*$ increasingly gains superiority over $\hat{\hat{y}}_{reg}$ for some initial values of p after which EB estimator loses its superiority and even becomes worse than $\widehat{\hat{y}}_{EB}^*$ when p exceeds n. Also, since $\mid \widehat{C}^+ - C \mid < \mid \widehat{C} - C \mid$

almost surely, $RSL(\xi; \widehat{\bar{y}}_{EB}^{+*}, \hat{\bar{y}}_{reg}) < RSL(\xi; \widehat{\bar{y}}_{EB}^{*}, \hat{\bar{y}}_{reg})$, meaning $\bar{y}_{EB}^{+*}$ is a more preceise predictor than $\hat{\bar{y}}_{EB}^{*}$.

The frequentist-risk performance of the estimators are then compared where the frequentist-risk is defined by

$$R(e) = E_p E'(e - \bar{y})^2$$

E' being the expectation with respect to model (i) of (3.7.1). It is shown that for $p > 2(1 + \alpha)$ for some real α and $\beta \in \mathcal{B}$, for all K such that

$$0 \leq K \leq \frac{4\alpha}{(n - p)(n - p + 2)},$$

$$R(\widehat{\bar{y}}_{EB}^{*}) < R(\hat{\bar{y}}_{reg})$$

when

$$\mathcal{B} = \{\beta \in R^p \mid \beta' X_s X_s \beta \leq \frac{p - 2(1 + \alpha)(p + 2)}{2n(1 + \alpha)(p - 2)} \sigma^2\}$$

Thus, for $\beta \in \mathcal{B}, \hat{\bar{y}}_{reg}$ is inadmissible. Also, if $p \geq 3$ and $\beta X_s' X_s \beta \leq \sigma^2$, then

$$R(\widehat{\bar{y}}_{EB(K)}^{+}) \leq R(\widehat{\bar{y}}_{EB(K)}) \leq R(\hat{\bar{y}}_{reg}).$$

3.7.1

We review in this subsection an application of the *generalised linear model* in estimation of strata means. First we recall the concept of univariate generalised linear modelling.

Consider $m + 1$ variables (y, x) where $x = (x_1, \ldots, x_m)$ is a vector of covariates or explainatory variables and y is the response or dependent variable. The data are given by $(y_i, x_i), i = 1, \ldots, n$ where $x_i = (x_{i1}, \ldots, x_{im}), x_{ij}$ being the value of x_j on unit i. The classical linear model is

$$y_i = z_i' \beta + \epsilon_i, \quad \epsilon_i \sim NID(0, \sigma^2) \tag{3.7.23}$$

where z_i, the design-vector is an appropriate function of the covariate vector x_i (often, $z_i = (1 \; x_i)$) and β is a vector of regression coefficients. We shall use the symbol

$$E(y_i \mid x_i) = \mu_i$$

The quantity $\eta_i = z_i' \beta$ is called the natural predictor of y_i. Thus, the classical linear model is

$$\mu_i = \eta_i \tag{3.7.24}$$

In the univariate generalised linear model, the assumptions of classical general linear model are relaxed as follows:

(i)It is assumed that given x_i, y_i are conditionally independently distributed with a distribution belonging to a simple exponential family (given below) with mean $E(y_i \mid x_i) = \mu_i$ and possibly a common scale parameter ϕ.

(ii) The mean μ_i is related to η_i by a *response function*

$$\mu_i = r(\eta_i) \tag{3.7.25}$$

instead of the relation (3.7.24). The corresponding *link function* is

$$\eta_i = g(\mu_i), \tag{3.7.26}$$

where g is the link function and is the inverse of r.

Univariate Exponential Family

The random variable y has a distribution belonging to a univariate exponential family if its density is

$$f(y \mid \theta_i, \phi, w_i) = exp\{\frac{y\theta_i - \psi(\theta_i)}{\phi}w_i + \rho(\theta_i, \phi, w_i)\} \tag{3.7.27}$$

where θ_i is called a *natural parameter*, ϕ is a scale parameter, $\psi(.)$ and $\rho(.)$ are specific functions corresponding to the type of exponential functions, w_i is a weight with $w_i = 1$ for ungrouped data $(i = 1, \ldots, n)$ and $w_i = n_i$ for grouped data $(i = 1, \ldots, h)$ if the average $\bar{y}_i$ of the ith group of n_i observations is considered as response (or $w_i = 1/n_i$ if the sum of the individual responses, $\sum_{j=1}^{n_i} y_{ij}$ is considered as response).

The natural parameter θ_i is a function of $\mu_i, \theta_i = \theta(\mu_i)$. The mean μ_i is given by

$$\mu_i = \psi'(\theta_i) = \frac{\partial \psi(\theta_i)}{\partial \theta_i}$$

Also, the variance is

$$V(y_i \mid x_i) = \sigma^2(\mu_i) = \frac{\phi v(\mu_i)}{w_i}$$

where

$$v(\mu) = \psi''(\theta) = \frac{\partial^2 \psi(\theta)}{\partial \theta^2}$$

For each exponential family there exists a natural or *canonical* link (also response) function. This is given by

$$\eta = \theta = \theta(\mu)$$

i.e.
$$z\beta = \eta = g(\mu) = \theta(\mu)$$

The canonical link function links the natural parameter directly to the linear predictor $z'\beta$.

Bernoulli distribution, Binomial distribution, Poisson distribution, Normal distribution, gamma distribution, inverse-Gaussian distributions are examples of univariate exponential distributions.

EXAMPLE 3.7.1

Suppose $y \sim N(\mu, \sigma^2)$. Here

$$f(y \mid \mu, \sigma^2) = exp\{\frac{y\mu - \mu^2/2}{\sigma^2} + \rho(y, \sigma^2)\}$$

The natural parameter $\theta = \mu; \psi(\theta) = \psi(\mu) = \mu^2/2, \phi = \sigma^2$. Also,

$$V(y_i \mid x_i) = \frac{\phi v(\mu)}{w_i} = \frac{\sigma^2}{w_i}$$

since, $v(\mu_i) = \psi''(\mu_i) = 1$. Natural response function is

$$\mu = \eta = z'\beta$$

Sometimes, a non-linear relationship, e.g.,

$$\mu = \eta^2, \ \mu = \ \log \eta, \ \mu = e^\eta$$

are more appropriate.

For further details, the reader may refer to Fahrmeir and Tutz (1994), among many others.

Ghosh et al (1998) used hierarchical Bayes estimators based on generalised linear models for simultaneous estimation of strata means. Suppose there are m strata and a sample of size n_i is drawn by srs from the ith stratum. Let y_{ik} denote the minimal sufficient statistic (discrete or continuous) for the kth unit within the ith stratum. The y_{ik}'s are assumed to be conditionally independent with

$$f(y_{ik} \mid \theta_{ik}, \phi_{ik}) = exp[\frac{y_{ik}\theta_{ik} - \psi(\theta_{ik})}{\phi_{ik}} + \rho(y_{ik}, \phi_{ik})] \tag{3.7.28}$$

$(k = 1, \ldots, n_i; i = 1, \ldots, m)$. The density (3.7.28) is parametrised with respect to the canonical parameters θ_{ik} and the scale parameters $\phi_{ik}(> 0)$

(supposed to be known) and ψ and ρ are functions specific for different models.

The natural parameters θ_{ik} are first modelled as

$$h(\theta_{ik}) = x'_{ik}\beta + u_i + \epsilon_{ik} \qquad (3.7.29)$$

where h is a strictly increasing function, the $x_{ik}(p \times 1)$ are known design vectors, $\beta(p \times 1)$ is the unknown regression coefficient, u_i are random effects due to strata and the ϵ_{ik} are random errors. It is assumed that u_i and ϵ_{ik} are mutually independent with $u_i \, ind \, N(0, \sigma_u^2)$ and $\epsilon \, ind \, N(0, \sigma^2)$.

Let $R_u = \sigma_u^{-2}, R = \sigma^{-2}, \theta = (\theta_{11}, \theta_{12}, \dots, \theta_{mn_m})', u = (u_1, \dots, u_m)'$. Then the hierarchical model is given by the following:

(I) Conditional on $(\theta, \beta, u, R_u = r_u, R = r), y_{ik}$ are independent with density given in (3.7.23).

(II) Conditional on $(\beta, u, R_u = r_u, R = r), h(\theta_{ik}) \sim NID(x'_{ik}\beta + u_i, r^{-1})$

(III) Conditional on $(\beta, R_u = r_u, R = r), u_i \sim NID(0, r_u^{-1})$.

Ghosh et al assigned the following priors to β, R_u, R.

(IV) β, R_u and R are mutually independent with $\beta \sim$ Uniform $(R^p)(p < m), R_u \sim$ gamma $(a/2, b/2), R \sim$ gamma $(c/2, d/2)$. (The variable $Z \sim$ gamma (α, β) if f(z) $\propto exp(-\alpha z)z^{\beta-1}I_{(0,\infty)}(z))$.

Our interest is in finding the joint posterior distribution of $g(\theta_{ik})$'s where g is a strictly increasing function given the data $\mathbf{y} = (y_{11}, \dots, y_{mn_m})'$, in particular, in finding the posterior mean, variances, covariances of these parameters. In typical applications $g(\theta_{ik}) = \psi'(\theta_{ik}) = E(y_{ik} \mid \theta_{ik})$.

Direct evaluation of the joint posterior distribution of the $g(\theta_{ik})$'s given $\mathbf{y}$ involve high-dimensional numerical integration and is not computationally tractable. The authors used Gibbs sampler techniques to evaluate the posterior distributions. The models in (3.7.28) were extended to the multicategorical data situation and were analysed with reference to an health hazards data set.

In the next section we review applications of some of the results in sections 3.2 - 3.7 in small area estimation. It will be clear that many results developed in all these sections can be applied to the problems of estimation in small areas.

3.8 APPLICATIONS IN SMALL AREA ESTIMATION

A sample survey is often designed to produce estimates for the whole geographical area (country or state) as well as those for some broad well-defined geographical areas (eg. districts). In some situations it is required to produce statistics for local areas (eg. blocks of villages, groups of wards in a city). These are called small areas. The sample is usually selected from the whole population or from different subpopulations (strata, which generally consist of a number of small areas) and as such sample size for a small area is a random variable. Clearly, sample sizes for small areas would be small, even zeros for some areas, because the overall sample size in a survey is usually determined to provide specific accuracy at a much higher level of aggregration than that of small areas. Usually survey estimates based on such small areas would involve formidable standard errors leading to unacceptably wide confidence intervals. Special methods are, therefore, needed to provide reliable estimates for small areas.

Let the finite population $\mathcal{P}$ be divided into A non-overlapping and exhaustive small areas $\mathcal{P}_a$ of sizes $N_a(a = 1, \ldots, A; \cup_a \mathcal{P}_a = \mathcal{P}; \sum_{a=1}^{A} N_a = N)$. Apart from small areas, we shall, consider G mutually exclusive and exhaustive domains or strata $D_g(g = 1, \ldots, G); \cup_g D_g = \mathcal{P}$. The boundaries of the domain may cut across the small areas. In socio-economic surveys, the domains are, generally, socio-economic strata, such as age-sex-religion-occupation-classification such that the units within the same domain are homogeneous in respect of the characteristic of interest and boundaries of these domains cut across the small areas. A domain is generally much larger in size than a small area. It is often assumed that the units belonging to a particular (small area, domain)-cell has the same character as the units belonging to that particular domain, irrespective of any particular small area. Thus when an estimate is required for a small area we use data from the region outside this small area but belonging to different domains which overlap this small (local) area. This is known as "borrowing strength" from similar regions and such estimators are called 'synthetic estimators'.

let y_{agk} be the value of the characteristic 'y' on the unit k belonging to the domain g and area a, i.e. to the cell $(a, g), k = 1, \ldots, N_{ag}$:

$$\sum_{g=1}^{G} N_{ag} = N_{a0} = N_a; \quad \sum_{a=1}^{A} N_{ag} = N_{0g}, \quad \sum_{a=1}^{A} \sum_{g=1}^{G} N_{ag} = N$$

A sample s of size n is selected by a sampling design $p(s)$ from $\mathcal{P}$ and let s_{ag} be the part of the sample of size n_{ag} that falls in the cell (a, g). Let

$s_a = \cup_g s_{ag}$, the part of the sample of size n_a that falls in area a; $s_{og} = \cup_a s_{ag}$ the part of the sample of size n_{og} that falls in domain g. Clearly, $s = \cup_a s_a = \cup_g s_{og}$. Also, n_{ag} is a random variable, whose value may be zero for some (a, g); $n_a = n_{a0} = \sum_g n_{ag}$; $n_{0g} = \sum_a n_{ag}$; $\sum_g \sum_a n_{ag} = n$. We shall denote by $T_a = \sum_{k \in \mathcal{P}_a} y_k$, $\bar{y}_a = \sum_{k \in \mathcal{P}_a} y_k / N_a = T_a / N_a$, $\bar{y}_{s_a} = \sum_{k \in s_a} y_k / n_a$, $\bar{y}_{s_{og}} = \sum_{k \in s_{og}} y_k / n_{0g}$, $\bar{y}_{s_{ag}} = \sum_{k \in s_{ag}} y_k / n_{ag}$, the population total, population mean, the sample mean for small area a, the sample mean for domain g and the sample mean for the cell (a, g), respectively, where y_k denotes the value of y on unit k in the population. Also let $\bar{s}_a = \mathcal{P} - s_a$, $\bar{s}_{ag} = \mathcal{P}_{ag} - s_{ag}$, $\mathcal{P}_{ag} = \mathcal{P} \cap D_g$.

We shall now consider Bayes prediction of small area totals for some simple models. Consider the simple one-way classified random effects model

$$y_{agk} = \alpha_a + \epsilon_{agk} \qquad (3.8.1)$$

where α_a is the effect due to area a and

$$\epsilon_{agk} \sim NID(0, \sigma^2)$$

It is assumed that α_a is a random variable having distribution,

$$\alpha_a \sim NID(\alpha_a^*, \sigma_\alpha^2)$$

$$\alpha_a \underset{\sim}{ind} \epsilon_{agk} \qquad (3.8.2)$$

The models (3.8.1), (3.8.2) can be used when one believes that the value of y in an area depends solely on the characteristic of the area, while this area-effect itself is a random sample from the distribution of such area-effects. Bayes estimator of small area population total T_a, obtained by Theorem 3.3.1 is

$$\hat{T}_{a;B}^A = \sum_{k \in s_a} y_k + \sum_{k \in \bar{s}_a} [(1 - \lambda_a)\alpha_a^* + \lambda_a \bar{y}_{sa}] \qquad (3.8.3)$$

where

$$\lambda_a = \frac{n_a M}{n_a M + 1}, \quad M = \frac{\sigma_\alpha^2}{\sigma^2} \qquad (3.8.4)$$

In (3.8.3) an estimate of $y_k (k \in \bar{s})$ is a weighted average of the prior mean α_a* and the sample mean $\bar{y}_{sa}$. As $M \to \infty$ (the prior information is very much inaccurate relative to the current) $\hat{T}_{a;B}^A$ tends to be the simple expansion estimator $N_a \bar{y}_{s_a}$. The prediction-variance of $\hat{T}_{a:B}^A$ under frequentist approach, with respect to model (3.8.1) is

$$V(\hat{T}_{a;B}^A - T_a) = (N_a - n_a)\sigma^2 + (N_a - n_a)\lambda_a^2 \sigma^2 / n_a \qquad (3.8.5)$$

Note that since λ_a is an increasing function of M, larger is the value of M, greater is the prediction variance (3.8.5).

If the α_a^*'s are not known, the maximum likelihood estimate *(mle)* of α_a^* under model (3.8.1), (3.8.2) is given by $\hat{\alpha}_a^* = \bar{y}_{s_a}$. Substituting this in (3.8.3) one gets the EB predictor of T_a,

$$\hat{T}_{a;EB}^A = N_a \bar{y}_{s_a} \tag{3.8.6}$$

which is the simple expansion predictor. Its prediction-variance is

$$V(\hat{T}_{a;EB}^A - T_a) = \frac{N_a(N_a - n_a)\sigma^2}{n_a} \tag{3.8.7}$$

When $\alpha_a^* = \alpha_0^* \ \forall \ a = 1, \ldots, A$ (i.e. the area effects are samples from a common distribution $N(\alpha_0^*, \sigma_\alpha^2)$), Bayes predictor of population total T_a, $\hat{T}_{a;B}^1$ is obtained by substititing α_0^* for α_a^* in (3.8.3). Clearly, as $M \to \infty$, $\hat{T}_{a;B}^1$ also tends to the simple expansion predictor $N_a \bar{y}_{sa}$.

Similarly, if α_0^* is not known, substituting

$$\hat{\alpha}_0^* = \sum_a \lambda_a \bar{y}_{sa} / \sum_a \lambda_a$$

into $\hat{T}_{a;B}^1$ we get $\hat{T}_{a;EB}^1$ with prediction-variance under frequentist approach, with respect to model (3.8.1),

$$V(\hat{T}_{a;EB}^1 - T_a) = (N_a - n_a)\sigma^2 + (N_a - n_a)^2 \lambda_a \sigma^2 / n_a$$

$$+ (N_a - n_a)^2 (1 - \lambda_a)^2 \sigma_\alpha^2 / \sum_a \lambda_a \tag{3.8.8}$$

This quantity can be shown to be smaller than (3.8.7). Thus when α_a^*'s are nearly equal but unknown, one may prefer $\hat{T}_{a;EB}^1$ in the smaller mean square error(mse)- sense to $\hat{T}_{a;EB}^A$, even though $\hat{T}_{a;EB}^1$ has a bias under (3.8.1) when α_a's are unknown.

Consider now the domain-dependent model

$$y_{agk} = \beta_g + \epsilon_{agk}, \quad \epsilon_{agk} \sim NID(0, \sigma^2) \tag{3.8.9}$$

Assume a normal prior for β_g,

$$\beta_g \sim NID(\beta_g^*, \sigma_B^2), \quad \beta_g \ ind \ \underset{\sim}{\epsilon}_{agk} \tag{3.8.10}$$

The Bayes predictor of T_a by application of Theorem 3.3.1 is

$$\hat{T}_{a;B}^{(G)} = \sum_g \sum_{k \in s_{ag}} y_k + \sum_g \sum_{k \in \bar{s}_{ag}} \{(1 - \mu_g)\beta_g^* + \mu_g \bar{y}_{s0g}\} \qquad (3.8.11)$$

where

$$\mu_g = \frac{n_{0g}K}{n_{0g}K + 1}, \quad K = \frac{\sigma_B^2}{\sigma^2}$$

with prediction-variance (under frequentist approach)

$$V(\hat{T}_{a;B}^{(G)} - T_a) = \sum_g (N_{ag} - n_{ag})\sigma^2 + \sum_g (N_{ag} - n_{ag})\mu_g^2 \sigma^2 / n_{0g} \qquad (3.8.12)$$

When $\beta_g^* = \beta_0^* \ \forall \ g$ we get the predictor $\hat{T}_{a;B}^{(1)}$ by substituting β_0^* for β_g^*. Its prediction-variance is also given by (3.8.12).

If β_g^* is unknown, replacing β_g^* by its *mle* (UMVU estimator) $\bar{y}_{s0g}$ we get the EB estimator

$$\begin{aligned}
\hat{T}_{a;EB}^{(G)} &= \sum_g \sum_{k \in s_{ag}} y_k + \sum_g \sum_{k \in \bar{s}_{ag}} \bar{y}_{s0g} \\
&= \sum_g \sum_{k \in s_{ag}} y_k + \sum_g (N_{ag} - n_{ag})\bar{y}_{s0g} \qquad (3.8.13) \\
&= \hat{T}_a^{*(G)}
\end{aligned}$$

which is identical to the least squares estimator. Similarly, if β_0^* is unknown, we substitute

$$\hat{\beta}_0^* = \sum_g \mu_g \bar{y}_{0g} / \sum_g \mu_g$$

for β_0^* in $\hat{T}_{a;B}^{(1)}$ to get the EB-estimator $\hat{T}_{a;EB}^{(1)}$.

Note that $\hat{T}_{a;B}^{(G)}(\hat{T}_{a;EB}^{(G)}), \hat{T}_{a;B}^{(1)}(\hat{T}_{a;EB}^{(1)})$ can be calculated even if some $n_{ag} = 0$. It is only needed that $n_{0g} > 0 \ \forall \ g$.

If the normality assumption is relaxed, $\hat{T}_{a;B}^{(G)}, \hat{T}_{a;B}^{(1)}$ become best linear unbiased estimators under the models (3.8.9),(3.8.10) (without normality).

The UMVU estimator of σ^2 when β_g^* is unknown is

$$\hat{\sigma}^2 = \sum_{s_{ag}} (y_k - \bar{y}_{0g})^2 / (n - G) \qquad (3.8.14)$$

If K is known, the UMVU estimator of σ_B^2 is $K\hat{\sigma}^2$.

Ghosh and Meeden (1986), Mukhopadhyay (1998 c) discussed methods for estimating $\dot{K}$ and this has been reviewed in section 3.6. Often one can provide a guessed value of K, say, K_f. It can be shown that so long as $K_f > K/2$, $\mathrm{MSE}(\hat{T}_{a;B}^{(G)})$ and $\mathrm{MSE}(\hat{T}_{a;B}^{(1)})$, using $K = K_f$, are less than that of the least square estimator $\hat{T}_a^{*(G)}$. Also, an incorrect value of K_f does not bias $\hat{T}_{a;B}^{(G)}$. The condition $K_f > K/2$ is sufficient to guarantee that $\hat{T}_{a;B}^{(G)}$ and $\hat{T}_{a:B}^{(1)}$ are superior to $\hat{T}_a^{*(G)}$ with repect to mse. Also, $\hat{T}_{a;EB}^{(1)}$ is superior to $\hat{T}_a^{*(G)}$ if $K_f > K$.

The models (3.8.1),(3.8.2) and (3.8.9),(3.8.10) are special cases of the model

$$y_{agk} = \mu_{ag} + \epsilon_{agk}, \ \epsilon_{agk} \sim NID(0, \sigma^2) \tag{3.8.15}$$

where we assume that μ_{ag} has one of the following priors:

$$(i)\mu_{ag} \sim NID(\mu_{a0}^*, \sigma_\mu^2), \tag{3.8.16}$$

$$(ii)\mu_{ag} \sim NID(\mu_{0g}^*, \sigma_\mu^2), \tag{3.8.17}$$

$$(iii)\mu_{ag} \sim NID(\mu_0^*, \sigma_\mu^2), \tag{3.8.18}$$

$$(iv)\mu_{ag} \sim NID(\mu_{ag}^*, \sigma_\mu^2) \tag{3.8.19}$$

with

$$\mu_{ag} \underset{\sim}{ind} \epsilon_{agk}$$

The cases (3.8.16) and (3.8.17) (together with (3.8.15)) coincide with (3.8.1), (3.8.2) and (3.8.9),(3.8.10) respectively. The case (3.8.18) states that there is an overall-effect throughout the survey population and there is no specific area- effect or domain-effect. The model (3.8.19) states that for each (small area, domain) cell there is a specific effect. In (3.8.19) the area-effects and domain-effects may or may not be additive.

Application of Theorem 3.3.1 gives the general form of the Bayes predictor

$$\hat{T}_{a;B}' = \sum_g \sum_{k \in s_{ag}} y_k + \sum_g \sum_{k \in \bar{s}_{ag}} \{(1 - \lambda_{ag})O_{ag} + \lambda_{ag}\bar{y}_{ag}\} \tag{3.8.20}$$

where $O_{ag} = E(\mu_{ag})$, the known prior mean and

$$\lambda_{ag} = \frac{n_{ag}L}{n_{ag}L + 1}, L = \frac{\sigma_\mu^2}{\sigma^2} \tag{3.8.21}$$

In $\hat{T}'_{a;B}$ a weighted average of prior information and the current data is used to predict the unobserved part $\sum_{\bar{s}_a} y_k$. This estimator becomes the post-stratified direct estimator $\sum_g N_{ag}\bar{y}_{ag}$ when $L \to \infty$.

When $n_{ag} = 0$, we use O_{ag} to predict the non-sampled units in the cell (a,g). When $n_{ag} > 0$, we use the weighted average of O_{ag} and the cell means $\bar{y}_{ag}$ for the same.

When O_{ag} is not known, its least square estimates under models (3.8.16)-(3.8.18) are, respectively,

$$\hat{\mu}^*_{a0} = \sum_g \lambda_{ag}\bar{y}_{ag} / \sum_g \lambda_{ag} \tag{3.8.22}$$

$$\hat{\mu}^*_{0g} = \sum_a \lambda_{ag}\bar{y}_{ag} / \sum_a \lambda_{ag} \tag{3.8.23}$$

$$\hat{\mu}^*_0 = \sum_a \sum_g \lambda_{ag}\bar{y}_{ag} / \sum_a \sum_g \lambda_{ag} \tag{3.8.24}$$

When $\sigma^2_\mu \simeq 0$, and n_{ag} values are such that $\lambda_{ag} \simeq 0 \ \forall \ (a,g)$ and prior means are unknown, estimate of $\mu^*_{a0}, \tilde{\mu}^*_{a0} = (\bar{y}_{sa})$ produces the simple expansion estimator $\hat{T}_{aE} = N_a\bar{y}_{sa}$, estimate of $\mu^*_{0g}, \tilde{\mu}^*_{0g}(= \bar{y}_{s0g})$, the synthetic estimator $\hat{T}^{*(G)}_a$ and estimate of $\mu^*_0, \tilde{\mu}^*_0(= \bar{y}_s)$, the EB-estimator $\sum_{k\in s_a} y_k + (N_a - n_a)\bar{y}_s$.

When $\sigma^2_\mu = \infty$ all these EB-estimators (including the one for the model (3.8.19)) produce the simple direct estimator $\hat{T}_a = \sum_g N_{ag}\bar{y}_{sag}$.

EXAMPLE 3.8.1.

A survey was carried out in the district Hugli, West Bengal, India in 1997-98 under the support of Indian Statistical Institute, Calcutta to estimate the population size and some household characteristics (like, distribution of persons according to age-groups, religion, educational status, occupational category, employment status) in different municipal towns (small areas). The urban area of Hugli district was divided into two strata - stratum 1 consisting of 24 single-ward towns and stratum 2 consisting of 12 towns, each having more than one ward. Two towns from the first stratum and nine towns from the second stratum were selected with probability proportional to the number of households in the 1991 census. For each selected ward a list of residential houses was prepared from the assessment registrars of the municipality and samples of these houses were selected using linear systematic sampling.

The results of the survey were mainly confined to the sample municipal areas in stratum 2. For municipalities in stratum 1, list of residential houses was not available with the municipal or Gram-Panchayet offices. Even the boundaries of these non-municipal towns were not clearly defined. As such the data that were collected for these areas could not be used to make any reliable estimate.

The estimates of population counts depended heavily on the estimation of number of households in these areas. Although the municipalities maintained some figures, they were found unreliable on the basis of earlier census estimates for the same. The number of households for 1998 for these areas were estimated on the basis of the census figures for 1971, 1981 and 1991 and an exponential growth.

In the synthetic estimation techniques we intend to obtain some synthesizing characters whose values remain independent of the municipal areas i.e. remain approximately constant over all municipalities. With the help of such characteristics we can then obtain estimates of population sizes for different small areas. The characteristics tried were: average household (hh) size by four occupation groups (x_1), average hh size by three education groups (x_2), average hh size by two types of houses (x_3), average hh size by two types of religion (x_4), average hh size by the number of living rooms (x_5), average hh size by possession of TV/Scooter/Refrigerator/Telephone (x_6), average hh size by type of ownership of houses (x_7), average hh size by possession of agricultural land (x_8). For each of these characteristics, estimated average household size for different municipalities and for all the sampled municipalities taken together were calculated. It was found that for each of the characteristics x_i, the sample estimate $S_j^i = \{S_{j1}^i, \ldots, S_{jm_i}^i\}$ of average hh size for the m_i groups of x_i, for the jth municipal town was of the same order as the overall easimate $S_0^i = \{S_{01}^i, \ldots, S_{om}^i\}$. The population size for each group for the characteristics x_i for each municipality and hence the population size for each municipality were calculated on the basis of the synthesizing character $x_i (i = 1, \ldots, 8)$. These were compared with the corresponding census estimates for 1991. It was found that x_5 produced reliable estimates of population sizes in the sense that the estimates obtained showed greater consistency with the census estimates than those obtained using other synthesizing characters. For details on synthetic method of estimation the reader may refer to Mukhopadhyay (1998 e).

Another series of estimates based on the linear regression of population size (y) on the number of households in the sample (x) was also considered.

The model was

$$Y_{as} = \alpha + \beta x_{as} + e_a, \ a = 1, \ldots, 9$$

where Y_{as} denotes the number of persons in the sample, x_{as} the number of households in the sample in municipal area a and e_a are independently distributed errors with mean zero and variance σ^2. The estimated number of persons for the area a is

$$\hat{Y}_a = \hat{\alpha} + \hat{\beta} X_a$$

where X_a is the estimated number of households in area a. Empirical Bayes estimation of population size was also considered. Consider the following normal theory model:

$$Y_{as} = \beta_a X_{as} + e_a, \ e_a \sim NID(0, \sigma^2)$$

$$\beta_a = B_a + u_a, \ u_a \sim N(B_a, \tau^2), \ u_a \ ind \ e_a, \ a = 1, \ldots, 9$$

Therefore, the posterior distribution of β_a given Y_{as} is independent normal with mean β^* and posterior variance τ^* where

$$\beta^* = \frac{Y_{as} X_{as} \tau^2 + B_a \sigma^2}{X_{as} \tau^2 + \sigma^2}, \ \tau^* = \frac{\sigma^2 \tau^2}{X_{as}^2 \tau^2 + \sigma^2}$$

An estimate of B_a is

$$\hat{B}_a = \frac{1}{4} \left[\frac{Y_a(71)}{X_a(71)} + \frac{Y_a(81)}{X_a(81)} + \frac{Y_a(91)}{X_a(91)} + \frac{Y_{as}}{X_{as}} \right], a = 1, \ldots, 9$$

where $Y_a(71)(X_a(71))$ denotes the total population (number of households) in area a in 1971 and similarly for the other notations. Also,

$$\hat{\sigma}^2 = \frac{1}{8} \sum_{a=1}^{9} (Y_{as} - \hat{\beta} X_{as})^2, \ \ \hat{\beta} = \sum_{a=1}^{9} Y_{as} / \sum_{a=1}^{9} X_{as}$$

$$\hat{\tau}^2 = \frac{1}{8} (r_{as} - \bar{r}_s)^2$$

where $r_{as} = Y_{as}/X_{as}, \bar{r}_s = \sum_{a=1}^{9} r_{as}/9$. An empirical Bayes estimate of population size for area a for 1997-98 is

$$\hat{Y}_a^B = \hat{\beta}_a^* X_a$$

Also, the 95% confidence interval for Y_a is

$$(\hat{Y}_a^B - 1.96 \hat{\sigma}(\hat{Y}_a^B), \hat{Y}_a^B + 1.96 \hat{\sigma}(\hat{Y}_a^B))$$

where

$$\hat{\sigma}(\hat{Y}_a^B) = \sqrt{X_a^2 \hat{V}(\hat{\beta}_a)}.$$

The estimated population are given in table 3.1. The empirical Bayes estimates seem to be most consistent with the rate of population growth in the region. In recent times the next decennial census for the year 2001 has been initiated by the Registrar General of India for the whole country. Its results, when available, will throw further light on the performane of these estimators. Further details are available in Mukhopadhyay (1998 e, g).

Table 3.1 Estimates of Population of Some Municipal Areas of Hugli District, West Bengal

Municipal Town	1991 Census	Estimated Number of Persons in 1997-98			
		Synthetic Estimate	Regression Estimate	EB Estimate $\hat{Y}_a^B$	$\hat{\sigma}(\hat{Y}_a^B)$
Arambagh	45211	56084	46693	55678	6449.1
Bansberia	93520	118520	125286	124366	12139.8
Hoogly-Chinsurah	151806	173305	159773	163406	10611.8
Bhadreswar	72474	79725	77852	81628	7812.3
Chandannagar	120378	148384	132081	158355	11238.1
Baidyabati	90081	105223	104073	105544	10143.0
Srirampur	137028	141921	146770	140961	14990.3
Rishra	102815	114882	115894	111058	11427.6
Konnagar	62200	80087	73891	66838	7285.8

In the next subsection we consider application of Carter-Ralph (1974) modification of James-Stein estimators discussed in Section 3.5 in small area estimation.

3.8.1 FAY-HERRIOT ESTIMATES

In US, statistics on population, per capita income (PCI) and adjusted taxes, among other characteristis, are used for determining the allocation of funds to the state governements and subsequently to the local governments. The PCI were estimated from the 1970-census. However, for small areas, with population less than 500, the sampling error of the census estimates were large and were replaced by the respective county averages. Fay and Herriot (1979) used James-Stein estimates based on auxiliary data related to PCI-data available from the Internal Revenue Service (IRS) and the 1970-census to improve upon estimates of allocations for local areas. In deriving their estimates Fay and Herriot (1979) considered extension of (3.5.23), (3.5.25) and (3.5.26) to the linear regression case. Consider

$$y_a \, ind \, N(\theta_a, D_a), \theta_a \, ind \, N(x_a'\beta, B) \qquad (3.8.25)$$

where x_a' is a p-dimensional row vector and β has a uniform (improper) prior distribution. The row vector x_a' and sampling variance D_a are both known, but β and B are to be estimated from the data. Now, assuming B is known, the weighted regression estimate

$$y_a^* = x_a'(X'V^{-1}X)^{-1}X'V^{-1}\mathbf{y} \qquad (3.8.26)$$

where $V = \text{Diag}(V_{aa}, a = 1, \ldots, k), V_{aa} = D_a + B$, gives the minimum variance unbiased estimate of $x_a'\beta$. Over the same joint distribution,

$$E(\sum_a \frac{(y_a^* - y_a)^2}{B + D_a}) = k - p \qquad (3.8.27)$$

Following Carter and Ralph (1974), Fay and Herriot estimated B^* as the unique solution to the constrained equation

$$B > 0, \quad \sum_a \frac{(y_a - y_a^*)^2}{B + D_a} = k - p \qquad (3.8.28)$$

They considered $B^* = 0$ when no positive solution is found. The estimator of θ_a is then

$$\delta_{aCR}^{(1)} = \frac{B^*}{B^* + D_a} y_a + \frac{D_a}{B^* + D_a} y_a^* \qquad (3.8.29)$$

If B is known, $\delta_{aCR}^{(1)}$ is the classical Bayes estimator (compare with (3.5.19)).

An alternative estimator for this problem based on a maximum likelihood approach to fitting the model

$$\theta_a \, ind \, N(x_a'\beta, BD_a^\alpha)$$

when β, B and α may be jointly estimated from the data has also been discussed by Fay and Herriot. Earlier, Ericksen (1973, 1974) explored the use of sample data to determine regression estimates. For further details on prediction for small areas in finite population sampling the reader may refer to Ghosh and Rao (1994), Chaudhuri (1994) and Mukhopadhyay (1998 e).

3.9 BAYES PREDICTION UNDER RANDOM ERROR VARIANCE MODEL

Butar and Lahiri (2000) considered empirical Bayes estimation of several (infinite) population means and variances under random error variance model. Suppose there are m populations. A random sample of size n_i is drawn from the ith population. Let y_{ij} be the value of y on the jth sampled unit in the ith population $(j = 1, \ldots, n_i; i = 1, \ldots, m), \bar{y}_{is} = \sum_{j=1}^{m_i} y_{ij}/n_i, s_i^2 = \sum_{j=1}^{n_i}(y_{ij} - \bar{y}_i)^2/(n_i - 1), y_{is} = (y_{i1}, \ldots, y_{n_i})$. Consider the following model

$$y_{ij} \mid \theta_i, \sigma_i^2 \sim NID(\theta_i, \sigma_i^2)$$

$$\theta_i \sim NID(x_i'\beta, \tau^2) \quad (i = 1, \ldots, m; j = 1, \ldots, n_i) \tag{3.9.1}$$

$$\sigma_i^2 \underset{\sim}{ind} IG\{\eta, (\eta - 1)\xi\}$$

where the Inverse Gamma (IG) density is given by

$$f(\sigma_i^2) = \{(\eta - 1)\xi\}^\eta (1/\sigma_i^2)^{\eta+1}$$

$$e^{-(\eta-1)\xi/\sigma_i^2}/\Gamma(\eta), \sigma^2 > 0 \tag{3.9.2}$$

Here $x_i = (x_{i1}, \ldots, x_{ip})$ is a $p \times 1$ vector of known and fixed observations x_{ij}'s on auxiliary variables $x = (x_1, \ldots, x_p)$ on unit i and β is a $p \times 1$ vector of regression coefficients. The authors considered Bayes estimation of θ_i under squared error loss function. The model (3.9.1) is an extention of the random error variance model considered by Kleffe and Rao (1992) who, however, obtained empirical best linear estimators of θ_i.

The posterior density of σ_i^2 is

$$f(\sigma_i^2 \mid y_i, \Psi) \propto (\sigma_i^2)^{-(\frac{n_i+1}{2}+\eta)}(n_i\tau^2 + \sigma_i^2)^{-1/2}$$

$$exp\{-\frac{(n_i - 1)s_i^2}{2\sigma_i^2} - \frac{n_i(\bar{y}_i - x_i'\beta)^2}{2(n_i\tau^2 + \sigma_i^2)} - \frac{(\eta - 1)\xi}{\sigma_i^2}\} \tag{3.9.3}$$

where $\Psi = (\beta, \tau^2, \eta, \xi)$. Hence, Bayes estimator of $b(\sigma_i^2)$, a function of σ_i^2 is

$$\hat{b}^B(\sigma_i^2) = \int_0^\infty b(\sigma_i^2) f(\sigma_i^2 \mid y_i, \Psi) d\sigma_i^2$$

Now, the posterior distribution of θ_i given y_i and σ_i^2 is normal with mean $(1 - B_i)\bar{y}_i + Bx_i'\beta$ and variance $\tau^2 B_i$ where $B_i = \sigma_i^2/(\sigma_i^2 + n_i\tau^2)$. Hence, Bayes estimator of θ_i when Ψ is known, is

$$\hat{\theta}_i^B = E(\theta_i \mid y_i, \Psi) = E[E\{\theta_i \mid y_i, \Psi, \sigma_i^2\} \mid y_i, \Psi]$$

$$= (1 - w_i)\bar{y}_i + w_i x_i'\beta \qquad (3.9.4)$$

where

$$w_i = E(B_i \mid y_i, \Psi) = \int_0^\infty B_i f(\sigma_i^2 \mid y_i, \Psi) d\sigma_i^2 \qquad (3.9.5)$$

The measure of uncertainty of $\hat{\theta}_i^B$ is

$$Var(\theta_i \mid y_i, \Psi) = \tau^2 w_i + (\bar{y}_i - x_i'\beta)^2 Var[B_i \mid y_i, \Psi]$$

In practice, the parameters in Ψ need to be estimated from the available data. The authors considered empirical Bayes estimators of $b(\sigma_i^2)$ and θ_i by using the ANOVA-type estimators of the parameters in Ψ as proposed by Arora et al (1997).

Following Tierney et al (1989) and Kass and Steffey (1989) they also proposed suitable approximations to the Bayes estimators of θ_i and $b(\sigma_i^2)$. It is known that the posterior probability density is proportional to the product of the likelihood and the prior. Hence, Bayes estimator of $b(\theta)$ is of the type

$$E(b(\theta) \mid y) = \frac{\int b(\theta) L(\theta) \pi(\theta) d(\theta)}{\int L(\theta) \pi(\theta) d\theta} \qquad (3.9.6)$$

where $b(\theta)$ is some function of θ. The expression (3.9.6) can be derived approximately by using the following lemma.

LEMMA 3.9.1 Let $h(\theta)$ be some smooth function of a m-dimensional vector θ, having first five derivatives and having a minimum at $\hat{\theta}$ and b be some other smooth function of θ having first three derivatives. Then under suitable conditions

$$E[b(\theta)] \simeq b(\hat{\theta}) + \frac{1}{2n}\{b''(\hat{\theta})[h''(\hat{\theta})]^{-1} - b'(\hat{\theta})h'''(\hat{\theta})[h''(\hat{\theta})]^{-2}\} \qquad (3.9.7)$$

where $h(\theta) = -n^{-1}logL(\theta)\pi(\theta)$.

The authors used the transformation $\sigma_i^2 = exp(-\rho_i)$ and calculated the posterior density $f(\rho_i \mid y_i)$ and hence $h(\rho_i)$. Using (3.9.7) they obtained approximate expressions for $E[b(\rho_i) \mid y_i, \hat{\psi}]$ where $B_i = b(\rho_i)$. They extended the resampling method of Laird and Louis (1987) to measure uncertainty of estimators $\hat{\theta}_i^{EB}$ and $\hat{b}^{EB}(\sigma_i^2)$ and made a numerical study to check the accuracy of Laplace approximation and compared the performance of empirical Bayes estimators of θ_i and σ_i^2. They (1999) also considered the problem of estimation of a finite population variance.

3.10 EXERCISES

1. Consider the Binomial superpopulation model: y_k's are independent with $P_\theta(y_k = 1) = \theta = 1 - P_\theta(y_k = 0), k = 1, \ldots, N$. Suppose the quantity to be predicted is the population distribution function $F_N(t) = \frac{1}{N} \sum_{i=1}^{N} \Delta(t - y_i)$ where $\Delta(z) = 1(0)$ when $z \geq 0$ (elsewhere). Using a $B(a, b)$ prior for θ, as in example 3.2.1, find a Bayes predictor and hence a minimax predictor of $F_N(t)$.

(Bolfarine, 1987)

2. Consider the Binomial superpopulation model and the $B(a, b)$ prior for θ as in example 3.2.1. Using the Linex loss function (3.4.1) find the Bayes predictor of the population total T and its Bayes risk.

(Bolfarine, 1989)

3. Consider p independent random variables $X_i \sim N(\theta_i, \sigma^2), i = 1, \ldots, p$. The maximum likelihood estimate of θ_i is $\hat{\theta}_i = X_i$. Suppose $\theta_i \sim N(\mu, \tau^2)$. Then posterior distribution of θ_i is

$$\pi(\theta_i \mid X_i) \sim N[\delta^B(X_i), \frac{\sigma^2 \tau^2}{\sigma^2 + \tau^2}], \ i = 1, \ldots, p$$

where Bayes estimate of θ_i is

$$\delta^B(X_i) = \frac{\sigma^2}{\sigma^2 + \tau^2}\mu + \frac{\tau^2}{\sigma^2 + \tau^2}X_i$$

In EB estimation μ, τ^2 are estimated from the marginal distribution of $X_i : f(X_i) \sim N(\mu, \sigma^2 + \tau^2), i = 1, \ldots, p$. Hence,

$$E(\bar{X}) = \mu, \ E[\frac{(p - 3)\sigma^2}{\sum_i(X_i - \bar{X})^2}] = \frac{\sigma^2}{\sigma^2 + \tau^2}$$

Therefore,

$$\hat{\mu} = \bar{X}, \ \hat{\tau}^2 = \sigma^2 [\frac{\sum_i (X_i - \bar{X})^2}{(p-3)\sigma^2} - 1]$$

Hence, show that an EB-estimator of θ_i is

$$\delta_i^E(X) = [\frac{(p-3)\sigma^2}{\sum_i (X_i - \bar{X})^2}]\bar{X} + [1 - \frac{(p-3)\sigma^2}{\sum_i (X_i - \bar{X})^2}]X_i$$

Also, if $p \geq 4$,

$$E[\sum_{i=1}^{p} (\theta_i - \delta_i^E(X))^2] < E[\sum_{i=1}^{p} (\theta_i - X_i)^2] \ \forall \ \theta_i$$

where the expectation is taken over the distribution of X_i given θ_i.

(Effron and Morris, 1973)

4. Let $Y = (Y_1, \ldots, Y_n)'$ have a joint distribution conditional on $\theta(= (\theta_1, \ldots, \theta_s))'$ with

$$E_\theta(Y_i) = \mu(\theta), i = 1, \ldots, n$$

Let the prior distribution of θ be such that

$$E\{\mu(\theta)\} = m$$

and $V\{\mu(\theta)\} < \infty$. Let $\bar{Y} = \sum_{i=1}^{n} Y_i/n$ and assume $V_\theta(\bar{Y}) < \infty$. Prove that if the posterior expectation

$$E\{\mu(\theta) \mid Y = y\} = \alpha\bar{y} + \beta$$

where α, β are independent y, then

$$E\{\mu(\theta)\} \mid Y\} = [V\{\mu(\theta)\} + E_\theta\{V(\bar{Y} \mid \theta)\}]^{-1}$$

$$[\bar{y}V\{\mu(\theta)\} + mE_\theta V\{\bar{Y} \mid \theta)]$$

The above model is satisfied for the following simple cases: Y_i are condition-ally independent normal, binomial, negative binomial, gamma, exponential and their unknown parameters follow the natural conjugate priors.

(Ericson, 1969 b)

5. Let $X = (X_1, \ldots, X_N)'$ [the random variable X_i having the finite popu-lation value x_i] have a prior distribution $\xi_X(x)$ generated by the assumption

that conditional on $\theta = (\theta_1, \ldots, \theta_m)'$, X_i's are iid with density $f(z \mid \theta)$ and θ has a distribution $g(\theta)$. Thus

$$\xi(x) = \int_\Theta \prod_{i=1}^N f(x_i \mid \theta) g(\theta) d\theta.$$

Let $\mu(\theta) = E(X_i \mid \theta), m = E(X_i), \mu = \sum_{i=1}^N X_i/N, x_s = \{(i, x_i), i \in s\}$. Let G_f be a class of distributions of θ having density $g(\theta \mid x', n', y')$ with the property that if x be any observation on X then the posterior distribution of θ is

$$g(\theta \mid x' + x, n' + 1, y'') \in G_f$$

Assume that for every $g(\theta \mid x', n', y') \in G_f$,

$$m = \frac{x' + a}{n' + b}$$

where a, b are constants. Show that

$$E[\mu(\theta) \mid x_s] = \frac{\bar{x}_s V(\mu(\theta)) + m E_\theta V(\bar{X}_s \mid \theta)}{V(\mu(\theta)) + E_\theta V(\bar{X}_s \mid \theta)}$$

$$E[\mu \mid x_s] = \frac{\bar{x}_s V(\mu) + m E_\mu V(\bar{X}_s \mid \mu)}{V(\mu) + E_\mu V(\bar{X}_s \mid \mu)}$$

$$V(\mu \mid x_s) = \frac{N - n}{N} \{V(\mu(\theta) \mid x_s)/V(\mu(\theta))\} V(\mu)$$

where $\bar{x}_s = \sum_{i \in s} x_i/n$, $\bar{X}_s = \sum_{i \in s} X_i/n$ and n is the size of the sample s.

6. Consider the model of example 3.2.2 and suppose that σ^2 is unknown. Assuming a non-informative prior for (β, σ^2) as given in (3.3.12), show that the ratio predictor $\hat{T}_R = (\bar{y}_s/\bar{x}_s)X$ is the Bayes predictor of T with Bayes prediction risk

$$E(\hat{T}_R - T)^2 = \frac{n - 1}{n - 3} \frac{\sum_{\bar{s}} x_i}{\sum_s x_i} X \hat{\sigma}_s^2$$

where $\hat{\sigma}_s^2$ is given in (3.3.15).

(Bolfarine, 1987; Bolfarine and Zacks, 1991)

7. Suppose the population is divided into K clusters of size $N_i (i = 1, \ldots, K)$. In the first stage a sample s of size k clusters is selected. In the second stage, a sample s_i of size n_i is selected from the ith cluster selected at the

first stage ($i = 1, \ldots, k$). Let y_{ij} be the y-value associated with the j-th unit in the i-th cluster. Suppose the model is

$$y_{ij} = \mu_i + e_{ij}, e_{ij} \sim NID(0, \sigma_i^2)$$

$$\mu_i = \nu + v_i, \quad v_i \sim NID(0, \sigma_v^2), j = 1, \ldots, N_i; i = 1. \ldots, K$$

$$e_{ij} \underset{\sim}{ind} v_i$$

Let $y_s = \{y_{ij}, j \in s_i; i \in s\}$ denote the observed data. Show that the posterior distribution of $\mu = (\mu_1, \ldots, \mu_K)$ given y_s is multivariate normal with mean vector $\hat{\mu}$, where

$$\hat{\mu}_i = \begin{cases} (1 - \lambda_i)\bar{y}_s + \lambda_i \bar{y}_{si} & i \in s \\ \bar{y}_s & i \in \bar{s}, \end{cases}$$

where $\lambda_i = \sigma_v^2/(\sigma_v^2 + \sigma_i^2/n_i)$, $\bar{y}_{si} = \sum_{j \in s_i} y_{ij}/n_i$, $\bar{y}_s = \sum_{i \in s} \lambda_i \bar{y}_{si} / \sum_{i \in s} \lambda_i$ and posterior covariance matrix with diagonal and off-diagonal elements given by

$$C_{ij} = \begin{cases} (1 - \lambda_i)^2 v^2 + (1 - \lambda_i)\sigma_v^2; & i = j \\ (1 - \lambda_i)(1 - \lambda_j)v^2; & i \neq j, \end{cases}$$

where $v^2 = [\sum_{i \in s}(\sigma_v^2 + \sigma_i^2/n_i)^{-1}]^{-1}$. Hence observing that the population total T is given by

$$T = \sum_{i \in s}\sum_{j \in s_i} y_{ij} + \sum_{i \in s}\sum_{j \notin s_i} y_{ij} + \sum_{i \notin s}\sum_{j=1}^{N_i} y_{ij}$$

find Bayes predictor of T along with its Bayes risk. Also find Bayes predictor of T and its Bayes risk with respect to the Linex loss function.

(Scott and Smith, 1969; Bolfarine, 1989)

Chapter 4

Modifications of Bayes Procedures

4.1 Introduction

This chapter considers different modifications of Bayes procedures and their applications in finite population sampling. Section 4.2 reviews Bayes least squares prediction or Linear Bayes prediction. Section 4.3 addresses restricted Bayes least squares prediction. The problem of Constrained Bayes prediction and Limited Translation Bayes prediction have been considered in the next section. Applications of these procedures in finite population sampling have been illustrated in various stages. Section 4.5 considers the robustness of a Bayesian predictor derived under a working model with respect to a class of alternative models as developed by Bolfarine et al (1987). Robust Bayes estimation of a finite population mean under a class of contaminated priors as advocated by Ghosh and Kim (1993, 1997) has been addressed in the last section.

4.2 Linear Bayes Prediction

Bayesian analysis requires full specification of the prior distribution of parameters which may often be large in number. But, in practice, one may not have full knowledge of the prior distribution but firmly believes that the prior distribution belongs to a class of distributions with specified first and second order moments. A Bayesian procedure, *Linear Bayes Procedure*, which is applicable in such circumstances was proposed by Hartigan

(1969). The procedure only requires the specification of first two moments and not the full knowledge of the distribution of the prior. The resulting estimator has the property that it minimises the posterior expected squared loss among all the estimators that are linear in the data and thus can be regarded as an approximation to the posterior mean. In certain sitations, a posterior mean is itself linear in the data (eg. Ericson (1969 a, b), Jewell (1974), Diaconis and Yalvisaker (1979), Goel and DeGroot (1980)) so that the Linear Bayes estimate is an exact Bayes estimate under squared error loss function.

Hartigan's procedure has similarity with the ordinary least squares procedure and as such may also be termed as 'Bayesian Least Squares' method.

The linear Bayes (LB) estimation theory is as follows. Suppose the data Z has likelihood function $f(Z \mid \theta)$ while θ has prior $g(\theta), \theta \in \Theta$. Under squared error loss function Bayes estimate of θ is $\hat{\theta}_B = E(\theta \mid Z)$. In the linear Bayes estimation we do not specify the density function $f(Z \mid \theta)$ and $g(\theta)$ but only their first two moments.

DEFINITION 4.2.1 Let $\Psi = (U, W_1, \ldots W_q)'$ be random variables with finite variances and covariances defined over a common probability space. The linear expectation of U given $W = (W_1, \ldots, W_q)'$ is defined as

$$E_L(U \mid W) = a_0 + \sum_{i=1}^{q} a_i W_i \qquad (4.2.1)$$

where $a_0, a_1, \ldots a_q$ are suitable constants determined by minimising

$$E(U - a_0 - \sum_{i=1}^{q} a_i W_i)^2 \qquad (4.2.2)$$

the expectation being taken with respect to the joint distribution of Ψ. The linear variance of U given W is defined as

$$V_L(U \mid W) = E(U - a_0 - \sum_{i=1}^{q} a_i W_i)^2 \qquad (4.2.3)$$

where a_j's $(j = 0, 1, \ldots, q)$ are determined as above.

The idea is that the true regression of U on $W_1, \ldots, W_q$ which is given by $E(U \mid W_1, \ldots, W_q)$ may be a complicated function of W. Instead, we consider a linear function

$$a_0 + \sum_{i=1}^{q} a_i W_i$$

which gives the best predictor of U in the sense that it minimises (4.2.2).
If the true regression is linear, the minimisation of (4.2.2) gives the true
regression, otherwise it gives the best fit linear regression. The quantity
(4.2.1) is the linear regression of U on W. The linear expectation $E_L(U \mid W)$
may, therefore, be considered as an approximation to $E(U \mid W)$. If Ψ has
$(q+1)$-variate normal distribution, then $E_L(U \mid W) = E(U \mid W)$ and hence
$V_L(U \mid W) = V(U \mid W)$.

DEFINITION 4.2.2 The linear Bayes estimate of θ given the data Z is

$$\hat{\theta}_{LB} = E_L(\theta \mid Z) \qquad (4.2.4)$$

Suppose linear expectation and linear variance of distribution of $g(\theta), f(Z \mid \theta)$ are given respectively by $E_L(\theta)(= E(\theta)), V_L(\theta) = (V(\theta)), E_L(Z \mid \theta), V_L(Z \mid \theta)$. If (Z, θ) are jointly normal, then $E(\theta), V(\theta), E(Z \mid \theta), V(Z \mid \theta)$ coincide
with the corresponding linear expectations and linear variance respectively.

In LB inference we assume that the relationship which hold among $E(\theta), V(\theta)$,
$E(Z \mid \theta), V(Z \mid \theta), E(\theta \mid Z), V(\theta \mid Z)$ in case (Z, θ) follow jointly normal distribution, also extend to the corresponding linear expectations and
variances, $E_L(\theta), V_L(\theta), E_L(Z \mid \theta)$, etc. The relations are: If

$$E_L(Z \mid \theta) = c\theta + d \qquad (4.2.5)$$

then

$$V_L^{-1}(\theta \mid Z) = c^2 V_L^{-1}(Z \mid \theta) + V_L^{-1}(\theta) \qquad (4.2.6.1)$$

and

$$V_L^{-1}(\theta \mid Z) E_L(\theta \mid Z) = c V_L^{-1}(Z \mid \theta)(Z - d)$$
$$+ V_L^{-1}(\theta) E_L(\theta) \qquad (4.2.6.2)$$

The LB estimate of $\theta, E_L(\theta \mid Z)$ and its linear variance $V_L(\theta \mid Z)$ are
calculated from the relations (4.2.6.1), (4.2.6.2).

More generally, suppose we have a prior distribution for Y given $X = (X_1, \ldots, X_n)'$ with linear expectation $E_L(Y \mid X)$ and linear variance $V_L(Y \mid X)$. A new data X_{n+1} is obtained and the likelihood of X_{n+1} given (Y, X)
has linear expectation

$$\begin{aligned} E_L(X_{n+1} \mid Y, X) &= cY + \sum_{i=1}^{n} \alpha_i X_i + \alpha \\ &= cY + d \text{ (say)} \end{aligned} \qquad (4.2.7)$$

where $d = \sum_{i=1}^{n} \alpha_i X_i + \alpha$ and linear variance $V_L(X_{n+1} \mid Y, X)$. Then the
following relations hold:

$$V_L^{-1}(Y \mid X, X_{n+1}) = c^2 V_L^{-1}(X_{n+1} \mid Y, X) + V_L^{-1}(Y \mid X) \qquad (4.2.8.1)$$

$$V_L^{-1}(Y \mid X, X_{n+1})E_L(Y \mid X, X_{n+1}) = cV_L^{-1}(X_{n+1} \mid Y, X)$$

$$(X_{n+1} - d) + V_L^{-1}(Y \mid X)E_L(Y \mid X) \qquad (4.2.8.2)$$

The LB estimation does not assume any loss function. Under squared error loss the procedure gives classical Bayes estimates under normality assumptions and approximately classical Bayes estimates when normality is not assumed.

EXAMPLE 4.2.1

Let $X_1, \ldots, X_n$ be independently and identically distributed random variables with mean μ and variance σ^2. Let μ have a distribution with mean μ_0 and variance σ_0^2. Here, $E_L(\mu) = E(\mu) = \mu_0, V_L(\mu) = V(\mu) = \sigma_0^2$. Also $E_L(X_i \mid \mu) = E(X_i \mid \mu) = \mu, V_L(X_i \mid \mu) = V(X_i \mid \mu) = \sigma^2 (i = 1, \ldots, N)$. Let $\bar{X} = \sum_{i=1}^{n} X_i/n$. Then $E_L(\bar{X} \mid \mu) = \mu(c = 1, d = 0), V_L(\bar{X} \mid \mu) = \sigma^2/n$. Here

$$V_L^{-1}(\mu \mid \bar{X}) = V_L^{-1}(\bar{X} \mid \mu) + V_L^{-1}(\mu) = n/\sigma^2 + 1/\sigma_0^2 \qquad (i)$$

and

$$(n/\sigma^2 + 1/\sigma_0^2)E_L(\mu \mid \bar{X}) = \frac{n}{\sigma^2}\bar{X} + \frac{\mu_0}{\sigma_0^2} \qquad (ii)$$

Therefore, LB estimate of μ is

$$E_L(\mu \mid \bar{X}) = (\frac{n\bar{X}}{\sigma^2} + \frac{\mu_0}{\sigma_0^2})/(\frac{n}{\sigma^2} + \frac{1}{\sigma_0^2}) \qquad (iii)$$

which coincides with the ordinary Bayes estimate of μ. If $\sigma_0 \to \infty, E_L(\mu \mid \bar{X}) = \bar{X}$ and $V_L(\mu \mid \bar{X}) = \sigma^2/n$.

If a further observation X_{n+1} is obtained, the prior has linear expectation $E_L(\mu \mid X) = E_L(\mu \mid \bar{X})$ (where $X = (X_1, \ldots, X_n)'$) given in (iii) and linear variance $V_L(\mu \mid X) = V_L(\mu \mid \bar{X})$ given in (i) (Before observing X_{n+1} these were the linear posterior expectation and linear posterior variance, respectively.) Also, $E_L(X_{n+1} \mid X, \mu) = \mu$ (so that c=1, d=0), $V_L(X_{n+1} \mid X, \mu) = \sigma^2$. Hence, by (4.2.8.1) and (4.2.8.2)

$$V_L^{-1}(\mu \mid X, X_{n+1}) = 1/\sigma^2 + (n/\sigma^2 + 1/\sigma_0^2)$$

$$V_L^{-1}(\mu \mid X, X_{n+1})E_L(\mu \mid X, X_{n+1}) = \frac{(n+1)\bar{X}_{n+1}}{\sigma^2} + \frac{\mu_0}{\sigma_0^2}$$

so that

$$E_L(\mu \mid X, X_{n+1}) = \frac{(n+1)\bar{X}_{n+1}\sigma_0^2 + \mu_0\sigma^2}{(n+1)\sigma_0^2 + \sigma^2}$$

where $\bar{X}_{n+1} = \sum_{i=1}^{n+1} X_i/(n+1)$. Note that the normality is not assumed anywhere.

Multivariate generalisaton

Consider now $Y, X_1, \ldots, X_n$, all vector random variables, $Y = (Y_1, \ldots, Y_m)'$, $X_i = (X_{i1}, \ldots, X_{ip})', (i = 1, \ldots, n)$. Define the linear expectation of the prior distribution of Y on $X_1, \ldots, X_n$ as

$$E_L(Y \mid X_1, \ldots, X_n) = \hat{Y} = (\hat{Y}_1, \ldots, \hat{Y}_m)'$$

and the linear variance

$$V_L(Y \mid X_1, \ldots, X_n) = E(Y - \hat{Y})(Y - \hat{Y})'$$

Suppose that the new data Z given $X = (X_1, \ldots, X_n)$ and Y have linear expectations

$$E_L(Z \mid X, Y) = CY + D$$

where

$$D = \alpha_0 + AX \tag{4.2.9}$$

Then

$$V_L(Z \mid X, Y) = E[(Z - CY - D)(Z - CY - D)' \mid X, Y]$$

The following relations now hold:

$$\begin{array}{rcl} V_L^{-1}(Y \mid Z, X) & = & C'V_L^{-1}(Z \mid Y, X)C + V_L^{-1}(Y \mid X) \\ V_L^{-1}(Y \mid Z, X)E_L(Y \mid Z, X) & = & C'V_L^{-1}(Z \mid Y, X)(Z - D) \end{array}$$

$$+V_L^{-1}(Y \mid X)E_L(Y \mid X) \tag{4.2.10}$$

Brunk(1980) has given the following results on LB-estimation. Consider $\theta = (\theta_1, \ldots, \theta_m)'$, a vector of parameters and $Z = (Z_1, \ldots, Z_n)'$, a sample of n observations. The LB-estimator of θ given Z is

$$\hat{\theta}_{LB} = E_L(\theta \mid Z) \tag{4.2.11}$$

The linear dispersion matrix of θ given Z is

$$V_L(\theta \mid Z) = E[(\theta - \hat{\theta}_{LB})(\theta - \hat{\theta}_{LB})'] \tag{4.2.12}$$

The following results hold:

$$\begin{array}{rcl} E_L(\theta \mid Z) & = & E(\theta) + Cov(\theta, Z)[Cov(Z)]^{-1}(Z - E(Z)) \\ V_L(\theta \mid Z) & = & Cov(\theta) - Cov(\theta, Z)[Cov(Z)]^{-1}Cov(Z, \theta) \end{array} \tag{4.2.13}$$

4.2.1   LINEAR BAYES ESTIMATION IN FINITE POP-ULATION SAMPLING

In the finite population set up, the problem is to estimate a linear function $b(\mathbf{y}) = \sum_{k=1}^{N} b_k y_k$ where $b = (b_1, \ldots, b_N)'$ is a known vector. Let, as before, $\mathbf{y} = (y_s', y_{\bar{s}}')'$, $b = (b_s', \ b_{\bar{s}}')'$ so that $b(\mathbf{y}) = b_s' y_s + b_{\bar{s}}' y_{\bar{s}}$, when s is a sample and other symbols have obvious meanings. Smouse (1984) considered LB estimation of $b(\mathbf{y})$ in sampling from a finite population.

Assume $\mathbf{y}$ to be a random vector having *pdf* ξ on R^N. Using Bayesian principle for making inference, the posterior distribution of $\mathbf{y}$ given the data $d = \{i, y_i; i \in s\}$ is

$$\xi_{\mathbf{y}|d}^* = \xi(\mathbf{y}) / \int \cdots \int_{\Omega_d} \xi(\mathbf{y}) d\mathbf{y} \qquad (4.2.14)$$

where $\Omega_d = \{\mathbf{y} : \mathbf{y} \text{ is consistent with } d\}$. If $\xi(\mathbf{y})$ is completely specified one can find ξ^* and hence Bayes estimator $\hat{b}_B(\mathbf{y})$ of $b(\mathbf{y})$.

The LB-estimate of $b'(\mathbf{y})$ is $b'\hat{\mathbf{y}}_{LB}$ where $\hat{\mathbf{y}}_{LB}$ is the linear expectation of $\mathbf{y}$ given the data. Let

$$
\begin{aligned}
E(\mathbf{y}) &= E(y_s', \ y_{\bar{s}}')' = (\mu_s', \ \mu_{\bar{s}}')', \\
D(\mathbf{y}) &= \begin{bmatrix} \Sigma_s & \Sigma_{sr} \\ \Sigma_{rs} & \Sigma_{\bar{s}} \end{bmatrix}
\end{aligned}
\qquad (4.2.15)
$$

Considering Brunk's results in (4.2.11)-(4.2.12.2), $y_s = Z, \mathbf{y} = \theta$. Hence,

$$Cov(\theta, Z) = Cov(\mathbf{y}, y_s) = \begin{bmatrix} \Sigma_s \\ \Sigma_{rs} \end{bmatrix} \qquad (4.2.16)$$

Therefore, from (4.2.12.1),

$$\hat{\mathbf{y}}_{LB} = \mu + (\Sigma_s, \Sigma_{sr})' \Sigma_s^{-1}(y_s - \mu_s)$$

i.e. LB estimate of y_s is

$$\hat{y}_{sLB} = y_s \qquad (4.2.17.1)$$

and similarly, LB estimate of $y_{\bar{s}}$ is

$$\hat{y}_{\bar{s}LB} = \mu_{\bar{s}} + \Sigma_{rs} \Sigma_s^{-1}(y_s - \mu_s) \qquad (4.2.17.2)$$

From (4.2.13)

$$
\begin{aligned}
V_L(\mathbf{y} \mid y_s) &= \Sigma - [\Sigma_s \quad \Sigma_{sr}]' \Sigma_s^{-1}[\Sigma_s \quad \Sigma_{sr}] \\
&= \begin{bmatrix} 0 & 0 \\ 0 & \Sigma_r - \Sigma_{rs} \Sigma_s^{-1} \Sigma_{sr} \end{bmatrix}
\end{aligned}
\qquad (4.2.18)
$$

In case μ and Σ are known, (4.2.17), (4.2.18) gives LB-estimator $\tilde{b}(\mathbf{y})$ and its variance. In case μ is not known precisely, we, generally, assume a completely specified prior distribution for μ. For LB-approach it is enough to know the mean and covariance matrix of μ. Suppose

$$
\begin{aligned}
E(\mu) &= \nu = (\nu_s, \ \nu_{\bar{s}})' \\
D(\mu) &= \Omega = \begin{bmatrix} \Omega_s & \Omega_{sr} \\ \Omega_{rs} & \Omega_r \end{bmatrix}
\end{aligned} \tag{4.2.19}
$$

Then,

$$
\begin{aligned}
E(\mathbf{y}) &= \nu \\
Cov(\mathbf{y}) &= \Sigma + \Omega \\
Cov(\mathbf{y}, y_s) &= [(\Sigma_s + \Omega_s), (\Sigma_{sr} + \Omega_{sr})]',
\end{aligned} \tag{4.2.20}
$$

The LB-estimator of $\mathbf{y}$ and its linear variance can ,therefore, be obtained by replacing μ by λ and Σ by $\Sigma + \Omega$ in (4.2.17) and (4.2.18).

Cocchi and Mouchart (1986) considered LB estimation in finite population with a categorical auxiliary variable. O'Hagan (1987) considered Bayes linear estimation for randomized response models. Mukhopadhyay (19998 b) considered linear Bayes prediction of a finite population total under measurement error models. Godambe (1999) investigated linear Bayes estimation procedure in the light of estimating functions.

4.3 RESTRICTED LINEAR BAYES PREDICTION

Rodrigues (1989) considered a different kind of Bayesian estimation of finite population parameters. His procedure also does not require full specification of the distribution of the prior. We consider two relevant definitions.

DEFINITION 4.3.1 A predictor $\hat{\theta} = a + t'y_s$, where a is a constant and t is a $n \times 1$ real vector is said to be a Restricted Bayes Least Squares Predictor (RBLSP) or Restricted Linear Bayes Predictor (RLBP) of θ if

$$
E(\theta - a - t'y_s) = 0 \tag{4.3.1}
$$

where the expectation is taken with respect to the predictive distribution of y_s. The corresponding class of predictors satisfying (4.3.1) is denoted as $\mathcal{L}$.

DEFINITION 4.3.2 A predictor $\hat{\theta}^* = a^* + t^{*'}y_s$ is said to be the best RBLSP of θ if $\hat{\theta}^* \in \mathcal{L}$ and

$$E(\hat{\theta}^* - \theta)^2 \leq E(\hat{\theta} - \theta)^2 \; \forall \; \hat{\theta} \in \mathcal{L} \qquad (4.3.2)$$

and for all parameters involved in the predictive distribution of $\mathbf{y}$ with strict inequality holding for at least one $\hat{\theta}$.

In RBLSP we are restricting ourselves to the linear unbiased predictors where unbiasedness is with respect to the predictive distribution of the data y_s. The unbiasedness in (4.3.1) is a generalisation of model-unbiasedness defined in Section 2.2, where the unbiasedness is with respect to the super-population model and no prior distribution is assumed for the parameters involved in the model. The concept can be extended to the quadratic unbiased Bayesian estimation which may be useful in estimating quadratic functions of finite population values, like population variance, design-variance of a predictor.

The RBLSP method was used by La Motte (1978) for estimating super-population parameters.

We now recall some results on least squares theory when the parameters are random variables (vide Rao, 1973, p.234). Consider the linear model

$$E(\mathbf{y} \mid \beta) = X\beta, D(\mathbf{y} \mid \beta) = V \qquad (4.3.3)$$

where β itself is a random vector with

$$E(\beta \mid \nu) = \nu, \; D(\beta \mid \nu) = R \qquad (4.3.4)$$

Our problem is to find the best linear unbiased estimator (BLUE) of $P'\beta$. where P is a vector of constants. Here,

$$E(\mathbf{y}) = X\nu, \; D(\mathbf{y}) = V + XRX'$$

$$
\begin{aligned}
C(\mathbf{y}, P'\beta) &= E[C(\mathbf{y}, P'\beta) \mid \beta] + C[E(\mathbf{y} \mid \beta), P'\beta)] \\
&= XRP
\end{aligned} \qquad (4.3.5)
$$

where C denotes model-covariance. We find a linear function $a + L'\mathbf{y}$ such that

$$E(P'\beta - a - L'\mathbf{y}) = 0 \qquad (4.3.6)$$

and

$$V(P'\beta - a - L'\mathbf{y}) \qquad (4.3.7)$$

is minimum among all functions satisfying (4.3.6).

Case 1. ν known. The optimum choice of L and a are

$$L^* = (V + XRX')^{-1}XRP = V^{-1}X(R^{-1} + X'V^{-1}X)^{-1}P \qquad (4.3.8)$$

$$a^* = \nu'P - \nu'X'L^*$$ (4.3.9)

and the prediction variance is

$$
\begin{aligned}
V(P'\beta - a^* - L^{*'}\mathbf{y}) &= P'RP - P'RX'L^* \\
&= P'(R^{-1} + X'V^{-1}X)^{-1}P
\end{aligned}
$$ (4.3.10)

Case 2. ν unknown. The optimum choice of L and a are

$$L^* = V^{-1}X(X'V^{-1}X)^{-1}P$$ (4.3.11)

$$a^* = 0$$ (4.3.12)

provided that there exists an L such that $X'L = P$ and the prediction variance is

$$V(P'\beta - L^{*'}\mathbf{y}) = P'(X'V^{-1}X)^{-1}P$$ (4.3.13)

Rodrigues (1989) obtained RBLSP of a population function $\theta = q'\mathbf{y}$ for a known vector $q = (q_1, \ldots, q_N)'$ of constants under the random regression coefficients model (4.3.3), (4.3.4). This model is called $M(V, R)$. We denote

$$
\Omega = \begin{bmatrix} \Omega_s & \Omega_{sr} \\ \Omega_{rs} & \Omega_r \end{bmatrix} = \begin{bmatrix} X_s R X'_s & X_s R X'_r \\ X_r R X'_s & X_r R X'_r \end{bmatrix}
$$

$$
V^* = \begin{bmatrix} V^*_s & V^*_{sr} \\ V^*_{rs} & V^*_s \end{bmatrix} = \begin{bmatrix} V_s + \Omega_s & V_{sr} + \Omega_{sr} \\ V_{rs} + \Omega_{rs} & V_r + \Omega_r \end{bmatrix}
$$

$$= V + \Omega$$ (4.3.14)

$$K = [V_s \quad V_{sr}], \quad K^* = [V^*_s \quad V^*_{sr}]$$

The following lemma can be proved by standard computations.

LEMMA 4.3.1 Under the model $M(V, R)$

$$\hat{\theta} = a + t'y_s \in \mathcal{L} \text{ iff } E[a + f'y_s - u'\beta] = 0$$

where

$$f' = t' - q'K'V_s^{-1}$$

and

$$u' = q'(X - K'V_s^{-1}X_s)$$

LEMMA 4.3.2 Under the model $M(V, R)$, for all a and t',

$$E(\hat{\theta} - \theta)^2 = V(a + f'y_s - u'\beta) + q'Vq - q'K'V_s^{-1}Kq$$

$$+[a + (t'X_s - q'X)\nu]^2$$

Lemma 4.3.2 corresponds to a result in Tam (1986) under frequentist approach. The above lemmas reduce the problem of predicting $\theta = q'y$ into the problem of predicting $u'\beta$.

It is clear from lemmas 4.3.1 and 4.3.2 that the problem of finding $\hat{\theta}^*$ is equivalent to the determination of $a + f'y_s$ such that

$$E(a + f'y_s - u'\beta) = 0$$

and

$$V(a + f'y_s - u'\beta)$$

is minimum. The problem can be solved in a straight-forward manner by using Rao's result stated in (4.3.6)-(4.3.13).

Case(i) ν known. Here

$$\begin{aligned} f^* &= (V_s + X_s R X_s')^{-1} X_s R u \\ &= V_s^{-1} X_s (R^{-1} + X_s' V_s^{-1} X_s)^{-1} u \end{aligned} \tag{4.3.15.1}$$

$$a^* = \nu'u - \nu'X_s'f^* \tag{4.3.15.2}$$

$$V(a^* + f^{*'}y_s - u'\beta) = u'(R^{-1} + X_s'V_s^{-1}X_s)^{-1}u \tag{4.3.15.3}$$

This gives

$$\begin{aligned} \hat{\theta}^* &= a^* + t^{*'}y_s \\ &= q_s'y_s + q_r'(X_r\tilde{\beta} + V_{rs}V_s^{-1}(y_s - X_s\tilde{\beta})) \end{aligned} \tag{4.3.16.1}$$

where

$$\begin{aligned} \tilde{\beta} &= C\tilde{\beta}_V + (I - C)\nu, \\ \tilde{\beta}_V &= (X_s'V_s^{-1}X_s)^{-1}X_s'V_s^{-1}y_s \\ C &= (R^{-1} + X_s'V_s^{-1}X_s)^{-1}(X_s'V_s^{-1}X_s) \end{aligned}$$

and

$$E(\hat{\theta}^* - \theta)^2 = u'(R^{-1} + X_s'V_s^{-1}X_s)^{-1}u + q'Vq - q'K'V_s^{-1}Kq, \tag{4.3.16.2}$$

Case (ii) ν unknown and fixed. Here

$$f^* = V_s^{-1}X_s(X_s'V_s^{-1}X_s)^{-1}u, \quad a^* = 0, \tag{4.3.17.1}$$

$$V(f^{*'}y_s - u'\beta) = u'(X_s'V_s^{-1}X_s)^{-1}u \tag{4.3.17.2}$$

This gives

$$\hat{\theta}^* = q'_s y_s + q'_r [X_r \tilde{\beta}_v + V_{rs} V_s^{-1}(y_s - X_s \tilde{\beta}_v)] \qquad (4.3.18.1)$$

and

$$E(\hat{\theta}^* - \theta)^2 = u'(X'_s V_s^{-1} X_s)^{-1} u + q' V q - q' K' V_s^{-1} K q \qquad (4.3.18.2)$$

The predictor $\hat{\theta}^*$ given in (4.3.18.1) was obtained by Royall and Pfeffermann (1982) by using the multivariate normal distribution for **y** given β and diffuse prior for β. The results (4.3.16.1),(4.3.16.2) were also obtained by Bolfarine et al (1987) and Malec and Sedransk (1985) by using normality assumptions.

EXAMPLE 4.3.2

Model $M(V, R)$ with $X = (1, \ldots, 1)', V = \sigma^2 I, R = \sigma_0^2, q' = (1/N, \ldots, 1/N)'$. Here BRBLSP of $\bar{y}$ is, using formula (4.3.16.1),

$$\hat{\bar{y}}^* = \hat{\theta}^* = (\frac{\sigma^2/N + \sigma_0^2}{\sigma^2/n + \sigma_0^2})\bar{y}_s + (1 - \frac{\sigma^2/N + \sigma_0^2}{\sigma^2/n + \sigma_0^2})\nu, \bar{y}_s = \sum_{i \in s} y_i/n,$$

and

$$E(\hat{\bar{y}}^* - \bar{y})^2 = (1 - f)\sigma^2/n[\frac{(N + n)\sigma_0^2 + \sigma^2}{N(\sigma^2/n + \sigma_0^2)}], f = n/N$$

Goldstein (1975) with the purpose of estimating the superpopulation parameter β considered the Bayes linear predictor of $\beta = E(\bar{y})$, which is actually a predictor of $\bar{y}$,

$$\hat{\theta}_G = \frac{\sigma_0^2}{\sigma^2/n + \sigma_0^2}\bar{y}_s + \frac{\sigma^2/n}{\sigma^2/n + \sigma_0^2}\nu$$

with

$$E(\hat{\theta}_G - \beta)^2 = (n/\sigma^2 + 1/\sigma_0^2)^{-1}$$

For further details the reader may refer to Rodrigues (1989).

4.4 CONSTRAINED BAYES PREDICTION

Suppose we have m parameters $\theta_1, \ldots \theta_m$ with corresponding estimates $\hat{\theta}_1, \ldots, \hat{\theta}_m$. Sometimes, it is desirable to produce an ensemble of parameter estimates whose histogram resemble the histogram of population parameters in some sense. This occurs, for example, in subgroup analysis when the problem is not only to estimate the different components of a vector, but also to identify the parameters whose values are above and below a certain cut-off point (see Ghosh and Maiti (1999) for further examples in this area). Louis (1984) wanted to modify the Bayes estimates to satisfy this property. He attempted to match the first two moments from the histogram of Bayes estimates with the corresponding moments from the histogram of parameters in a normal theory set up. Specifically, let $\theta_i \, ind \, N(\mu, \tau^2), i = 1, \ldots, m$ and $X_i \mid \theta_i \, ind \, N(\theta_i, 1)$. Then Bayes estimate of θ_i under the summed squared loss [SSEL $= \sum_{i=1}^{m} (\hat{\theta}_i - \theta_i)^2$] function is

$$\hat{\theta}_i^B = \mu + D(x_i - \mu), i = 1, \ldots, m \qquad (4.4.1)$$

where
$$D = \tau^2 / (1 + \tau^2) \qquad (4.4.2)$$

Letting
$$\bar{\theta} = \sum_i \theta_i / m, \bar{\hat{\theta}}^B = \sum_i \hat{\theta}_i^B / m \qquad (4.4.3)$$

Louis proved that
(i)
$$E(\bar{\theta} \mid x) = \bar{\hat{\theta}}^B$$

but
(ii)
$$E[\sum_{i=1}^{m} (\theta_i - \bar{\theta})^2 \mid x] \geq \sum_{i=1}^{m} (\hat{\theta}_i^B - \bar{\hat{\theta}}^B)^2 \qquad (4.4.4)$$

where $x = (x_1, \ldots, x_m)'$, the sample observations. Thus, for any given x, the mean of two histograms, of estimates and of posterior expected values of θ's coincide , while the variance of histogram of estimates is only a fraction of the posterior expected values of the variance of histogram of parameters. Louis pointed out that this phenomenon was due to overshrinking of the observed estimates towards the prior means (exercise 1).

EXERCISE 4.4.1 Suppose $X_i \mid \theta_i$ are independent $N(\theta_i, 18)$ and θ_i are *iid* $N(0, 9)$. Then the Bayes estimate of θ is $\hat{\theta}^B(x) = (\frac{1}{3} X_1, \ldots, \frac{1}{3} X_m)'$. Also,

$$E[(m-1)^{-1} \sum_{i=1}^{m} (\hat{\theta}_i^B(x) - \bar{\hat{\theta}}^B(x))]^2 = 3$$

$$E[(m-1)^{-1}\sum_{i=1}^{m}(\theta_i - \bar{\theta})^2 \mid x] = 9$$

so that the Bayes estimates underestimate the posterior expected value of the variance of the parameters by a factor $2/3$.

Ghosh (1992) proved that (4.4.4) holds true in a more general set up. Suppose that $\theta_1, \ldots, \theta_m$ are the m parameters of interest and $e_1^B(x), \ldots e_m^B(x)$ are the corresponding Bayes estimates for any likelihood function of x and any prior of $\theta = (\theta_1, \ldots, \theta_m)'$ under any quadratic loss function. Assume that

(A) not all $\theta_1 - \bar{\theta}, \ldots, \theta_m - \bar{\theta}$ have degenerate posterior distributions.

The assumption A is much weaker than the assumption that $V(\theta \mid x)$ is positive definite.

THEOREM 4.4.1 Under assumptions (A)

$$E[\sum_{i=1}^{m}(\theta_i - \bar{\theta})^2 \mid x] > \sum_{i=1}^{m}(e_i^B - \bar{e}^B)^2 \qquad (4.4.5)$$

where

$$\bar{e}^B = \bar{e}^B(x) = \sum_{i=1}^{m} e_i^B(x)/m$$

Proof.

$$E[\sum_{i=1}^{m}(\theta_i - \bar{\theta})^2 \mid x] = E[\theta'(I_m - m^{-1}J_m)\theta \mid x]$$

$$= E(\theta' \mid x)(I_m - m^{-1}J_m)E(\theta \mid x) + \mathrm{tr}[(I_m - m^{-1}J_m)V(\theta \mid x)]$$

$$= \sum_{i=1}^{m}(E(\theta_i \mid x) - E(\bar{\theta} \mid x))^2 + \mathrm{tr}[V((\theta - \bar{\theta}1_m) \mid x] \qquad (4.4.6)$$

$$> \sum_{i=1}^{m}(e_i^B(x) - \bar{e}^B(x))^2 \qquad (4.4.7)$$

since $tr(B) > 0$, where $J_m = 1_m 1'_m$ and B is a positive semi-definite matrix.

Taking as a desirable criterion the property of invariance of first two moments of the histogram of parameters and their estimates, we have the following definition.

DEFINITION 4.4.1 A set of estimators

$$e^{CB}(x) = (e_1^{CB}(x), \ldots, e_m^{CB}(x))'$$

of θ is said to be a set of constrained Bayes (CB) estimators of θ if $e^{CB}(x)$ minimises

$$E[\sum_{i=1}^{m}(\theta_i - t_i)^2 \mid x] \tag{4.4.8}$$

within the class of estimates $t(x) = (t_1(x), \ldots, t_m(x))'$ of θ which satisfy
(a)

$$E(\bar{\theta} \mid x) = m^{-1}\sum_{i=1}^{m}t_i(x) = \bar{t}(x) \tag{4.4.9}$$

(b)

$$E[\sum_{i=1}^{m}(\theta_i - \bar{\theta})^2 \mid x] = \sum_{i=1}^{m}(t_i(x) - \bar{t}(x))^2 \tag{4.4.10}$$

The Bayes estimate $e^B(x) = (e_1^B(x), \ldots, e_m^B(x))'$ satisfies (4.4.9) but not (4.4.10).

Let us write

$$H_1(x) = \mathrm{tr}[V(\theta - \bar{\theta}1_m \mid x)] \tag{4.4.11}$$

$$H_2(x) = \sum_{i=1}^{m}(e_i^B(x) - \bar{e}^B(x))^2 \tag{4.4.12}$$

THEOREM 4.4.2 Let $\mathcal{X}_0 = \{x : H_2(x) > 0\}$ and

$$e^B(x) = (e_1^B(x), \ldots, e_m^B(x))'$$

denote the Bayes estimate of θ under any quadratic loss function . Then, for any $x \in \mathcal{X}_0$,

$$e_i^{CB}(x) = ae_i^B(x) + (1-a)\bar{e}^B(x), i = 1, \ldots, m \tag{4.4.13}$$

where

$$a = a(x) = [1 + H_1(x)/H_2(x)]^{1/2} \tag{4.4.14}$$

Proof. We have

$$E[\sum_{i=1}^{m}(\theta_i - t_i)^2 \mid x]$$

$$= E[\sum_{i=1}^{m}(\theta_i - e_i^B(x))^2 \mid x] + \sum_{i=1}^{m}(e_i^B(x) - t_i)^2 \tag{4.4.15}$$

Now,

$$\sum_{i=1}^{m}(e_i^B(x) - t_i)^2 = \sum_{i}(t_i - \bar{t})^2 - 2\sum_{i}(e_i^B(x) - \bar{e}^B(x))(t_i - \bar{t})$$

$$+ \sum_{i=1}^{m} (e_i^B(x) - \bar{e}^B(x))^2$$

$$= m[V(Z_1) + V(Z_2) - 2Cov(Z_1, Z_2)] \tag{4.4.16}$$

where

$$P[Z_1 = e_i^B(\mathbf{x}), Z_2 = t_i] = 1/m, \ i = 1, \ldots, m.$$

Now, $V(Z_1) = H_2$ is a fixed quantity. Also, $V(Z_2) = \sum_{i=1}^{m}(t_i - \bar{t})^2/m = H_1 + H_2$ is a fixed quantity (because of the requirement (4.4.10)). Hence, minimum value of (4.4.16) is attined when the correlation $\rho(Z_1, Z_2) = 1$, i.e. when $Z_2 = aZ_1 + b$ with probability one for some constants $a(> 0)$ and b. Thus

$$t_i = t_i(x) = ae_i^B(x) + b, i = 1, \ldots, m \tag{4.4.17}$$

Now, (4.4.9) requires

$$\bar{e}^B(x) = \bar{t}(x) = a\bar{e}^B(x) + b$$

This gives

$$t_i(x) = ae_i^B(x) + (1 - a)\bar{e}^B(x)$$

By virtue of (4.4.6), (4.4.10) and (4.4.17),

$$H_1(x) + H_2(x) = \sum_{i=1}^{m}(t_i - \bar{t})^2 = a^2 H_2(x).$$

Hence, for $x \in \mathcal{X}_0$,

$$a = [1 + H_1(x)/H_2(x)]^{1/2}$$

EXAMPLE 6.4.2 Let $X_1, \ldots, X_m$ be m independent random variables, where x_i has *pdf* (with respect to some σ-finite measure),

$$f_{\phi_i}(x_i) = exp\{n\phi_i x_i - n\psi(\phi_i)\}, i = 1, \ldots, m \tag{i}$$

Each X_i can be viewed as an average of n *iid* random variables, each having a *pdf*, belonging to a one-parameter Exponential family. Assuming that $\psi(.)$ is twice differentiable in its argument, it is desired to estimate $\theta_i = E_{\phi_i}(X_i) = \psi'(\phi_i), i = 1, \ldots, m$. For conjugate prior

$$g(\phi_i) = exp(\nu\phi_i\mu - \nu\psi(\phi_i)) \tag{ii}$$

for ϕ_i, the Bayes estimate of θ_i for the squared error loss function is given by

$$e_i^B(x) = E(\theta_i \mid x) = (1 - B)x_i + B\mu \tag{iii}$$

where $B = \nu/(n+\nu)$. Also,

$$
\begin{aligned}
V(\theta_i \mid x) &= V(\psi'(\phi_i) \mid x_i) \\
&= E[\psi''(\phi_i) \mid x_i]/(n+\nu) = q_i \text{ (say)}
\end{aligned} \qquad (iv)
$$

It follows that

$$
H_1(x) = (1 - 1/m) \sum_{i=1}^{m} q_i
$$

$$
H_2(x) = (1 - B)^2 \sum_{i=1}^{m} (x_i - \bar{x})^2
$$

from which 'a' can be determined.

In particular, suppose that *pdf* of X_i belongs to the QVF (quadratic variance function) subfamily of the natural exponential family. Thus,

$$
\psi''(\phi_i) = \nu_0 + \nu_1\theta_i + \nu_2\theta_i^2, \ i = 1, \ldots, m \qquad (v)
$$

where ν_0, ν_1, ν_2 are not all zero and $\nu_2 < n + \nu$. It follows from (iv) and (v),

$$
q_i = [\nu_0 + \nu_1 e_i^B(x) + \nu_2(e_i^B(x))^2]/(n + \nu - \nu_2)
$$

$$
H_1(x) = (m-1)(n+\nu-\nu_2)^{-1}[\nu_0 + \nu_1\bar{e}^B(x) + \nu_2\{(\bar{e}^B(x))^2 + H_2(x)/m\}] \quad (vi)
$$

Therefore, for $x \in \chi_0$,

$$
a^2(x) = [1 + \nu_2(n + \nu - \nu_2)^{-1}(1 - 1/m)]
$$

$$
+(m - 1)(n + \nu - \nu_2)^{-1}[\nu_0 + \nu_1\bar{e}^B(x) + \nu_2(\bar{e}^B(x))^2]/H_2(x) \qquad (vii)
$$

When the X_i's are averages of *iid* Bernoulii variables, $\nu_0 = 0, \nu_1 = 1, \nu_2 = -1$. For the Poisson case, $\nu_0 = \nu_2 = 0, \nu_1 = 1$. For the normal case, $\nu_1 = \nu_2 = 0, \nu_0 = Var(X_i)$.

Note 4.4.1

Unlike the classical Bayes estimators, the CB estimators change if weighted squared error loss $(\hat{\theta} - \theta)'W(\hat{\theta} - \theta)$, where W is a $m \times m$ matrix of weights, is used instead of the Eucledian distance $\sum_{i=1}^{m}(\hat{\theta}_i - \theta_i)^2$.

The following theorem indicates the Bayes risk dominance of CB-estimator over the sample mean under a conjugate prior. Consider the model M:

$$
X \mid \theta \sim N(\theta, \sigma^2 I_m) \ (\sigma^2 \text{unknown})
$$

$$
\xi : \theta \sim N(0, \tau^2 I_m) \qquad (4.4.18)
$$

Here,
$$X = (X_1, \ldots, X_m), \quad a^2(X) = 1 + \sigma^2/(1 - B)S^2$$
where
$$S^2 = \sum_{i=1}^{m} (X_i - \bar{X})^2/(m - 1), (m \geq 2)$$
$$B = \sigma^2/(\sigma^2 + n\tau^2)$$

Hence,
$$e^{CB}(x) = (1 - B)[\bar{X}1_m + a(X)(X - \bar{X}1_m)] \qquad (4.4.19)$$

THEOREM 4.4.3 Let $r(\xi, e) = E\{\sum_{i=1}^{m}(e_i - \theta_i)^2 \mid x\}$ where $e = (e_1, \ldots, e_m)$, denote the Bayes risk of an estimator e of θ under the model M and SSEL. Then
$$r(\xi, e^{CB}) < r(\xi, X) \text{ for } m \geq 4.$$
Ghosh and Maiti (1999) considered generalisation when $\theta_1, \ldots, \theta_m$ are vectors of parameters.

Note 4.4.2

Efron and Morris (1971, 1972) (exercise 2) pointed out that Bayes estimators may perform well overall but do poorly (in frequentist sense) for estimating individual θ_i's with unusually large or small values. To overcome this problem they recommended the use of *limited translation* (LT) Bayes estimators of θ. They suggested a compromise, which consists of restricting the amount by which the Bayes estimator $\hat{\theta}_i^B$ differs from the *ml* estimator $\hat{\theta}_i^{ML}$ by some multiple of the standard error of X_i. In the model $X_i \sim N(\theta_i, \sigma^2), \theta_i \sim N(0, \tau^2), i = 1, \ldots, k$, the modified estimator is

$$\hat{\theta}_i^{LT}(X) = \begin{cases} \hat{\theta}_i^B & \text{if } |\hat{\theta}_i^{ML} - \hat{\theta}_i^B| < K\sigma \\ \hat{\theta}_i^{ML} - K\sigma & \text{if } \hat{\theta}_i^B < \hat{\theta}^{ML} - K\sigma \\ \hat{\theta}^{ML} + K\sigma & \text{if } \hat{\theta}_i^B > \hat{\theta}^{ML} + K\sigma \end{cases} \qquad (4.4.20)$$

where K is a suitable constant.

The estimator (4.4.20) compromises between limiting the maximum possible risk to any component $\hat{\theta}_i^B$ for any unusually large or small value of θ_i and preserving the average gain of $\hat{\theta}^B (= (\hat{\theta}_1^B, \ldots, \hat{\theta}_k^B))$. The choice $K = 1$, for example, ensures that $E(\hat{\theta}_i^{LT} - \theta_i)^2 < 2\sigma^2 \ \forall \ i$ while retaining more than 80 % of the average gain of $\hat{\theta}^B$ over $X = (X_1, \ldots, X_k)$.

The LT Bayes estimator does not seem to have received considerable attention in survey sampling.

4.4.1   APPLICATIONS IN FINITE POPULATION SAMPLING

EXAMPLE 4.4.3

Suppose there are m strata in the population, the ith stratum of size N_i having values $(y_{i_1}, \ldots, y_{iN_i})'$ of the study variable y on the units in the population $(y_{is}[= (y_{i1}, \ldots, y_{in_i})'],$ on the n_i sampled units) $(i = 1, \ldots, m)$. The objective is to predict

$$\gamma = (\gamma_1, \ldots, \gamma_m)' \text{ where } \gamma_i = \sum_{j=1}^{N_i} y_{ij}/N_i \qquad (4.4.21)$$

on the basis of $y_{is}(i = 1, \ldots, m)$. Denote

$$y_{i\bar{s}} = (y_{in_i+1}, \ldots, y_{iN_i})', \quad y_i = (y'_{is}, y'_{i\bar{s}})'$$

$$y_s = (y'_{1s}, \ldots, y'_{ms})'$$

Consider the following model:

(a)
$$y_i \mid \theta = (\theta_1, \ldots, \theta_m)' \, ind \, N(\theta_i 1_{N_i}, \sigma^2 I_{N_i})$$

(b)
$$\theta_i \, ind \, N(\mu, \tau^2)$$

It follows from Ghosh and Meeden (1986), Ghosh and Lahiri (1987 a) that

$$
\begin{aligned}
E(\gamma_i \mid y_s) &= (1 - f_i B_i)\bar{y}_{is} + f_i B_i \mu \\
V(\gamma_i \mid y_s) &= f_i \sigma^2 [N_i^{-1} + f_i n_i^{-1}(1 - B_i)], i = 1, \ldots, m \\
Cov(\gamma_i, \gamma_k \mid y_s) &= 0, i \neq k = 1, \ldots, m
\end{aligned}
$$

where

$$\bar{y}_i = \sum_{j=1}^{n_i} y_{ij}/n_i, B_i = \sigma^2/(\sigma^2 + n_i \tau^2), f_i = 1 - n_i/N$$

CB-predictors of γ, $\hat{\gamma}^{CB}$ are found by computing $H_1(\mathbf{y})$ and $H_2(\mathbf{y})$ and using formula (4.4.13).

Lahiri(1990) obtained constrained empirical Bayes (CEB) predictors of γ by finding estimates of μ, σ^2 and τ^2 and then substituting these in $\hat{\gamma}^{CB}$. He called these as 'adjusted EB-predictors'. Following Hartigan (1969), Ericson (1969), Goldstein (1975) and others, he also replaced the normality assumptions by a weaker assumption of 'posterior linearity', discussed in section 3.3.

EXAMPLE 4.4.4

Ghosh (1992) used CB-estimation to estimate the average wage and salaries of workers in a certain industry consisting of 114 units and spread over 16 small areas. He also wanted to identify the areas with very low or very high average wages. A simple random sample was taken and the sampled units were post-stratified into these small areas. It turned out that 3 of the 16 areas had no representation in the sample.

The following mixed effects model was considered:

$$y_{ij} = \beta_0 + \beta_1 x_{ij} + v_i + e_{ij}^* \sqrt{x_{ij}} \qquad (4.4.22)$$

where $y_{ij}(x_{ij})=$ average wage (gross business income) of unit j in area i, $v_i = $ a random area effect, v_i's and e_{ij}^*'s are independently distributed with $v_i \, ind \, N(0, (\lambda r)^{-1})$, $e_{ij}^* \, ind \, N(0, r^{-1})(i = 1, \ldots, 16; j = 1, \ldots, N_i; \sum N_i = 116), \beta_0, \beta_1$ are unknown regression coefficients. The model (4.4.22) gives the conditional likelihood of the data given the parameters $(\beta_0, \beta_1, \lambda, r)$. The following priors were assumed:

(i)

$$\beta = (\beta_0, \ \beta_1)' \sim Uniform(R^2)$$

(ii)

$$r \sim Gamma(a_0/2, g_0/2) \qquad (4.4.23)$$

(iii)

$$\lambda r \sim (a_1/2, g_1/2)$$

The authors used diffuse Gamma priors on r and λr with $a_0 = g_0 = g_1 = 0$ and $a_1 = 0.00005$ ($a_1 = 0$ could lead to an improper distribution). Small area models using mixed effects and Hierarchical Bayes estimation have been discussed in details in Mukhopadhyay (1998 e).

The model given in (4.4.22) and (4.4.23) is a special case of the model in Datta and Ghosh (1991). Using these results Ghosh found Hierarchical Bayes (HB) predictors $\hat{\gamma}_i^{HB} = E(\gamma_i \mid y_s)$ and $V(\gamma_i \mid y_s), H_1(y_s)$ and $H_2(y_s)$. He adjusted these HB-predictors to find constrained HB-predictors $\hat{\gamma}_i^{CHB}(y_s)$ of small area means $\gamma_i(i = 1, \ldots, 16)$. The three estimators, sample averages $\bar{y}_{is}, \hat{\gamma}_i^{HB}(\mathbf{y}), \hat{\gamma}_i^{CHB}(\mathbf{y})$ along with their associated standard errors were compared with reference to the average of the squared deviations of the estimates from the true means, $ASD = \sum_{i=1}^{A}(e_i - M_i)^2/A$, average bias, $AB = \sum_{i=1}^{A} \mid e_i - M_i \mid /A$ and average relative bias of the estimates

$ARB = \sum_{i=1}^{A} (\mid e_i - M_i \mid)/AM_i$, where M_i is the true mean of the small area i and A is the number of small areas for which estimates are available (A=13 for $\bar{y}_{is}$ and 16 otherwise). It was found that on an average the HB predictors $\hat{\gamma}_i^{HB}$ resulted in a 77.05% reduction in ASD and 60% reduction in ARB compared to sample means. The adjusted Bayes estimators $\hat{\gamma}_i^{CHB}$ had slight edge over $\hat{\gamma}_i^{HB}$ resulting in 79.7% reduction in ASD and 52.9% reduction in ARB compared to $\bar{y}_{is}$ (Ghosh and Maiti, 1999). The author observed that the CB estimators identified the areas with very high and low wages more successfully than the other two estimators. It has been recommended that in case of the dual problem of estimation and subgroup identification, one should use CB estimators in preference to usual Bayes estimators.

4.5   BAYESIAN ROBUSTNESS UNDER A CLASS OF ALTERNATIVE MODELS

In this section we shall consider the robustness of a Bayesian predictor derived under a working model with respect to a class of alternative models. Consider the model $M(V, R, \nu)$:

$$\mathbf{y} = X\beta + \epsilon, \ \epsilon \sim N_N(0, V) \tag{4.5.1}$$

$$\beta \sim N_p(\nu, R)$$

The model has been discussed in details in Theorem 3.3.1. We use here slightly different notations. Let

$$\Delta_s = Diag(\delta_k, k = 1, \ldots, N), \ \delta_k = 1(0) \text{ if } k \in (\notin)s$$

$$y_s = \Delta_s \mathbf{y}, y_r = (I - \Delta_s)\mathbf{y}$$

$$X_s = \Delta_s X, X_r = (I - \Delta_s)X$$

$$V_s = \Delta_s V \Delta_s, V_r = (I - \Delta_s)V(I - \Delta_s)$$

$$V_{sr} = V'_{rs} = \Delta_s V(I - \Delta_s)$$

Hence,

$$y'_s y_r = 0, \ V_S V_r = \Phi \tag{4.5.2}$$

where Φ is the square null matrix (here of order p). Note that both y_s and y_r are $N \times 1$ vectors. Suppose rank $(X_s) = p$. Let

$$
\begin{aligned}
A^{-1} &= X_s' V^- X_s \\
D &= (A^{-1} + R^{-1})^{-1} A^{-1} \\
D_0 &= (A^{-1} + R^{-1})^{-1} R^{-1} \\
\hat{\beta}^* &= A X_s' V_s^- y_s
\end{aligned}
$$

where G^- denotes the generalised inverse of a matrix G. Note that $D_0 + D$ is the identity matrix of order p.

THEOREM 4.5.1 In the model $M(V, R, \nu)$ the posterior distribution of y_r given y_s is multivariate normal with mean

$$E(y_r \mid y_s) = X_r \bar{\beta} + V_{rs} V_s^{-1} (y - X_s \bar{\beta}) \tag{4.5.3}$$

and conditional variance

$$D(y_r \mid y_s) = (V_r - V_{rs} V_s^- V_{sr}) + (X_r - V_{rs} V_s^- V_{sr})$$

$$D A (X_r - V_{rs} V_s^- V_{sr})' \tag{4.5.4}$$

where

$$\bar{\beta} = E(\beta \mid y_s) = D \hat{\beta}^* + D_0 \nu$$

is the Bayesian estmate of β.

Proof. Follows as in Theorem 3.3.1 or from Lindley and Smith (1972).

The Bayesian estimator $\bar{\beta}$ may be looked upon as a generalisation of the convex combination of the generalised least square estimator $\hat{\beta}^*$ and prior mean ν. In case prior distribution of β is non-informative, $R^{-1} = \Phi$ and $\bar{\beta} = \hat{\beta}^*$.

Consider an alternative model

$$\mathbf{y} = X^* \beta^* + \epsilon, \epsilon \sim N_N(0, V)$$

$$X^* = (X, \ Z), \beta^* = (\beta', \delta)'$$

$$\beta^* \sim N_{p+L}(\nu^* = (\nu', \lambda')', R^*)$$

$$R^* = \begin{bmatrix} R & \Omega_0 \\ \Omega_0' & \Omega \end{bmatrix}$$

where Z is a $N \times L$ matrix of values of additional L variables $x_{p+1}, \dots, x_{p+L}, \delta$ is a $L \times 1$ vector of corresponding regression coefficients and $\lambda, \Omega_0, \Omega$ have obvious interpretations. Call this model as $M^*(V, R^*, \nu^*)$. Let E^*, D^* denote, respectively, expectation and dispersion matrix (variance) with respect to M^*. Define

$$X_s^* = \Delta_s X^*, \quad X_r^* = (I - \Delta_s)X^*,$$

$$Z_s = \Delta_s Z, \quad Z_r = (I - \Delta_s)Z \qquad (4.5.5)$$

Thus

$$X_s^* = (X_s, \ Z_s), \quad X_r^* = (X_r, \ Z_r)$$

Bolfarine et al (1987) considered the conditions under which the posterior mean or posterior distribution of a linear function $q'y$ remains the same under a class of Bayesian models $\mathcal{B}$, i.e. under a class of combinations of likelihood functions and priors. They called these conditions as a set of conditions for robustness. Under these conditions Bayes predictors under squared error loss remains the same for all the models in $\mathcal{B}$. We have the following definition.

DEFINITION 4.5.1 *Weak Robustness (Strong robustness or simply Robustness)* A set of conditions $\mathcal{R}$ is a set of conditions for weak robustness (or strong robustness or simply, robustness) in relation to a linear function $q'y$ for a class $\mathcal{B}$ of Bayesians (i.e. Bayesian models), if under $\mathcal{R}$ the posterior expectation (distribution) of y_r [of $q'y$] given y_s remains the same for all elements of $\mathcal{B}$.

The following theorem considers a class of Bayesian models $\mathcal{B}$, for which expectation of y_r given y_s equal either to $E(y_r \mid y_s)$ or $E^*(y_r \mid y_s)$. Clearly, models M and M^* are members of $\mathcal{B}$. The authors found conditions under which the posterior expectation of $q'y$ given y_s remain the same under all models in $\mathcal{B}$.

THEOREM 4.5.2 (Bolfarine et al, 1987) For any class $\mathcal{B}$ of Bayesians whose posterior means of y_r given y_s are either equal to $E(y_r \mid y_s)$ or $E^*(y_r \mid y_s)$ the following set of conditions $\mathcal{R}$ form a 'weak robustness set' in relation to the linear function $q'y$. The condition $\mathcal{R}$ are:
(i) δ and β are independent
(ii)

$$q'Z_r = q'[(X_r - V_{rs}V_s^- X_s)(A^{-1} + R^{-1})^{-1}X_s'$$

$$+ V_{rs}]V_s^- Z_s \qquad (4.5.6)$$

For models M and $M^*, \mathcal{R}$ is actually a set of conditions for (strong) robustness in relation to the linear function $q'y$.

Example 4.5.1

Suppose V is a positive definite diagonal matrix and define $V_s^{-1} = \Delta_s V^{-1} \Delta_s$. Suppose $R^{-1} = \Phi$ and condition (i) of $\mathcal{R}$ holds. Then condition (4.5.6) reduces to

$$q'Z_r = q'X_r A X_s' V_s^- Z_s$$

$$\Leftrightarrow q'(I - \Delta_s)Z = q'(I - \Delta_s)X(X'\Delta_s V^{-1}\Delta_s X)^{-1}$$

$$X'\Delta_s V^{-1}Z \qquad (4.5.7)$$

Under these conditions, $q'y_r \mid y_s$ is normally distributed with mean

$$\begin{aligned} E\{q'y_r \mid y_s\} &= E^*\{q'y_r \mid y_s\} \\ &= q'(I - \Delta_s)X(X'\Delta_s V^{-1}X)^{-1}X'\Delta_s V^{-1}y_s \end{aligned}$$

and variance

$$\begin{aligned} D\{q'y_r \mid y_s\} &= D^*\{q'y_r \mid y_s\} \\ &= q'(I - \Delta_s)[V^{-1} + X(X'\Delta_s V^{-1}X)^{-1}X'] \end{aligned}$$

$$(I - \Delta_s)q$$

Condition (4.5.7) coincides with the condition of Pereira and Rodrigues (1983) for unbiasedness of Royall's BLUP $\hat{T}^*(X, V)$ under the model (X^*, V) where V is diagonal.

Example 4.5.2

Consider the model of example 3.2.2. Suppose $Z = 1_N$. Condition (4.5.6) reduces to

$$\bar{x}_{\bar{s}} = \bar{x}_s + \sigma^2/nR. \qquad (i)$$

This condition ensures robustness of the Bayesian predictor of population total

$$\hat{T}_B = T_1 + \frac{N - n}{n}\bar{x}_{\bar{s}}(\bar{x}_s + \sigma^2/nR)^{-1}(n\bar{y}_s + \sigma^2\nu/R) \qquad (ii)$$

where $T_1 = \sum_s y_i$, if β and β and δ are independently and normally distributed.

For a non-informative prior of $\beta, R \to \infty$ and $\hat{T}_B$ reduces to the ratio predictor which remains robust under both these models (model of example 3.2.2 and the present) if $\bar{x}_{\bar{s}} = \bar{x}_s$. A Bayesian, therefore, may use the following sampling rules. If one is sure that δ is not included in the model, one may use the optimal s.d. p^* to control the variance as suggested in example 3.2.2. However, if there is doubt that δ may appear in the model one should take a balanced sample for which $\bar{x}_s = \bar{x}_{\bar{s}}$ (or a sample for which the condition (i) is satisfied if the parameters R is known). In both the cases purposive samples are recommended.

EXAMPLE 4.5.3

Suppose $X = 1_N, V = \sigma^2 I, \nu$ is a finite real number, R is a positive real number,

$$
Z = \begin{bmatrix} x_1 & x_1^2 & \ldots x_1^L \\ \cdot & \cdot & \ldots \\ x_N & x_N^2 & \ldots x_N^L \end{bmatrix}, \quad \Omega_0 = (0, \ldots, 0)'
$$

The Bayes predictor of population total T under $M(V, R, \nu)$ is

$$
\hat{T}_B = T_1 + \frac{(N-n)}{n}(1 + \sigma^2/nR)^{-1}(T_1 + \sigma^2\nu/R).
$$

Condition (4.5.6) reduces to

$$
(1 + \sigma^2/nR)\bar{x}_{\bar{s}}^{(j)} = \bar{x}_s^{(j)}, j = 1, \ldots, L \qquad (4.5.8)
$$

where

$$
\bar{x}_s^{(j)} = \sum_s x_k^j/n, \quad \bar{x}_{\bar{s}}^{(j)} = \sum_{\bar{s}} x_k^j/(N-n)
$$

If $\sigma^2/R \simeq 0$, then condition (4.5.8) is close to the conditions of balanced sampling designs of Royall and Herson (1973) and $\hat{T}_B \simeq N\bar{y}_s$. If $\nu = 0, \hat{T}_B = [1 + \frac{N-n}{n}(1+\sigma^2/nR)^{-1}]T_1$, where $T_1 = \sum_s y_i$, which is similar to an estimator proposed by Lindley (1962). The factor σ^2/nR which uses the information about the current population relative to the prior may, therefore, be called a' shrinkage factor'.

In the next section we consider robust Bayesian estimation of finite population parameters under a class of contaminated priors.

4.6   ROBUST BAYES ESTIMATION UNDER CONTAMINATED PRIORS

Following Berger (1984), Berger and Berliner (1986), Sivaganeshan and Berger (1989), Ghosh and Kim (1993, 1997) considered robust Bayes estimation of a finite population mean $\gamma(\mathbf{y}) = \frac{1}{N}\sum_{i=1}^N y_i$ under a class of contaminated priors. We first review robust Bayesian view-point of Berger and Berliner (1986).

Let X denote an observable random vector having a *pdf* $f(x \mid \theta)$ indexed by a parameter vector $\theta \in \Theta$. Consider the class of priors π for θ,

$$\Gamma_Q = \{\pi : \pi = (1 - \epsilon)\pi_0 + \epsilon q, q \in Q\}, \epsilon \in [0, 1] \tag{4.6.1}$$

where π_0 is a particular well-specified distribution, q is another prior distribution, Q is a subset of $\mathcal{Q}$, the class of all prior distributions of θ on Θ. Clearly, the class Γ_Q is a broader class than the singletone class $\{\pi_0\}$ and thus considers errors in assessment of subjective prior π_0. Such priors have been used by Blum and Rosenblatt (1967), Hubler (1973), Merazzi (1985), Bickel (1984), Berger (1982, 1984), Berger and Berliner (1986), among others. The most commonly used method of selecting a robust prior in Γ_Q is to choose that prior π which maximises the (marginal) predictive density

$$m(x \mid \pi) = \int_\Theta m(x \mid \theta)\pi(d\theta) \tag{4.6.2}$$

$$= (1 - \epsilon)m(x \mid \pi_0) + \epsilon m(x \mid q)$$

over Q. This is equivalent to maximising $m(x \mid q)$ over Q. Assuming that the maximum of $m(x \mid q)$ is uniquely attained at $q = \hat{q}$, the estimated prior $\hat{\pi}$, called the *ml* (maximum likelihood)-II prior by Good (1963) is

$$\hat{\pi} = (1 - \epsilon)\pi_0 + \epsilon\hat{q} \tag{4.6.3}$$

For an arbitrary prior $q \in Q$, the posterior density of θ is

$$\pi(d\theta \mid x) = \lambda(x)\pi_0(d\theta \mid x) + (1 - \lambda(x))q(d\theta \mid x) \tag{4.6.4}$$

where $\lambda(x) \in [0, 1]$ and is given by

$$\lambda(x) = \frac{(1 - \epsilon)m(x \mid \pi_0)}{m(x \mid \pi)} \tag{4.6..5}$$

Further, the posterior mean δ^π and the posterior variance V^π of θ (when they exist) are given by

$$\delta^\pi(x) = \lambda(x)\delta^{\pi_0}(x) + (1 - \lambda(x))\delta^q(x) \tag{4.6.6}$$

$$V^\pi(x) = \lambda(x)V^{\pi_0}(x) + (1 - \lambda(x))V^q(x)$$
$$+ \lambda(x)(1 - \lambda(x))(\delta^{\pi_0}(x) - \delta^q(x))^2 \tag{4.6.7}$$

If C is a measurable subset of Θ, then the posterior distribution of C with respect to π is

$$P^\pi(\theta \in C) = \lambda(\mathbf{x})P^{\pi_0}(\theta \in C) + (1 - \lambda(x))P^q(\theta \in C)$$

When $Q = \mathcal{Q}$, assuming a unique *mle* $\hat{\theta}(x)$ exists for θ, the *ml*-II prior of θ in $\Gamma_{\mathcal{Q}}$ is given by

$$\hat{\pi}(d\theta) = (1 - \epsilon)\pi_0(\theta) + \epsilon\hat{q}_x(d\theta) \tag{4.6.8}$$

where $\hat{q}_x(.)$ is a degenerate prior of θ, which assigns probability one to $\theta = \hat{\theta}(x)$. The *ml*-II posterior of θ is then given from (4.6.4) as

$$\hat{\pi}(. \mid x) = \hat{\lambda}(x)\pi_0(. \mid x) + (1 - \hat{\lambda}(x))\hat{q}_x(.) \tag{4.6.9}$$

where

$$\hat{\lambda}(x) = (1 - \epsilon)m(x \mid \pi_0)/[(1 - \epsilon)m(x \mid \pi_0) + \epsilon f(x \mid \hat{\theta}(x))] \tag{4.6.10}$$

The *ml*-II posterior mean of θ is then

$$\delta^{\hat{\pi}}(x) = \hat{\lambda}(x)\delta^{\pi_0}(x) + (1 - \hat{\lambda}(x))\hat{\theta}(x) \tag{4.6.11}$$

and the posterior variance of θ is

$$V(\theta \mid \hat{\pi}(x)) = V^{\hat{\pi}}(x) = \hat{\lambda}(x)[V^{\pi_0}(x) + (1 - \hat{\lambda}(x))(\delta^{\pi_0}(x) - \hat{\theta}(x))^2] \tag{4.6.12}$$

When the data are consistent with $\pi_0, m(x \mid \pi_0)$ will be reasonably large and $\hat{\lambda}(x)$ will be close to one (for small ϵ), so that $\delta^{\hat{\pi}}$ will be essentially equal to δ^{π_0}. When the data and π_0 are incompatible, $m(x \mid \pi_0)$ will be small and $\hat{\lambda}(x)$ near zero; $\delta^{\hat{\pi}}$ will then be approximately equal to *mle* $\hat{\theta}$.

An interesting class of priors Γ_S involves symmetric modal contamination. Here

$$Q = \{ \text{ densities of the form } q(\mid \theta - \theta_0 \mid), q \text{ non-increasing}\}$$

Since any symmetric unimodal distribution is a mixture of symmetric uniform distributions (cf. Berger and Silken, 1987) it suffices to restrict q to

$$Q' = \{ \text{ Uniform } (\theta_0 - a, \theta + a) \text{ densities }, a \geq 0\} \tag{4.6.13}$$

where a is to be chosen optimally. For the class Γ_S, the *ml*-II prior is

$$\tilde{\pi} = (1 - \epsilon)\pi_0 + \epsilon\tilde{q}$$

where $\tilde{q}$ is uniform in $(\theta_0 - \hat{a}, \theta_0 + \hat{a})$, $\hat{a}$ being the value of a which minimises

$$m(x \mid a) = \begin{cases} (2a)^{-1} \int_{\theta_0 - a}^{\theta_0 + a} f(\mathbf{x} \mid \theta)d\theta, & a > 0 \\ f(x \mid \theta_0) & a = 0 \end{cases}$$

EXAMPLE 4.6.1

Let $X = (X_1, \ldots, X_p)' \sim N_p(\theta, \sigma^2 I_p)$, θ unknown, σ^2 known. Suppose the elicited prior π_0 for θ is $N_p(\mu, \tau^2 I_p)$. Since the usual *mle* of θ is $\hat{\theta}(x) = x = (x_1, \ldots, x_p)$,

$$\delta^{\pi_0}(x) = x - [\frac{\sigma^2}{\sigma^2 + \tau^2}](x - \mu)$$

$$\delta^{\hat{\pi}}(x) = [1 - \hat{\lambda}(x)\frac{\sigma^2}{\sigma^2 + \tau^2}](x - \mu) + \mu \quad \text{by}(4.6.11)$$

where

$$\hat{\lambda}(x) = [1 + (\frac{\epsilon}{1 - \epsilon})(1 + \frac{\tau^2}{\sigma^2})^{p/2}$$

$$\exp \{| x - \mu |^2 /2(\tau^2 + \sigma^2)\}]^{-1}$$

and $| z |$ denotes $\sqrt{\sum z_i^2}$. Note that $\lambda \to 0$ exponentially fast in $| x - \mu |^2$ so that $\delta^{\hat{\pi}}(x) \to x$ quite rapidly as $| x - \mu |^2$ gets large.

Another class of priors is unimodality preserving contaminations where we denote by θ_0 the mode of π_0, assumed to be unique. Here the class of priors is

$$\Gamma_U = \{\pi : \pi = (1 - \epsilon)\pi_0 + \epsilon q, q \in Q_U\} \quad (4.6.14)$$

where Q_U is the set of all probability densities for which π is unimodal with mode θ_0 (not necessarily unique) and $\pi(\theta_0) \leq (1 + \epsilon')\pi_0(\theta_0)$.

However, it may be noted that the *ml*-II technique is not fullproof and can produce bad results, specially, when Γ includes unreasonable distributions.

Sivaganeshan (1988) obtained the following result for the range of posterior mean of θ when $\theta \in \Gamma_Q$. Assume that the parameter space Θ is the real line R_1 and $f(x \mid \theta) > 0$ $\forall \theta \in R$.

THEOREM 4.6.1 Let $\Gamma_1 \in \Gamma_Q$ be defined by

$$\Gamma_1 = \{\pi : \pi = (1 - \epsilon)\pi_0 + \epsilon q; q \text{ is a point mass }\} \quad (4.6.15)$$

Then,

$$\sup{}_{\pi \in \Gamma_Q} \delta^\pi(x) = \sup{}_{\theta \in R_1} R(\theta)$$

and

$$\inf{}_{\pi \in \Gamma_Q} \delta^\pi(x) = \inf{}_{\pi \in \Gamma_1} \delta^\pi(x) = \inf{}_{\theta \in R_1} R(\theta)$$

where

$$R(\theta) = [a\delta^{\pi_0}(\mathbf{x}) + \theta f(\mathbf{x} \mid \theta)]/[a + f(x \mid \theta)]$$

and
$$a = (1 - \epsilon)m(x \mid \pi_0)/\epsilon$$

The problem has also been considered by Sivaganesan and Berger (1989). Clearly, the smaller is the range of $\delta^\pi(x)$, more robust is the Bayes estimate over priors in Γ_Q.

EXAMPLE 4.6.2

Suppose $x \mid \theta \sim N(\theta, \sigma^2), \sigma^2$ known and $\pi_0 = N(\theta_0, \tau^2)$ for given θ_0, τ^2. Then
$$R(\theta) = \lambda(\theta)\delta^{\pi_0}(x) + (1 - \lambda(\theta))\theta \qquad (i)$$

where
$$\lambda(\theta) = \frac{a}{a + f_x(\theta)},$$

$$a = \frac{1-\epsilon}{\epsilon} \frac{1}{\sqrt{2\pi(\sigma^2 + \tau^2)}} \exp\left[-\frac{(x-\theta_0)^2}{2(\sigma^2 + \tau^2)}\right]$$

$$\delta^{\pi_0}(x) = \frac{\sigma^2\theta_0 + \tau^2 x}{\sigma^2 + \tau^2}$$

The range of $\delta^\pi(x)$ for $\pi \in \Gamma_Q$ is given by
$$[\lambda_l\delta^{\pi_0}(x) + (1 - \lambda_l)\theta_l, \lambda_u\delta^{\pi_0}(x) + (1 - \lambda_u)\theta_u]$$

where θ_l is the value of θ in R which minimises $R(\theta)$ given in (i) and similarly for θ_u.

Robustness of inference based on posterior probability distribution of θ with respect to prior $\pi \in \Gamma_Q$ can be checked from the following result due to Hubler (1973).

THEOREM 4.6.2 Let C be a measurable subset of Θ and define β_0 to be the posterior probability of C under π_0 i.e.
$$\beta_0 = \int_C f(x \mid \theta)\pi_0(d\theta) = P^{\pi_0}[\theta \in C \mid X = x]$$

Then

$$\inf{}_{\pi \in \Gamma} P^\pi[\theta \in C \mid X = x] = \beta_0\{1 + \frac{\epsilon \sup{}_{\theta \in C} f(x \mid \theta)}{(1 - \epsilon)m(x \mid \pi_0)}\}^{-1}$$

$$\sup{}_{\pi \in \Gamma} P^\pi[\theta \in C \mid X = x] = \frac{(1 - \epsilon)m(x \mid \pi_0)\beta_0 + \epsilon \sup{}_{\theta \in C} f(x \mid \theta)}{(1 - \epsilon)m(x \mid \pi_0) + \epsilon \sup{}_{\theta \in C} f(\mathbf{x} \mid \theta)}$$

Thus robustness with respect to Γ will usually significantly depend on the observed x values. A lack of robustness may also be due to the fact that Γ is too large. Generally C is taken as the $100(1-\alpha)\%$ credibility interval of θ under π. Berger and Berliner (1983) determined the optimal $(1-\alpha)$ robust credible set, optimal in the sense of having smallest size (Lebesgue measure) subject to the posterior probability having at least $1-\alpha$ for all π in Γ.

4.6.1   APPLICATIONS IN FINITE POPULATION SAMPLING

A sample $s = (i_1, \ldots, i_n)$ is drawn from $\mathcal{P}$ using the sampling design $p(s)$ and let $y_r = \{y_i, i \in r = \bar{s}\}, \bar{y} = \frac{1}{n}\sum_{i \in s} y_i$. Consider the following super-population model

$$y_i \mid \theta \underset{\sim}{iid} N(0, \sigma^2), i = 1, \ldots, N \qquad (4.6.16)$$

where θ has a prior

$$\theta \sim N(\mu_o, \tau_0^2) = \pi_0 \text{ (say)} \qquad (4.6.17)$$

From Ericson (1969 a) it follows that

$$y_r \mid (s, y_s) \sim N(\{(1 - B_0)\bar{y}_s + B_0\mu_0\}1_{N-n}, \sigma^2(I_{N-n} + (M_0 + n)^{-1}J_{N-n})) \qquad (4.6.18)$$

where

$$M_0 = \sigma^2/\tau_0^2, \quad B_0 = \frac{M_0}{(M_0 + n)} \qquad (4.6.19)$$

Bayes estimate of $\gamma(\mathbf{y}) = \frac{1}{N}\sum_{i=1}^{N} y_i$ is, therefore,

$$\delta^{\mu_0, B_0}(s, y_s) = E[\gamma(\mathbf{y}) \mid s, y_s)]$$

$$= \bar{y} - (1 - f)B_0(\bar{y} - \mu_0) \qquad (4.6.20)$$

where $f = n/N$. Also, the posterior variance of $\gamma(\mathbf{y})$ is

$$V(\gamma(\mathbf{y}) \mid s, y_s) = N^{-2}(N - n)\sigma^2(M_0 + N)/(M_0 + n) \qquad (4.6.21)$$

The classical estimator of $\gamma(\mathbf{y})$ is

$$\delta^c(s, y_s) = \bar{y}_s \qquad (4.6.22)$$

which remains p-unbiased under srs and m-unbiased under any model which assumes that y_i's have a common mean.

Ghosh and Kim (1993) considered robust Bayes estimation of $\gamma(\mathbf{y})$ under the class of priors Γ_Q. The ml-II prior in this class is given by (4.6.8) as

$$\hat{\pi}_s(\theta) = (1 - \epsilon)\pi_0(\theta) + \epsilon\delta_{\bar{y}_s}(\theta) \qquad (4.6.23)$$

where $\delta_{\bar{y}_s}(\theta) = 1(0)$ for $\theta = \bar{y}_s$ otherwise .

THEOREM 4.6.3 Under prior $\hat{\pi}_s(\theta)$ (given in (4.6.8)), the posterior distribution of y_r is

$$\hat{\pi}_s(y_r \mid s, y_s) = \hat{\lambda}_{ML}(\bar{y}_s)N(\{(1 - B_0)\bar{y}_s + B_0\mu_0\}1_{N-n},$$

$$\sigma^2(I_{N-n} + (M_0 + n)^{-1}J_{N-n})) + (1 - \hat{\lambda}_{ML}\bar{y}_s)N(\bar{y}1_{N-n}, \sigma^2I_{N-n}) \quad (4.6.24)$$

where

$$\hat{\lambda}_{ML}^{-1}(\bar{y}_s) = 1 + \frac{\epsilon}{(1 - \epsilon)\sqrt{B_0}}exp\{nB_0(\bar{y} - \mu_0)^2/2\sigma^2\} \qquad (4.6.25)$$

The Bayes estimator of $\gamma(\mathbf{y})$ is

$$\delta^{RB}(s, y_s) = \bar{y}_s - (1 - f)\hat{\lambda}_{ML}(\bar{y}_s)B_0(\bar{y}_s - \mu_0) \qquad (4.6.26)$$

with posterior variance

$$V(\gamma(\mathbf{y}) \mid s, y_s) = N^{-2}[(N - n)\sigma^2 + (N - n)^{-2}$$

$$\{\sigma^2\frac{\hat{\lambda}_{ML}}{M_0 + n} + \hat{\lambda}_{ML}(1 - \hat{\lambda}_{ML})B_0^2(\bar{y}_s - \mu_0)^2\}] \qquad (4.6.27)$$

Proof The conditional *pdf* of y_r given (s, y_s) is

$$\hat{\pi}_s(y_r \mid s, y_s) = \int f(y_r \mid \theta)\hat{\pi}_s(\theta \mid s, y_s)d\theta \qquad (4.6.28)$$

The results are then obtained by using (4.6.10) - (4.6.12).

We note that for ϵ very close to zero, i.e. when one is very confident about the π_0- prior, (since $\hat{\lambda}_{ML}(\bar{y}) \simeq 1$), δ^{RB} is very close to δ^0. For ϵ close to one, δ^{RB} is close to δ^c.

For a given prior ξ, the posterior risk of an estimator $e(s, y_s)$ of $\gamma(\mathbf{y})$ is

$$\rho(\xi, (s, y_s), e) = E[\{e(s, y_s) - \gamma(\mathbf{y})\}^2 \mid s, y_s] \qquad (4.6.28)$$

DEFINITION 4.6.1 An estimator $e_0(s, y_s)$ is ψ-posterior-robust (POR) *wrt* priors in a class Γ if

$$POR(e_0) = sup_{\xi\in\Gamma} \mid \rho(\xi, (s, y_s), e_0) - \rho(\xi, (s, y_s), \delta^\xi \mid < \psi \qquad (4.6.29)$$

where $\delta^\xi = \delta^\xi(s, y_s)$ is Bayes estimator of $\gamma(\mathbf{y})$ under the prior ξ. The quantity $POR(e_0)$ is called the *posterior robustness index* of e_0. Taking

$$\Gamma_{\mu, B} = \{\pi_{\mu, B} = N(\mu, \sigma^2), \ \mu \in R, \sigma^2 > 0\}, \tag{4.6.30}$$

where B is defined as in (4.6.4), Bayes estimator under $\pi_{\mu, B}$ and its posterior variance $\rho(\pi, (s, y_s), \delta^{\mu, B})$ are given by (4.6.5) and (4.6.6). The following results hold.

$$\rho(\pi_{\mu, B}, (s, y_s), \delta^0) - \rho(\pi_{\mu, B}, (s, y_s), \delta^{\mu, B})$$
$$= (1 - f)^2 [B_0(\mu - \mu_0) + (B_0 - B)(\bar{y}_s - \mu)]^2 \tag{4.6.31}$$

$$\rho(\pi_{\mu, B}, (s, y_s), \delta^c) - \rho(\pi_{\mu, B}, (s, y_s, \delta^{\mu, B})$$
$$= (1 - f)^2 B^2 (\bar{y} - \mu)^2 \tag{4.6.32}$$

$$\rho(\pi_{\mu, B}, (s, y_s), \delta^{RB}) - \rho(\pi_{\mu, B}, (s, y_s), \delta^{\mu, B})$$
$$= (1 - f)^2 [B_0 \hat{\lambda}_{ML}(\bar{y}_s)(\bar{y}_s - \mu_0) - B(\bar{y}_s - \mu)]^2 \tag{4.6.33}$$

It follows from (4.6.31) - (4.6.32), therefore, that all the estimators δ^0, δ^c and δ^{RB} are POR-non-robust under $\Gamma_{\mu, B}$. This is because Γ is very large. If one confines to the narrower class $\Gamma_0 = \{\pi_{\mu_0, B} = N(\mu_0, \sigma^2), \sigma^2 > 0\}$, it follows from (4.6.31) - (4.6.33) that

$$POR(\delta^0) = (1 - f)^2 max[B_0^2, (1 - B_0)^2](\bar{y}_s - \mu_0)^2 \tag{4.6.34}$$

$$POR(\delta^c) = (1 - f)^2 (\bar{y}_s - \mu_0)^2 \tag{4.6.35}$$

$$POR(\delta^{RB}) = (1 - f)^2 max[B_0^2 \hat{\lambda}_{ML}(\bar{y}_s), (1 - B_0 \hat{\lambda}_{ML}(\bar{y}_s))^2](\bar{y}_s - \mu_0)^2 \tag{4.6.36}$$

Thus, given ψ and f, posterior robustness of the predictors depend on the closeness of $\bar{y}$ to μ_0. Also, both the subjective Bayes predictor δ^0 and robust Bayes predictor δ^{RB} are more posterior robust than δ^c under Γ_0. Again, δ^{RB} is more posterior robust than δ^0 if $B_0 \hat{\lambda}_{ML}(\bar{y}) > 1/2$. Defining

$$\gamma(\xi, e) = E[\rho(\xi, (s, \mathbf{y}_s)), e] \tag{4.6.37}$$

where expectation is taken with respect to marginal predictive distribution of y_s, as the *overall Bayes risk* of e, we consider

DEFINITION 4.6.2 An estimator $e_0(s, y_s)$ is said to be ψ- procedure robust with respect to Γ if

$$PR(e_0) = sup_{\xi \in \Gamma} \mid r(\xi, e_o) - r(\xi, \delta^\xi) \mid < \psi \tag{4.6.38}$$

PR(e_0) is called the *procedure robustness* of e_0. Considering the class Γ_0, and denoting by π_B the $N(\mu_0, \sigma^2)$ prior,

$$r(\pi_B, \delta^0) - r(\pi_B, \delta^B) = (1 - f)^2 (B_0 - B)^2 \sigma^2 / nB \qquad (4.6.39)$$

$$r(\pi_B, \delta^c) - r(\pi_B, \delta^B) = (1 - f)^2 B\sigma^2 / n \qquad (4.6.40)$$

$$r(\pi_B, \delta^{RB}) - r(\pi_B, \delta^B) = (1 - f)^2 E[(B_0\hat{\lambda}_{ML}(\bar{y}_s) - B)^2(\bar{y}_s - \mu_0)^2]$$

It follows, therefore,

$$PR(\delta^0) = \infty$$

$$PR(\delta^c) = (1 - f)^2 \sigma^2 / n$$

$$PR(\delta^{RB}) = (1 - f)^2 \sup_{0 < B < 1} E[(B_0\hat{\lambda}_{ML}(\bar{y}_s) - B)^2(\bar{y}_s - \mu_o)^2] \qquad (4.6.41)$$

$$= O(\sqrt{B})$$

Thus δ^0 is not procedure robust. For small B, δ^c is more procedure robust than δ^{RB}. This is what is to be expected since small B signifies small variance ratio δ^2/τ^2 which amounts to instability in the assessment of prior for θ. In this case, long-run performance of δ^c is expected to be better than that of δ^{RB}.

We note that subjective Bayes predictor δ^0 which is POR-robust (i.e. robust for a given sample) fails completely in long-run performance as measured by procedure robustness. The robust Bayes procedue δ^{RB} seems to achieve a balance between a frequentist and subjective Bayesian viewpoint.

Following Sivaganeshan and Berger (1989) the authors considered the range of posterior means of $\gamma(\mathbf{y})$ over $\pi \in \Gamma_Q$.

The authors extended the study to the symmetric class of unimodal contaminated priors to obtain the estimate δ^{SU}.

Ghosh and Kim (1997) considered robust Bayes competitors of ratio estimators. Under the superpopulation model

$$y_i = \beta x_i + e_i, \; i = 1, \dots, N$$

$$e_i \underset{\sim}{iid} N(0, \sigma^2 x_i) \qquad (4.6.42)$$

with β having a uniform prior over $(-\infty, \infty)$, Bayes estimate of $\gamma(\mathbf{y})$ is given by the ratio estimator $e_R = (\sum_s y_i / \sum_i x_i)\bar{X}$. Under model (4.6.42) and a $\pi_1 = N(\beta_0, \tau^2)$- prior for β, the Bayes estimator of $\gamma(\mathbf{y})$ is

$$\delta^1 = \delta^1(s, y_s) = f\bar{y}_s + (1 - f)\bar{x}_s[(1 - B_1)\bar{y}_s/\bar{x}_s + B_1\beta_0] \qquad (4.6.43)$$

where

$$B_1 = \frac{M_0}{M_0 + n\bar{x}}, \quad \bar{x}_s = \frac{1}{n}\sum_{i \in s} x_i$$

and M_0 is as defined in (4.6.19) with associated Bayes posterior variance

$$V_\pi = V(\gamma(\mathbf{y}) \mid s, y_s) = \sigma^2 N^{-1}[(N-n)\bar{x}_s$$

$$+ (N-n)^2 \bar{x}^2 / (M_0 + n\bar{x})] \tag{4.6.44}$$

Under the class of contaminated priors Γ_Q where $\pi_1 = \pi$, the authors found the ml-II prior and the robust Bayes predictor of $\gamma(\mathbf{y})$ under this prior as

$$\delta^{RB(1)}(s, y_s) = f\bar{y}_s + (1-f)\bar{x}_s\{(1 - \hat{\lambda}_{ML}(\bar{y}))B_1(s))\bar{y}_s/\bar{x}_s + \hat{\lambda}'_{ML}(\bar{y}_s)B_1\beta_0 \tag{4.6.45}$$

where

$$\hat{\lambda}_{ML}^{l-1} = 1 + \frac{\epsilon}{(1-\epsilon)}\sqrt{B_1}\,\exp\,(nB_1(s)(\bar{y}_s - \beta_0\bar{x}_s)^2/2\sigma^2\bar{x}_s) \tag{4.6.46}$$

Also its posterior variance is

$$V^{RB(1)} = V(\gamma(\mathbf{y}) \mid s, y_s) = N^{-2}[\sigma^2(N-n)\bar{x}_s + (N-n)^2\bar{x}_r^2$$

$$\{\sigma^2\frac{\hat{\lambda}'_{ML}}{M_0 + n\bar{x}_s} + \hat{\lambda}'_{ML}(1 - \hat{\lambda}'_{ML})B_1^2(\frac{\bar{y}_s}{\bar{x}_s} - \beta_0)^2\}] \tag{4.6.47}$$

where $\bar{x}_r = \sum_{i \in r} x_i/(N-n)$. The authors compared $\delta^{RB(1)}, \delta^1$ and e_R in terms of posterior risks as well as the overall Bayes risk under the class of priors $\{N(\beta_0, \tau^2), \tau^2 > 0\}$. It is found that both δ^1 and $\delta^{RB(1)}$ are superior to e_R in terms of posterior robustness. Also, δ^1 lacks procedure robustness, while e_R is quite procedure robust. It was found that for small values of δ^2/τ^2 (which amounts to greater instability in the assessment of the prior distribution of β relative to the superpopulation model) e_R is more procedure robust than $\delta^{RB(1)}$. This shows that in such circumstances it is safer to use e_R, if one is seriously concerned about the long-run performance of the estimator. The authors extended the study to symmetric unimodal contaminated class of priors to obtain the estimator $\delta^{SU(1)}$.

It may be noted that in models (4.6.1) [(4.6.42)], σ^2 may be unknown and one may consider normal gamma priors for $(\mu, \sigma^2)[(\beta, \sigma^2)]$ and derive ml-II priors under the contaminated class and undertake similar studies. Clearly, such studies can be extended to multiple regression models.

Ghosh and Kim (1997) considered 1970 population (y) and 1960 population (x) of 125 US cities with 1960-population between 10^5 and 10^6 for an empirical study of the ratio estimator and estimators of its variance. A 20% $srswor$ of cities was taken. σ^2 was assumed to be known. To elicit the basic prior π_0 for β data on 1950 and 1960 population was used and ϵ was chosen as 0.1. It was found that $\delta^{RB(1)}$ and $\delta^{SU(1)}$ were closer to $\gamma(\mathbf{y})$ than e_R, which was worst in terms of posterior robustness index. The range of posterior mean of $\gamma(\mathbf{y})$ was found to be small in both the cases Γ_Q and Γ_S so that if the true prior was π_0 one could use modelling via any of the contaminations to achieve the robust Bayesian analysis.

4.7   EXERCISES

1. Let $X_k \mid \theta_k$ be distributed independently as $N(\mu, \tau^2)$, while θ_k has independent prior distribution $N(\mu, \tau^2), k = 1, \ldots, n$. The posterior density of $\theta_k \mid X_k$ is, therefore, $N(\mu + D(x_k - \mu), D)$ where $D = \tau^2/(1 + \tau^2)$. The Bayes estimate of θ_k under SSEL are the posterior means

$$\hat{\theta}_k^B = \mu + D(x_k - \mu), k = 1, \ldots, n$$

Define $\bar{\theta} = \sum_{k=1}^{n} \theta_k/n, \ \bar{\hat{\theta}}^B = \sum_{k=1}^{n} \hat{\theta}_k^B/n, \ S^2 = \frac{1}{n-1}\sum_{k=1}^{n}(X_k - \bar{X})^2, \ \bar{X} = \sum_{k=1}^{n} X_k/n$. Note that $\bar{X} \to \mu$ and $S^2 \to 1 + \tau^2 = 1/(1-D)$ im probability. Hence, or otherwise show that

(i)
$$E(\bar{\hat{\theta}}^B) = E(\bar{\theta}),$$

(ii)
$$\sum_{i=1}^{n}(\hat{\theta}_i^B - \bar{\hat{\theta}}^B)^2/(n-1) = D^2 S^2 \to D\tau^2 \text{ as } n \to \infty$$

(iii)
$$E[\sum_{i=1}^{n}(\theta_i - \bar{\theta})^2/(n-1) \mid x] = D(1 + DS^2) \to \tau^2 \text{ as } n \to \infty$$

Therefore, histogram of $\hat{\theta}_i^B$ values are more concentrated about prior mean μ than the values of θ_i given $x = (x_1, \ldots, x_n)'$, since $0 \le D \le 1$.

Consider now the modified estimator

$$\hat{\theta}_k^L = \xi + A(X_k - \xi)$$

where

$$A = \sqrt{D}[\frac{1 + S^2 D}{S^2}]^{1/2}$$

$$\xi = \frac{(1 - D)\mu + \bar{X}(D - A)}{1 - A}$$

Show that for this estimator
   (iv)

$$E(\bar{\hat{\theta}}^L) = E(\bar{\theta})$$

   (v)

$$\sum_{k=1}^{n}(\hat{\theta}_k^L - \bar{\hat{\theta}}^L)^2/(n-1) = E[\sum_{k=1}^{n}(\theta_k - \bar{\theta})^2 \mid \mathbf{x}]/(n-1)]$$

where $\bar{\hat{\theta}}^L = \sum_{k=1}^{n} \hat{\theta}_k^L/n$

(Louis, 1984)

2. Let $X_j(j = 1, \ldots, n)$ be independently distributed $N(\theta, \sigma^2)$ and θ have the prior $\xi^* : \theta \sim N(\mu, \tau^2)$ where μ, τ, σ^2 are known and θ is an unobservable random quantity whose value we want to predict. Show that under the loss function $L(\theta, a)$ where $L(\theta, a)$ is an increasing function of $\mid \theta - a \mid$, the Bayes estimate of θ is

$$\delta_B(x) = \frac{\mu\sigma^2 + n\bar{x}_s\tau^2}{\sigma^2 + n\tau^2}$$

where $x = (x_1, \ldots, x_n)'$ and $\bar{x}_s = \sum_{j=1}^{n} x_j/n$.

For $L(\theta, a) = (\theta - a)^2$, show that the Bayes risk (expected risk) of δ_B with respect to ξ^* is

$$E(\delta_B(x) - \theta)^2 = R(\xi^*, \delta_B(x)) = \frac{\tau^2\sigma^2}{\sigma^2 + n\tau^2}$$

where expectation is taken with respect to predictive distribution of x for the given prior ξ^*.

Consider an alternative prior $\xi_1 : \theta \sim N(\mu_1, \tau_1^2), \tau_1^2 < \tau^2$. Show that under this prior expected risk of δ_B is

$$R(\xi_1, \delta_B(x)) = [(\frac{A}{A+1} - \frac{A_1}{A_1+1})^2(A_1 + 1) + \frac{A_1}{A_1+1}]\sigma^2/n$$

$$+ \frac{(\mu_1 - \mu)^2}{(A+1)^2}$$

$$= B\sigma^2/n + \frac{(\mu_1 - \mu)^2}{(A+1)} \text{ (say)}$$

where

$$A = \frac{n\tau^2}{\sigma^2}, \ A_1 = \frac{n\tau_1^2}{\sigma^2}.$$

Thus

$$R(\xi_1, \delta_B(x)) \lessgtr R(\xi_1, \bar{x}) = \sigma^2/n$$

according as

$$(\mu_1 - \mu)^2 \lessgtr \sigma^2(1 - B)(A+1)^2/n.$$

Thus for any fixed value of $\tau_1^2 < \tau^2$, $R(\xi_1, \delta_B(x))$ can be made arbitrarily large, by making $| \mu_1 - \mu |$ arbitrarily large.

In particular, let $\sigma^2 = 1, \mu = 0, n = 1$ when $\tau^2 = A$. Under the prior ξ^* which we now denote as ξ_A, Bayes estimate of θ is

$$\delta_A^* = \frac{Ax}{A+1}$$

with Bayes risk

$$R(\xi_a, \delta_A^*) = \frac{A}{A+1}$$

Saving in risk using δ_A^* over using the *mle* x is

$$R(\xi_A, x) - R(\xi_A, \delta_A^*) = 1/(A+1)$$

Again,

$$E_X(\delta_A^* - \theta)^2 = \frac{\theta + A^2}{(A+1)^2}$$

Therefore, an estimator which compromises between Bayes estimator δ_A^* (which has high risk $E_X(\delta_A^* - \theta)^2$ (in the frequentist sense) for high value of $| \theta |$) and the *mle* $\delta^0 (= x)$ (which has no savings in risk with respect to ξ_A but has minimax risk $R_\theta(\delta^0) = 1 \ \forall \ \theta$) is the Limited Translation (LT) Bayes estimator $\delta_{A,M}(x)$ defined as follows. For any A and $M(> 0)$, let $C = M(A+1)$. Then

$$\delta_{A;M}(x) = \begin{cases} \delta^0(x) + M = x + M \text{ for} & \delta_A^*(x) > x + M \\ \delta_A^*(x) = \frac{Ax}{(A+1)} \text{ for} & | \delta_A^*(x) - \delta^0(x) | < M \\ \delta^0(x) - M = x - M \text{ for} & \delta_A^*(x) < x - M \end{cases}$$

Show that relative savings loss (RSL) of $\delta_{A,M}$ with respect to δ^0 is

$$\frac{R(\xi_A, \delta_{A,M}) - R(\xi_A, \delta_A^*)}{R(\xi_A, \delta_0) - R(\xi_A, \delta_A^*)}$$

$$= (A+1)[R(\xi_A, \delta_{A,M}) - \frac{A}{A+1})]$$

(Effron and Morris, 1971)

Chapter 5

Estimation of Finite Population Variance, Regression Coefficient

5.1 INTRODUCTION

In this chapter we consider estimation of a finite population variance and regression coefficient. The estimation of population variance is of considerable importance in many circumstances. The geneticsts often classify their population according to population variance [Thompson and Thoday (1979)]. In allocating sample size in a stratified random sampling according to optimum allocation rules, the stratum standard deviations are required to be estimated. Sections 5.2 through 5.4 consider design-based, model-based and Bayes prediction of a finite population variance. Section 5.5 considers some asymptotic properties of a sample regression coefficient. The next section considers pm-unbiased prediction of the slope parameter in the linear regression model. The concluding section addresses optimal prediction of the finite population regression coefficient under multiple regression model.

5.2 DESIGN-BASED ESTIMATION OF A FINITE POPULATION VARIANCE

Liu (1974 a) first considered design-unbiased estimation of the finite population variance

$$\begin{aligned} V(\mathbf{y}) &= \frac{1}{N} \sum_{i=1}^{N} (y_i - \bar{y})^2 \\ &= a_1 \sum_{i=1}^{N} y_i^2 - a_2 \sum_{i \neq i'=1}^{N} y_i y_{i'} \end{aligned} \tag{5.2.1}$$

where

$$a_1 = \frac{1}{N}\left(1 - \frac{1}{N}\right), \quad a_2 = \frac{1}{N^2}$$

A homogeneous quadratic (*h.q.*) estimator

$$e_q(s, \mathbf{y}) = \sum_{k \in s} b(s, k) y_k^2 + \sum \sum_{k \neq k' \in s} b(s, k\ k') y_k y_{k'} \tag{5.2.2}$$

is unbiased for $V(\mathbf{y})$ *iff*

$$\sum_{s \ni k} b(s, k) p(s) = \frac{1}{N}\left(1 - \frac{1}{N}\right) \forall\ k = 1, \ldots, N \tag{5.2.3.1}$$

$$\sum_{s \ni (k,k')} b(s, k\ k') p(s) = -\frac{1}{N^2}\ \forall\ k \neq k' = 1, \ldots, N \tag{5.2.3.2}$$

It is clear that for a given *s.d.*, a necessary condition for the existance of an unbiased quadratic estimator e_q is $\pi_{ij} > 0\ \forall\ i \neq j = 1, \ldots, N$. The following *Horvitz-Thompson type* estimator, first considered by Liu (1974 a) is unbiased for V.

$$e_L(s, \mathbf{y}) = a_1 \sum_{i \in s} \frac{y_i^2}{\pi_i} - a_2 \sum \sum_{i \neq j \in s} \frac{y_i y_j}{\pi_{ij}} \tag{5.2.4}$$

For *srswor*, e_L reduces to

$$\frac{N-1}{N} s_y^2 = s_y'^2 \text{ (say)}$$

For *ppswr* sampling design, e_L reduces to

$$e_{Lp} = a_1 \sum_{i \in s} \frac{y_i^2}{1 - (1-p_i)^n} -$$

$$a_2 \sum \sum_{i \neq j \in s} \frac{y_i y_j}{1 - (1 - p_i)^n - (1 - p_j)^n + (1 - p_i - p_j)^n} \tag{5.2.5}$$

For *ppswr* sampling design, other unbiased estimators of V are

$$e_{Lt_1} = \frac{a_1}{n} \sum_{i \in s} \frac{t(s,i)}{p_i} y_i^2 - \frac{a_2}{n(n-1)} \sum \sum_{i \neq j \in s} \frac{t(s,i)t(s,j)}{p_i p_j} y_i y_j \tag{5.2.6}$$

$$e_{Lt_2} = \frac{a_2}{n} \sum_{i \in s} \frac{t^2(s,i)}{[1 + (n-1)p_i]p_i} y_i^2 - \frac{a_2}{n(n-1)} \sum \sum_{i \neq j \in s} \frac{t(s,i)t(s,j)}{p_i p_j} y_i y_j \tag{5.2.7}$$

where $t(s,i)$ is the number of times the unit i is drawn in s, $t(s,i) = 0, 1, \ldots, n$; $\sum_{i=1}^{N} t(s,i) = n$ for all samples with fixed size n. In particular, for *srswr* sampling design e_{Lp} reduces to

$$e_{Lp}^0 = \frac{1}{N^2} \Big[\frac{(N-1)}{1 - (1 - 1/N)^n} \sum_{i \in s} y_i^2 -$$

$$\frac{1}{1 - 2(1 - 1/N)^n + (1 - 2/N)^n} \sum \sum_{i \neq j \in s} y_i y_j \Big] \tag{5.2.8}$$

However, the commonly used estimator in this case is

$$s_y^2 = \frac{1}{n-1} \sum_{i \in s} t(s,i) \{ y_i - \frac{1}{n} \sum_{i \in s} t(s,i) y_i \}^2$$

Liu (1974 a) showed that for all s.d.'s with $\pi_{ij} > 0 \ \forall \ i \neq j$, the estimator e_L is admissible in the class of all unbiased estimators. He also considered the variance $V(e_L)$ and its quartic unbiased estimator $v(e_L)$ and showed it to be admissible in the class of all quartic unbiased estimators of $V(e_L)$. Strauss (1982) derived some other admissible estimators of V. Sengupta (1988) proved the admissibility of s_y^2 in the class of fixed size sampling designs with unbiased estimators for V.

Noting that both e_L and $v(e_L)$ can take negative values Chaudhuri (1978) considered several non-negative unbiased estimators of V for a *ppswor* s.d. First we note some of his notations:

$$\psi_{ij} = \sum_{s \ni (i,j)} \frac{1}{p(s)}$$

$$\psi_{ijkl} = \sum_{s \ni (i,j,k,l)} \frac{1}{p(s)}$$

$$I_{ij}(s) = 1(0) \text{ if } (i,j) \in s \text{ (otherwise)}$$

$$t_{ij} = \sum_{s \in S} I_{ij}(s)$$

$$= \text{ number of samples containing } (i,j).$$

$$l_{ij} = t_{ij}(t_{ij} - 1)$$

$$f_{ijkl} = \sum_{s \in S} m_{ij}(s)m_{kl}(s)$$

$$\gamma_{ijkl} = \sum_{s \neq s' \in S} \sum m_{ij}(s)m_{kl}(s')$$

$$d_{ij} = (y_i - y_j)^2$$

$$\alpha(s) = \sum_{i<j \in s} \sum \frac{d_{ij}}{t_{ij}}$$

$$\beta(s) = \sum_{i<j \in s} \sum \frac{d_{ij}^3}{t_{ij}^2} l_{ij}$$

$$\gamma(s) = \sum_{i<j \in s} \sum_{k<l \in s} \sum \frac{d_{ij}}{t_{ij}} \frac{d_{kl}}{t_{kl}} \frac{\gamma_{ijkl}}{f_{ijkl}}$$

$p(s \mid i,j) = $ conditional probability of selecting s according to the *s.d.* p when it is given that i and j have been chosen on the first two draws. The estimators considered are:

$$e_{C1} = \frac{1}{N^2} \sum_{i<j \in s} \sum \frac{d_{ij}}{\pi_{ij}}, \text{ assuming } \pi_{ij} > 0 \ \forall \ i \neq j$$

$$e_{C2} = \frac{1}{N^2} \frac{\alpha(s)}{p(s)}, \text{ assuming } p(s) > 0 \ \forall \ s \text{ and } t_{ij}, l_{ij} \geq 1$$

$$e_{C3} = \frac{1}{N^2} \frac{1}{p(s)} \sum_{i<j \in s} \sum \{d_{ij} p(s \mid i,j)\},$$

assuming $p(s) > 0 \ \forall \ s$. He considered conditions for non-negativity of variance estimators of these estimators.

We shall denote

$$s_y^2 = \frac{N}{N-1} V(\mathbf{y}) = \frac{1}{N-1} \sum_{i=1}^{N} (y_i - \bar{y})^2$$

$$= b_1 \sum_{i=1}^{N} y_i^2 - b_2 \sum_{i \neq j=1}^{N} y_i y_j \tag{5.2.9}$$

where

$$b_1 = \frac{1}{N}, \ b_2 = \frac{1}{N(N-1)}$$

For *ppswr* sampling design, Das and Tripathi (1978) considered the following unbiased estimator of S_y^2,

$$t_p = b_1 A_s - b_2 B_s \tag{5.2.10}$$

where

$$A_s = \sum_{i \in s} \frac{y_i^2}{np_i},$$

$$B_s = \frac{1}{n(n-1)} \sum_{i \neq j \in s} \{ \frac{y_i y_j}{p_i p_j} - \frac{1}{2} (\frac{y_i^2}{p_i} + \frac{y_j^2}{p_j}) \}$$

Bhattacharyya (1997) studied the properties of t_p.

Assuming *srswr*, Isaki (1983) considered ratio estimator and regression estimator of $V(\mathbf{y}) = \sigma_y^2$. Let x be an auxiliary variable closely related to y and assume all the values $x_k (k = 1, \ldots, N)$ are known. The ratio estimator of V is

$$\hat{\sigma}_{yR}^2 = \frac{s_y^2}{s_x^2} \sigma_x^2$$

It can be shown, retaining terms up to order n^{-1} in the Taylor series expansion , that the variance of s_y^2 is less than that of $\hat{\sigma}_{yR}^2$ *iff*

$$\text{Corr. Coeff. } (s_x^2, s_y^2) > \frac{1}{2} \frac{cv(s_x^2)}{cv(s_y^2)}$$

where *cv* denotes the coefficient of variation.

Following Olkin (1958) he extended the ratio estimator to the multivariate case (Exercise 1). Let $x = (x_1, \ldots, x_k)$ be a vector of k auxiliary variables whose values $x_{ij} (i = 1, \ldots, N; j = 1, \ldots, k)$ are known, Let $\sigma_i^2 (s_i^2 = s_{ii})$ denote the population (sample) variance of x_i. Under *srswr* he considered a multivariate difference estimator

$$\hat{\sigma}_{ydm}^2 = s_y^2 + \sum_{i=1}^{k} B_i (\sigma_i^2 - s_i^2) \tag{5.2.11}$$

where B_i's are known constants. Optimum values of B_i are obtained by minimising $\text{Var}\,(\hat{\sigma}^2_{ydm})$ with respect to $B_i(i = 1, \ldots, k)$. The equations are

$$AB = C \qquad (5.2.12)$$

where $A = ((a_{ij})), a_{ij} = s^2_{ij}, B = (B_1, \ldots, B_k)', C = (C_1, \ldots, C_K)', C_j = s^2_{0j},$ obtained by replacing terms like $\text{Var}\,(s^2_i)$ by s^4_i, $\text{Cov}\,(s^2_i, s^2_j)$ by s^2_{ij} in equations $\frac{\partial Var(\hat{\sigma}^2_{ydm})}{\partial B_i} = 0, i = 1, \ldots, k$. Let $\hat{B}^0_i = (\hat{B}^0_1, \ldots, \hat{B}^0_k)'$ be the solution of the equations (5.2.12). Assuming A^{-1} exists and using the results of Fuller (1976, Ch.5), we have

$$(\hat{B}^0 - B^0) = O_p(n^{-1/2})$$

where $B^0 = (B^0_1, \ldots, B^0_k)', B^0_j = \sigma_{yj}/\sigma^2_y$. The multivariate regression estimator is, therefore, defined as

$$\hat{\sigma}^2_{yrm} = s^2_y + \sum_{i=1}^{k} \hat{B}^0_i(\sigma^2_i - s^2_i) \qquad (5.2.13)$$

For the case $k = 1$,

$$\hat{\sigma}^2_{yrm} = s^2_y + \hat{\beta}^2(\sigma^2_1 - s^2_1)$$

where $\hat{\beta} = s_{y1}/s^2_1$. Under the specific multivariate model of exercise 1, it can be shown that $\text{Var}\,(\hat{\sigma}^2_{yrm})$ is minimised for

$$\hat{B}^0_i = \frac{\rho R^2_i}{1 + (k-1)\rho^2} \quad (i = 1, \ldots, k)$$

where $R_i = \bar{y}/\bar{x}_i$. In this case,

$$\text{Var}\,(\hat{\sigma}^2_{yrm}) \simeq \frac{2\sigma^2_y}{n-1}[1 - (1 + (k-1)\rho^2)^{-1}k\rho^4]$$

omitting terms of order $O_p(n^{-3/2})$.

Bhattacharyya (1997) considered non-negative unbiased estimator of S^2_y. Writing $M_i = \binom{N-i}{n-i}$, a non-negative unbiased estimator of S^2_y, for any *s.d.* $p \in \rho_n$ is

$$t = b_1 \sum_{i \in s} \frac{y^2_i}{M_1 p(s)} - b_2 \sum_{i \neq j \in s} \sum \frac{y_i y_j}{M_2 p(s)}$$

$$= \frac{s^2_y}{M_0 p(s)} \qquad (5.2.14)$$

Variance of t is

$$V(t) = \frac{1}{M_0^2} \sum_{s \in S} \frac{s_y^4}{p(s)} - S_y^4 \qquad (5.2.15)$$

For Midzuno's (1952) $s.d., p_M$ (say) $, p(s) = q_s/M_1$, where $q_s = \sum_{i \in s} p_i, p_i = x_i/X, X = \sum_{i=1}^N x_i$, x_i being the value of an auxiliary variable x on unit i. Here, t reduces to

$$t_M = \frac{n s_y^2}{N q_s} \qquad (5.2.16)$$

Under the sampling scheme due to Singh and Srivastava (1980), p_S (say), $p(s) = s_x^2/(M_o S_x^2)$. Here, t reduces to the ratio type estimator

$$t_R = \frac{s_y^2}{s_x^2} S_x^2 \qquad (5.2.17)$$

Bhattacharyya (1997) studied the properties of the strategies $H_0 = (p_0, s_y^2)$, $H_1 = (p_M, t_M), H_2 = (p_S, t_R), H_3 = (p_0, t_R), H_4 = (ppswr, t_p)$ where p_0 denotes $srswr$.

She also considered the ratio estimator t_R under a class of controlled sampling designs. As discussed in section 1.3, such designs use only a fraction of the total number of all possible M_0 samples and thus many samples, specially the "non-preferred" ones are left out of the scope of survey. Consider a balanced incomplete block design (BIBD) with parameters v, b, r, k, λ, where v is the number of varieties, b the number of blocks, r, the number of replications of each unit, k, the size of a block, and λ is the number of blocks in which every pair (i, j) of elements occur together, $i \neq j = 1, \ldots, v$. Each element is identified as a unit in the population and each block as a sample. Therefore, $N = v, n = k$. Samples are selected with probability

$$p(s) = \frac{r s_y^2}{r N S_x^2} \qquad (5.2.17)$$

For this $s.d.$ p_c, (say) $, t_R$ is unbioased fopr S_y^2.

The performance of the strategies $H_i(i = 0, \ldots, 7)$ where $H_5 = (p_c, t_R), H_6 = (p_M, e'), H_7(p_S, e')$ where $e' = \frac{N}{N-1} e_{C1}$ were studied numerically and also under a superpopulation model.

5.3   MODEL-BASED PREDICTION OF V

Mukhopadhyay (1978) considered the superpopulation model $\eta : y_1, \ldots, y_N$ are random variables such that the conditional distribution of y_k given x_k

(known value of an auxiliary variable x) is normal with

$$\mathcal{E}(y_k \mid x_k) = 0$$

$$\mathcal{E}(y_k^2 \mid x_k) = \sigma^2 w(x_k) \tag{5.3.1}$$

where σ^2 is an unknown constant and $w(x_k)$ is a known function of x_k. Now,

$$V(\mathbf{y}) = A(s) + B(s)$$

where

$$A(s) = a_1 \sum_{k \in s} y_k^2 - a_2 \sum \sum_{k \neq k' \in s} y_k y_{k'}$$

$$B(s) = a_1 \sum_{k \in s} y_k^2 - a_2 [2 \sum_{k \in s} \sum_{k' \in \bar{s}} y_k y_{k'} + \sum \sum_{k \neq k' \in \bar{s}} y_k y_{k'}] \tag{5.3.2}$$

For a given sample s, $A(s)$ is completely known. A predictor of V is, therefore, $\nu(s, \mathbf{y}) = A(s) + C(s)$, where $C(s)$ is a predictor of $V(\mathbf{y})$. The predictor ν is η-unbiased for V if

$$\mathcal{E}[C(s)] = \mathcal{E}[B(s)] \quad \forall \; s : p(s) > 0$$

and $\forall \; \sigma^2 > 0$. Now,

$$\mathcal{E}[B(s)] = a_1 \sigma^2 \sum_{k \in s} w(x_k)$$

The best unbiased predictor of V is, therefore,

$$\nu^*(s, \mathbf{y}) = A(s) + C^*(s) \tag{5.3.3}$$

where

$$C^*(s) = a_1 \hat{\sigma}_*^2 \sum_{k \in \bar{s}} w(x_k),$$

$\hat{\sigma}_*^2$ being the best unbiased predictor of σ^2 in the sense

$$\mathcal{E}\hat{\sigma}_*^2 = \sigma^2 \tag{5.3.4.1}$$

$$\mathcal{E}[\hat{\sigma}_*^2] \leq \mathcal{E}[\hat{\sigma}^2 - \sigma^2]^2 \tag{5.3.4.2}$$

where $\hat{\sigma}^2$ is any predictor satisfying (5.3.4.1). We shall consider here the quadratic unbiased predictors $Q(\sigma^2)$ of σ^2 and hence the quadratic unbiased predictors $Q(V)$ of V. Clearly, each member of $Q(\sigma^2)$ gives a unique member of $Q(V)$. We shall, therefore, derive $\nu_q^*(s, \mathbf{y})$, the best quadratic unbiased predictor in the class $Q(V)$, denoting the corresponding BQUP's as c_q^* and $\hat{\sigma}_q^2*$, respectively.

Consider the quadratic predictor

$$c_q(s, \mathbf{y}) = \sum_{k \in s} b(s, k\ k)y_k^2 + \sum_{k \neq k' \in s} \sum b(s, k\ k')y_k y_{k'} \tag{5.3.5}$$

of $B(s)$. We have

$$\mathcal{E}[\nu_q(s, \mathbf{y})] = \sigma^2 \sum_{k \in s} b(s, k\ k)w(x_k)$$

$$= \mathcal{E}[B(s)]$$

if

$$\sum_{k \in s} b(s, k\ k)w(x_k) = a_1 \sum_{\bar{s}} w(x_k)\ \forall s : p(s) > 0 \tag{5.3.6}$$

Hence, a q-unbiased predictor of σ^2 is

$$\hat{\sigma}^2 = \frac{c_q(s, \mathbf{y})}{a_1 \sum_{k \in \bar{s}} w(x_k)} \tag{5.3.7}$$

provided (5.3.6) holds. Our problem is, therefore, to minimise $\mathcal{E}[\{\hat{\sigma}^2\}^2]$ subject to the condition (5.3.6). Minimising

$$\mathcal{E}\{\frac{\nu_q(s, \mathbf{y})}{a_1^2(\sum_{\bar{s}} w(x_k))^2}\} + \mu[a(1) \sum_{\bar{s}} w(x_k) - \sum_{k \in s} b(s, k\ k)w(x_k)]$$

where μ is a Lagrangian multiplier, with respect to $b(s, k\ k), b(s, k\ k')$ gives the solution:

$$b(s, k\ k) = \frac{a_1 \sum_{l \in \bar{s}} w(x_l)}{nw(x_k)}$$

$$b(s, k\ k') = 0$$

Hence,

$$\hat{\sigma}_{*q}^2 = \sum_{k \in s} \frac{y_k^2}{nw(x_k)}$$

and

$$\nu_q^*(s, \mathbf{y}) = A(s) + \frac{a_1}{n} \sum_{l \in \bar{s}} w(x_l) \sum_{k \in s} \frac{y_k^2}{w(x_k)} \tag{5.3.7}$$

When in particular, $w(x_k) = 1\ \forall\ k = 1, \ldots, n$,

$$\nu_q^*(s, \mathbf{y}) = \nu_q^{*0} \text{ (say) } = A(s) + \frac{(N-n)(N-2)}{nN^2} \sum_{k \in s} y_k^2$$

$$= \frac{N-1}{nN} \sum_{k \in s} y_k^2 - \frac{1}{N^2} \sum_{k \neq k' \in s} \sum y_k y_{k'} \qquad (5.3.8)$$

The conventional predictor of V,

$$\frac{N-1}{N} s_y^2 = s_y'^2$$

is also η-unbiased for V. It follows, therefore, for all s with $p(s) > 0$,

$$\mathcal{E}[\nu_q^{*0} - V]^2 \leq \mathcal{E}[s_y'^2 - V]^2$$

Now,

$$\mathcal{E}[\nu_q^* - v]^2$$

$$= 2\sigma^4 [a_3 \sum_{k \in \bar{s}} (w(x_k))^2 + \sum_{\bar{s}} w(x_k) \{ \frac{a_1^2}{n} \sum_{\bar{s}} w(x_k) + 2a(2)^2 W(x) \}] \quad (5.3.9)$$

where

$$W(x) = \sum_{k=1}^{N} w(x_k)$$

and

$$a_3 = a_1^2 - 2a_2^2 = \frac{1}{N^4} [(N-1)^2 - 2] > 0$$

for $N > 3$. It follows, therefore, that in the class of all $p \in \rho_n$, the best sampling plan to use $\nu_q^*(s, \mathbf{y})$ is a $s.d.$ p^*, where p^* is such that

$$p^*(s) = \begin{cases} 1 & \text{for } s = s^* \\ 0 & \text{otherwise} \end{cases}$$

where s^* is such that

$$\sum_{k \in s^*} w(x_k) = \max_{s \in S_n} \sum_{k \in s} w(x_k)$$

Mukhopadhyay (1982) considered optimal p-unbiased, m-unbiased, pm-unbiased predictors of V under the above model with two different measures of uncertainty (Exercise 2). He (1984) also derived optimal predictors of V within a general class of quadratic predictors under a class of generalised random permutation models. Vijayan's (1975) class of non-negative unbiased polynomial predictors of variance was also examined and an optimal predictor within that class was derived under the model. Mukhopadhyay and Bhattacharyya (1989) considered a slightly different (from (5.3.1)) model and obtained optimal predictors (Exercise 3). They (1991) also obtained optimality results under some general linear models with exchangeable errors (Exercise 5).

Following Cassel et al (1976), Mukhopadhyay (1990) suggested a generalised predictor of a finite population variance. Suppose auxiliary variables x_j with its value x_{ij} on unit i is available for the population ($i = 1, \ldots, N; j = 1, \ldots, k$). Let I_i, I_{ij} denote indicator random variables with $I_i = 1(0)$ according as $i \in (\notin)s$ and $I_{ij} = 1(0)$ acording as the pair $(i,j) \in (\notin)s$. The predictor proposed for $V(\mathbf{y})$ is

$$v_G(\mathbf{y}) = a_1 \sum_{i=1}^{N} \frac{I_i y_i^2}{\pi_i} - a_2 \sum_{i \neq i'=1}^{N} \sum \frac{I_{ii'} y_i y_{i'}}{\pi_{ii'}}$$

$$+ \sum_{j=1}^{k} \hat{\beta}_j \{ a_1 \sum_{i=1}^{N} (\frac{I_i}{\pi_i} - 1) x_{ij}^2 - a_2 \sum_{i \neq i'=1}^{N} \sum (\frac{I_{ii'}}{\pi_{ii'}} - 1) x_{ij} x_{i'j}) \} \qquad (5.3.10)$$

Here $\hat{\beta}_j$ is a function of $I = (I_1, \ldots, I_N), \mathbf{y}$ and $\mathbf{x} = ((x_{ij}))$ an $N \times k$ matrix such that $\hat{\beta}_j$ when suitably assigned is computable given the data stated above. Following Isaki and Fuller (1982) and Robinson and Sarndal (1983), the author showed that $v_G(\mathbf{y})$ is asymptotically design unbiased and consistent for V under conditions which do not require modelling. He (1986) derived a lower bound to the asymptotic variance of v_G under certain regularity conditions on the *s.d.* and superpopulation model. The generalised estimator v_G was further studied by Shah and Patel (1995). Further researches in this area is welcome.

5.4 BAYES PREDICTION OF $V(\mathbf{y})$

We have
$$V(\mathbf{y}) = \frac{n}{N} s_y^2 + (1 - \frac{n}{N})[s_{ry}^2 + \frac{n}{N}(\bar{y}_s - \bar{y}_r)^2] \qquad (5.4.1)$$

where $s_y^2 = \sum_{i=1}^{n}(y_i - \bar{y})^2/(n-1), s_{ry}^2 = \sum_{i \in r}(y_i - \bar{y}_r)^2/(N-n), \bar{y}_r = \sum_{i \in r} y_i/(N-n)$.

Under model ψ_R defined in Section 3.3 the Bayes predictive distribution of $\bar{y}_r$ given y_s is normal with mean

$$h(y_s) = \frac{1}{N-n} 1_r' \eta_r(y_s) \qquad (5.4.2)$$

and variance
$$D_r^2 = \frac{1}{(N-n)^2} 1_r' \Sigma_r 1_r \qquad (5.4.3)$$

where $\eta_r(y_s)$ and Σ_r are defined in (3.3.4) and (3.3.5) respectively. It follows that Bayes predictive distribution of $(n/N)(\bar{y}_r - \bar{y}_s)^2$, given y_s, is

$$\frac{n}{N}D_r^2\chi^2(1;\lambda) \tag{5.4.4}$$

when $\lambda = \frac{(h(y_s)-\bar{y}_s)^2}{2D_r^2}$ is the non-centrality parameter of χ^2 distribution with one d.f. Therefore,

$$E_{\psi_R}[\frac{n}{N}(\bar{y}_r - \bar{y}_s)^2 \mid y_s] = \frac{n}{N}(D_r^2 + (h(y_s) - \bar{y}_s)^2) \tag{5.4.5}$$

Again,

$$E[s_{ry}^2 \mid y_s] = \text{tr } (E_r\Sigma_r) + \eta_r(y_s)E_r\eta(y_s) \tag{5.4.6}$$

where

$$E_r = \frac{1}{N-n}(I_{N-n} - \frac{1}{N-n}J_r),$$

$J_r = 1_r 1_r'$. The Bayes predictor of S_y^2 under model ψ_R and squared error loss is, therefore,

$$\hat{S}_y^2 = E_{\psi_R}[S_y^2 \mid y_s]$$

$$= \frac{n}{N}s_y^2 + (1 - \frac{n}{N})\{ \text{tr } (E_r\Sigma_r)+$$

$$\eta_r(y_s)'E_r\eta_r(y_s) + \frac{n}{N}(D_r^2 + (h(y_s) - \bar{y}_s)^2\} \tag{5.4.7}$$

EXAMPLE 5.4.1

Consider the model (3.3.1), (3.3.2) with $X = 1_N, V = \sigma^2 I(\sigma^2 \text{ (unknown)}, \beta \sim N(\nu, R)$. The Bayes estimator of β is

$$\hat{\beta}_R = \frac{\sum_s y_i/\sigma^2 + \nu/R)}{n/\sigma^2 + 1/R}$$

In this case,

$$\eta_r(y_s) = E(y_r \mid y_s) = \hat{\beta}_R 1_r$$

$$\Sigma_r = V(y_r \mid y_s) = \sigma^2(I_r + \frac{R}{nR + \sigma^2}J_r)$$

Moreover,

$$h(y_s) = \frac{1}{N-n}1_r'\eta_r(y_s) = \hat{\beta}_R$$

$$D_r^2 = \frac{1}{(N-n)^2}1_r'\Sigma_r 1_r = \frac{\sigma^2}{N-n}\frac{NR + \sigma^2}{nR + \sigma^2}$$

$$\text{tr } (E_r \Sigma_r) = \sigma^2 \frac{N - n - 1}{N - n}$$

$$\eta_r(y_s)' E_r \eta_r(y_s) = 0$$

Substituting these in (5.4.7), Bayes estimator of S_y^2 is

$$\hat{S}_{By}^2 = \frac{n}{N} s_y^2 + (1 - \frac{n}{N}) \sigma^2 [1 - \frac{\sigma^2}{(N - n)R}$$

$$(\frac{1}{n} - \frac{1}{N}) \frac{nR}{\sigma^2 + nR} + \frac{n}{N} (\frac{\hat{\beta}_R - \bar{y}_s}{\sigma})^2] \qquad (i)$$

In case $R \to \infty, \hat{S}_{yB}^2$ reduces to

$$\hat{S}_{My}^2 = \frac{n}{N} s_y^2 + (1 - \frac{n}{N}) \sigma^2 \qquad (ii)$$

In addition, if σ^2 is also unknown, the non-informative prior (3.3.12) yields the Bayes predictor

$$\hat{S}_{My}^2 = \frac{(N - 3)n}{N(n - 3)} s_y^2 \qquad (iii)$$

This predictor was also derived by Ericson (1969 a) and Zacks and Solomon (1981).

We shall show that the predictor (ii) is minimax for the squared error loss. Under this model, the unknown parameter is $\beta(\sigma^2$ is known). The Bayes prediction risk of $\hat{S}_{By}^2$ is

$$\rho(\hat{S}_{By}^2; \nu, R) = E_{\psi_n} \{Var_{\psi_n}[(1 - \frac{n}{N})[S_{ry}^2 + \frac{n}{N}(\bar{y}_r - \bar{y}_s)^2 \mid y_s]]\}$$

Let $y'Ay$ be a symmetric quadratic form and $l'y$ be a linear form. We have

(1) $y'Ay \sim \chi^2(p, \lambda)$ where $p = \text{rank } (A), \lambda = \mu' \Sigma \mu$, iff $A\Sigma$ is idempotent.

(ii) If $A\Sigma 1 = 0$, then $y'Ay$ and $l'y$ are independent.

Now,

$$S_{ry}^2 = y_r' E_r y_r / (N - n) \text{ with } E_r = I_r - \frac{J_r}{N - n}$$

and since

$$E_r \Sigma_r / \sigma^2 = (I_r - J_r / (N - n))(I_r + \frac{\nu J_r}{Rn + \sigma^2})$$

is idempotent of rank $(N - n)$, the Bayes predictive distribution of S_{ry}^2, given y_s is $(\sigma^2 / (N - n))\chi^2(N - n - 1)$. In fact

$$\lambda = \hat{\beta}_R 1_r' E_r 1_r / \sigma^2 = 0$$

Hence,

$$Var\psi_R[S_{ry}^2 \mid y_s] = \frac{2\sigma^2}{(N-n)^2}(N-n-1)$$

Moreover, since $\bar{y}_r = 1'_r y_r/(N-n)$ and $E_r \Sigma_r 1_r = 0, S_{ry}^2$ and y_r are conditionally independent, given y_s. Thus

$$Cov_{\psi_R}[S_{ry}^2, (\bar{y}_s - \bar{y}_r)^2 \mid y_s] = 0$$

Now, the Bayes predictive distribution of $\bar{y}_r - \bar{y}_s$ given y_s is normal with mean $\hat{\beta}_R - \bar{y}_s$ and variance

$$\frac{\sigma^2}{N-n} + \frac{R\sigma^2}{nR + \sigma^2}$$

Hence,

$$(\bar{y}_r - \bar{y}_s)^2 \mid y_s \sim \sigma^2(\frac{1}{N-n} + \frac{R}{Rn + \sigma^2})$$

$$\chi^2[1, (\hat{\beta}_R - \bar{y}_s)^2/\{2\sigma^2(\frac{1}{N-n} + \frac{R}{nR + \sigma^2})\}]$$

It follws that

$$Var_{\psi_R}[(\bar{y}_r - \bar{y}_s)^2 \mid y_s] = 2\sigma^4(\frac{1}{N-n} + \frac{R}{Rn + \sigma^2})^2$$

$$(1 + 2(\hat{\beta}_R - \bar{y}_s)^2)/[\sigma^2(\frac{1}{N-n} + \frac{R}{nR + \sigma^2})] \qquad (iii)$$

Taking the expected value of (iii) with respect to the marginal distribution of S_y^2 we obtain the Bayes prediction risk of $\hat{S}_{By}^2$ as

$$\rho(\hat{S}_{By}^2; \nu, R) = (1 - \frac{n}{N})^2 \frac{2\sigma^4}{(N-n)^2}\{N - n - 1 + (\frac{n}{N})^2(\frac{\sigma^2 + NR}{\sigma^2 + nR})^2$$

$$(1 + (2(N-n))/[\sigma^2(\frac{\sigma^2 + NR}{\sigma^2 + nR})(\sigma^2/n + R)])\}$$

Hence,

$$lim_{R\to\infty}\rho(\hat{S}_{By}^2; \nu, R) = \frac{2\sigma^4}{N^2} \qquad (iv)$$

Again, the righthand side of (iv) is the risk function of the predictor (ii) and is independent of β. Hence, by Theorem 3.2.2, $\hat{S}_{My}^2$ is a minimax predictor of S_y^2.

Under the superpopulation model, $y_1, \ldots y_N$ are independent with $\mathcal{E}(y_i) = 0 \, \forall \, i = 1, \ldots, N$ Liu (1974 b) obtained a lower bound to the Bayes risk of a design unbiased predictor of a finite population variance (Exercise 4).

In the next two sections we shall study the large sample properties of sample regression coefficient and estimation of finite population regression coefficient in survey sampling.

5.5 ASYMPTOTIC PROPERTIES OF SAMPLE REGRESSION COEFFICIENT

Consider the model

$$y_i = \sum_{j=0}^{p} \beta_j x_{ij} + e_i, \ i = 1, \ldots, N \tag{5.5.1}$$

$$e_i \underset{\sim}{iid} \ (0, \sigma^2)$$

$x_{i0} = 1 \ \forall \ i$. A sample $s = (1, \ldots, n)$ (say) of size n is drawn from the population by *srswor*. Define the finite population vector of regression coefficients

$$B = Q_N^{-1} H_N \tag{5.5.2}$$

and the infinite population vector of coefficients

$$\beta = Q^{-1} H \tag{5.5.3}$$

where

$$Q_N = ((q_{Nrs})), \ H_N = ((h_{Nr})), \ Q = ((q_{rs})), \ H = ((h_{rs}))$$

$$q_{Nrs} = N^{-1} \sum_{i=1}^{N} x_{ir} x_{is}, \ q_{rs} = E(x_r x_s)$$

$$h_{Nr} = N^{-1} \sum_{i=1}^{N} x_{ir} y_i, \ h_r = E(x_r y) \tag{5.5.4}$$

The sample estimator of β based on s is

$$b = Q_n^{-1} H_n \tag{5.5.5}$$

Let

$$G = ((g_{rs})), \ g_{rs} = E(x_r x_s e^2) \tag{5.5.6}$$

Fuller (1975) proved that under an asymptotic framework $\sqrt{n}(b - B)$ converges in law to a normal distributuion.

THEOREM 5.5.1 Let $\{\mathcal{P}_n, n = 1, 2, \ldots\}$ be a sequence of finite populations of size $N_n(N_n > N_{n-1})$ drawn from the superpopulation (5.5.1) having finite fourth order moments and a positive definite covariance matrix. Let a srs of size n be selected from $\mathcal{P}_n(n = 1, 2, \ldots)$ and let $f_n = n/N_n \to f$ as $n \to \infty$. Then

$$\sqrt{n}(b - B) \to^L N(0, (1 - f)Q^{-1}GQ^{-1}) \qquad (5.5.7)$$

Also,

$$\sqrt{n}(b - B) \to^L N(0, Q^{-1}GQ^{-1})$$

Proof

$$b - \beta = Q_n^{-1}H_n - Q^{-1}H \simeq Q_n^{-1}(H_n - H) = Q_n^{-1}R_n \text{ (say)}$$

where

$$R_n = \frac{1}{n}(\sum_{i=1}^n e_i, \sum_{i=1}^n e_i x_{i1}, \ldots, \sum_{i=1}^n x_{ip}e_i)'$$

Similarly,

$$b - \beta = Q_{N_n}^{-1}R_{N_n}$$

Since the elements of Q_n are sample moments with variance of order $\frac{1}{n}$, we have

$$Q_n - Q = O_p(\frac{1}{\sqrt{n}}), \quad Q_{N_n} - Q = O_p(\frac{1}{\sqrt{N_n}})$$

Hence,

$$\sqrt{n}(b - B) = \sqrt{n}[(b - \beta) - (B - \beta)] = \sqrt{n}[Q_n^{-1}R_n - Q_{N_n}^{-1}R_{N_n}]$$

$$= \sqrt{n}Q^{-1}(R_n - R_{N_n}) + O_p(n^{-1/2})$$

Now,

$$\sqrt{n}(R_n - R_{N_n}) = n^{-1/2}(1 - f_n)\begin{bmatrix} \sum_{i=1}^n e_i \\ \cdot \\ \cdot \\ \sum_{i=1}^n x_{ip}e_i \end{bmatrix}$$

$$- \sqrt{f_n(1 - f_n)}(N - n)^{-1/2}\begin{bmatrix} \sum_{i=n+1}^N e_i \\ \cdot \\ \cdot \\ \sum_{i=n+1}^N x_{ip}e_i \end{bmatrix}$$

Now, $E(x_j e_i) = 0$. Also, $(e_i, x_{i1}e_i, \ldots, x_{ip}e_i)'(i = 1, 2, \ldots)$ are iid with mean vector 0 and dispersion matrix G. Hence, by Lindeberg Central Limit Theorem,

$$\sqrt{n_t}Q^{-1}(R_{n_t} - \mathbf{R}_{N_t}) \to^L N(0, (1 - f)Q^{-1}GQ^{-1}).$$

Hence the first part. The second part follows similarly.

The results were extended to the case of regression coefficients estimated from stratified two-stage sampling and to the situations when the observations contain measurement errors.

5.6 PM-UNBIASED ESTIMATION OF SLOPE PARAMETER IN THE LINEAR REGRESSION MODEL

Consider the model

$$y_i = \beta_0 + \beta x_i + e_i, \ e_i \sim (0, \sigma^2) \tag{5.6.1}$$

A homogeneous linear estimator

$$\hat{\beta}_s = \sum_{i \in s} b_{si} y_i$$

is pm-unbiased for β iff

$$E_p E_m(\hat{\beta}_s) = \beta$$

i.e. if

$$\sum_{s \in S} p(s) \sum_{i \in s} b_{si} x_i = 1 \tag{5.6.2a}$$

$$\sum_{s \in S} p(s) \sum_{i \in s} b_{si} = 0 \tag{5.6.2b}$$

Also,

$$E_p E_m(\hat{\beta}_s - \beta)^2 = \sigma^2 \sum_s p(s) \sum_{i \in s} b_{si}^2 + \sum_s p(s)$$

$$\{\sum_{i \in s} b_{si}(\beta_0 + \beta x_i)\}^2 - \beta^2$$

$$= \psi(b) \text{ (say)} \tag{5.6.3}$$

THEOREM 5.6.1 (Thomsen, 1978) Under assumptions that there are two samples s_1, s_2 with $s_1 \cap s_2 \neq \phi$, and

$$\sum_{i \in s_1} x_i - \tilde{x} \neq \sum_{i \in s_2} x_i - \tilde{x} \tag{5.6.4}$$

where

$$\tilde{x} = \sum_{i=1}^{N} x_i \pi_i / \sum_{i=1}^{N} \pi_i,$$

there does not exist any linear pm-unbiased estimator of β_o and β.

Proof. The minimising equation for $\psi(b)$ (wrt b_{si})

$$\sigma^2 b_{si} + \{\sum_{j \in s} b_{sj}(\beta_0 + \beta x_j)\}(\beta_0 + \beta x_i) - \mu x_i - \gamma = 0 \qquad (5.6.5)$$

where μ and γ are Lagrangian multipliers should be satisfied for all values of β_0 and β for any given value of $\sigma^2(> 0)$. Putting $\beta_0 = 0$ and $\beta = 0$ and $\sigma = 1$ in (5.6.5),

$$b_{si} = b_i = \mu x_i + \gamma \qquad (5.6.6)$$

From (5.6.2a), (5.6.2b) and (5.6.6) it follows that

$$b_i = \frac{x_i - \tilde{x}}{\tilde{S}_x^2} \qquad (5.6.7)$$

where

$$\tilde{S}_x^2 = \sum_{i=1}^{N} x_i^2 \pi - \frac{(\sum_1^N x_i \pi_i)^2}{\sum_1^N \pi_i}$$

For $\beta_0 = 1, \beta = 0$ and $\sigma = 1$, (5.6.5) gives

$$b_i + (\sum_{k \in s} b_k) - \mu x_i - \gamma = 0 \forall s \text{ and } \forall i \in s \qquad (5.6.8)$$

Hence, for two samples s_1, s_2 with at least one unit in common,

$$\sum_{i \in s_1} b_i = \sum_{i \in s_2} b_i \qquad (5.6.9)$$

i.e.

$$\sum_{i \in s_1} x_i - \tilde{x} = \sum_{i \in s_2} x_i - \tilde{x}$$

(by (5.6.7)). This contradicts the assumption (5.6.4). Hence, the result for estimation of β. The case for β_0 can be proved similarly.

THEOREM 5.6.2 Under model (5.6.1), optimal m-unbiased estimator of β (that minimises (5.6.3))is

$$\hat{\beta}_s^* = \sum_{i \in s}(x_i - \bar{x}_s)(y_i - \bar{y}_s) / \sum_{i \in s}(x_i - \bar{x}_s)^2 \qquad (5.6.10)$$

Also,

$$E_p E_m(\hat{\beta}_s^* - \beta)^2 = \sigma^2 \sum_s p(s) / \sum_{i \in s}(x_i - \bar{x}_s)^2$$

An optimal sampling design to use $\hat{\beta}_s^*$ is, therefore, $p^*(s)$ where $p^*(s) = 1(0)$ for $s = s^*$ (otherwise) , s^* is the set of samples which minimises $\sum_s p(s) / \sum_{i \in s}(x_i - \bar{x}_s)^2$ among all samples s in $\mathcal{S}$.

COROLLARY 5.6.1 If the sample is a *srswor* of size n,

$$Var(\hat{\beta}_s^*) \geq \frac{\sigma^2}{(n-1)S_x^2}$$

Proof By Jensen's inequality,

$$E[\frac{\sigma^2}{\sum_{i \in s}(x_i - \bar{x}_s)^2}] \geq \frac{\sigma^2}{E \sum_{i \in s}(x_i - \bar{x}_s)^2} = \frac{\sigma^2}{(n-1)S_x^2}$$

THEOREM 5.6.3 Under model (5.6.1), optimal m-unbiased estimator of β_0 (that minimises $E_p E_m(\hat{\beta}_{0s} - \beta)^2$) is

$$\hat{\beta}_{0s}^* = \bar{y}_s - \hat{\beta}_s^* \bar{x}_s \qquad (5.6.11)$$

Also,

$$E_p E_m(\hat{\beta}_{0s}^* - \beta_0)^2 = \sigma^2 \sum_s p(s) \{\sum_{i \in s}[\frac{1}{\nu(s)} - \frac{(x_i - \bar{x}_s)\bar{x}_s}{\sum_{k \in s}(x_k - \bar{x}_s)}]^2\}$$

where $\nu(s)$ is the number of distinct units in s.

COROLLARY 5.6.2 For any design with

$$p(s) > 0 \Rightarrow \sum_{i \in s}(x_i - \bar{x}_s)^2 > 0 \qquad (5.6.12)$$

$$Cov(\hat{\beta}_{0s}^*, \hat{\beta}_s^*) = \sigma^2 \sum_s p(s)\bar{x}_s$$

It follows that

$$E(Q_{0s}) = (\bar{n} - 2)\sigma^2$$

where

$$Q_{0s} = \sum_{i \in s}(y_i - \hat{\beta}_{0s}^* - \hat{\beta}_s^* x_i)^2, \quad \bar{n} = \sum_s n(s)p(s),$$

provided (5.6.12) holds.

5.7 Optimal Prediction of Finite Population Regression Coefficient under Multiple Regression Model

Consider the multiple regression model

$$\mathbf{y} = X\beta + e$$

$$e \sim N(0, V)$$

denoted as $\psi(\beta, V)$, described in (3.3.1). We shall now consider optimal prediction of finite population regression coefficient

$$B_N = (X'V^{-1}X)^{-1}(X'V^{-1}\mathbf{y}) \qquad (5.7.1)$$

under model $\psi(\beta, V)$. We assume that the population model $\psi(\beta, V)$ also holds for the sample, i.e. there is no selection bias (see Krieger and Pfeffermann (1992) for discussion on the effects of sample selection). Bolfarine et al (1994) considered the following cases.

Case (a) Diagonal Covariance Matrix

When $V_{rs} = 0$, we can write

$$B_N = A_s\hat{\beta}_s + A_r\beta_r \qquad (5.7.2)$$

where

$$
\begin{aligned}
\hat{\beta}_s &= B_s\mathbf{y}_s \\
B_s &= (X'_sV_s^{-1}X_s)^{-1}X'_sV_s^{-1} \\
A_s &= (X'V^{-1}X)^{-1}X'_sV_s^{-1}X_s \\
A_r &= (X'V^{-1}X)^{-1}X'_rV_r^{-1}X_r \\
\beta_r &= (X'_rV_r^{-1}X_r)^{-1}X'_rV_r^{-1}y_r
\end{aligned}
\qquad (5.7.3)
$$

Note that $A_r + A_s = I_p$.

A predictor of B_N is, therefore,

$$\hat{B}_N = A_s\hat{\beta}_s + A_r\hat{\beta}_r$$

where $\hat{\beta}_r$ is a predictor of β_r based on y_s.

By definition 2.2.1, a predictor $\hat{B}_N$ of B_N is unbiased iff

$$E_\psi[\hat{B}_N - B_N] = 0 \ \forall \ \psi$$

Note that $\hat{B}_N$ is unbiased for B_N iff $\hat{\beta}_r$ is so for β_r. Also, $\hat{\beta}_s$ is an unbiased predictor of B_N.

DEFINITION 5.7.1 The generalised prediction mean square error (GMSE) of a predictor $\hat{B}_N$ of B_N is

$$GMSE_\psi[\hat{B}_N] = E_\psi[\lambda'(\hat{B}_N - B_N)(\hat{B}_N - \beta)'\lambda]$$

for any real $p \times 1$ vector λ.

DEFINITION 5.7.2 $\hat{B}_{BUN}$ is the best unbiased predictor (BUP)of B_N if $\hat{B}_{BUN}$ is unbiased and

$$GMSE_\psi[\hat{B}_{BUN}] \leq GMSE[\hat{B}_N]$$

for any other unbiased predictor $\hat{B}_N$ and all ψ.

THEOREM 5.7.1 Under the superpopulation model $\psi(\beta, V)$ the best unbiased predictor of B_N is

$$\hat{B}_{BUN} = \hat{\beta}_s \tag{5.7.4}$$

Furthermore,

$$GMSE[\hat{B}_{BUN}] = \lambda' A_r[(X_s' V_s^{-1} X_s)^{-1} + (X_r V_r^{-1} X_r)^{-1}]A_r \lambda \tag{5.7.5}$$

Proof Let $\hat{B}_{UN} = A_s\hat{\beta}_s + A_r\hat{\beta}_{ur}$ be any unbiased predictor of B_N. According to Arnold (1981), if V is known, $\hat{\beta}_s$ is a complete and sufficient statistic. Moreover, since $V_{rs} = 0$, y_s is independent of y_r, which implies by definition of Rodrigues et al (1985) (stated in definition 6.9.1) that $\hat{\beta}_s$ is totally sufficient. We may then write

$$E_\psi[\lambda'(\hat{B}_{UN} - B_N)(\hat{B}_{UN} - B_N)'\lambda]$$

$$= Var_\psi[\lambda'(B_{UN} - B_N)]$$

$$\geq Var_\psi[\lambda' E_\psi\{(\hat{B}_{UN} - B_N) \mid y_s, \hat{\beta}_s\}]$$

$$= \lambda' A_r Var[E_\psi\{\hat{\beta}_{ur} \mid y_s, \hat{\beta}_s\}]A_r' \lambda$$

Since, $\hat{\beta}_s$ is totally sufficient,

$$E_\psi\{\hat{\beta}_{ur} \mid y_r, \hat{\beta}_s\} = \hat{\beta}_s$$

Therefore,

$$\hat{B}_{BUN} = A_s\hat{\beta}_s + A_r\hat{\beta}_s = \hat{\beta}_s$$

Uniqueness follows from the completeness of $\hat{\beta}_s$. Again,

$$\mathrm{GMSE}_\psi[\hat{\beta}_s] = \lambda' A_r \, \mathrm{Var}\,[\hat{\beta}_s - \beta_r] A_r \lambda$$

Now,

$$Var_\psi[\hat{\beta}_s - \beta_r] = (X'_s V_s^{-1} X_s)^{-1} + (X'_r V_r^{-1} X_r)^{-1}$$

Hence the result.

Case (b) Covariance Matrix not necessarily diagonal

Here we can write

$$X'V^{-1}X - H$$

$$X'V^{-1} = (Bc^{-1} \; DE^{-1})$$

so that

$$B_N = H^{-1}BC^{-1}y_s + H^{-1}DE^{-1}y_r \qquad (5.7.6)$$

where

$$\begin{aligned} B &= X'_s - X'_r V_r^{-1} V_{rs} \\ C &= V_s - V_{sr} V_r^{-1} V_{rs} \\ D &= X'_r - X'_s V_s^{-1} V_{sr} \\ E &= V_r - V_{rs} V_s^{-1} V_{sr} \\ H &= BC^{-1} X_s + DE^{-1} X_r \end{aligned} \qquad (5.7.7)$$

Let

$$M_s = H^{-1}BC^{-1}, \; M_r = H^{-1}DE^{-1}, \; M = (M_s, \; M_r)$$

THEOREM 5.7.2 Under the model $\psi(\beta, V)$ where V is not necessarily diagonal, the BUP of B_N is given by

$$\hat{B}_{BUN} = M_s y_s + M_r[X_r \hat{\beta}_s + V_{rs} V_s^{-1}(y_s - X_r \hat{\beta}_s)] \qquad (5.7.8)$$

Furthermore, the GMSE of $\hat{B}_{BUN}$ is

$$\mathrm{GMSE}\,(\hat{B}_{BUN}) = \lambda' H^{-1}DE^{-1}D'H^{-1}\lambda + \lambda' M_r D'(X'_s V_s^{-1} X_s)^{-1} DQ'_r \lambda$$

$$= \lambda' Q_r [E + D'(X'_s V_s^{-1} X_s)^{-1}D]M'_r \lambda \qquad (5.7.9)$$

We now consider the BLUP of B_N when the data follow model $\psi(\beta, V)$ under Gauss-Markoff (GM) set up i.e. under the assumption $e \sim (0, V)$ (without assuming normality). Let $\hat{\beta}_{LN} = R'y_s$ be any linear predictor of B_N, where R is nay $p \times n$ matrix of known entries. It follows that $\hat{\beta}_{LN}$ is unbiased for β_N iff $R'X_s = I_p$.

LEMMA 5.7.2 Under GM -set up as above, for any $p \times 1$ vector λ, and any $n \times p$ matrix R,

$$E_\psi\{[\lambda'R'y_s - \lambda'My]^2\} = Var_\psi[\lambda'(R' - M_rV_{rs}V_s^{-1})y_s]$$

$$+\lambda'M_rV_rM_r'\lambda - \lambda'M_rV_{rs}V_s^{-1}V_{sr}M_r'\lambda + \{\lambda'(M'X_s - I_p)\beta\}^2 \qquad (5.7.10)$$

THEOREM 5.7.3 Under the GM-assumptions above, the unbiased linear predictor with minimum GMSE, $\hat{B}_{BLN}$ is as given in (5.7.8) with GMSE as given in (5.7.9).

Proof Since $\hat{\beta}_{LN}$ is unbiased ,

$$E(\lambda'\hat{B}_{LN} - \lambda'B_N) = \lambda'(R'X_s\beta - MX\beta) = 0 \forall \lambda, \; \beta$$

so that $R'X_s = MX = I_p$ and the last term in the expression (5.7.10) is zero. Therefore, it follows from (5.7.10) that to find the linear unbiased predictor of B_N with minimum GMSE, it is equivalent to find a predictor which is unbiased for

$$E_\psi[\lambda'R'y_s - \lambda'M_rV_{rs}V_s^{-1}y_s]$$

$$= (\lambda'MX - \lambda'M_rV_{rs}V_s^{-1}X_s)\beta \qquad (5.7.11)$$

and has minimum variance in the class of all linear unbiased predictors of (5.7.11). Hence, by GM theorem, it follows that the best linear unbiased estimator of the expected value (5.7.11) is given by

$$\lambda'R_*'y_s - \lambda'M_rV_{rs}V_s^{-1}y_s = \lambda'(MX - M_rV_{rs}V_s^{-1}X_s)\hat{\beta}_s \qquad (5.7.12)$$

where $\hat{\beta}_s$ is the usual least square estimator of β given in (5.7.4). From (5.7.12) it follows that

$$\lambda'R_*'y_s = \lambda'M_rV_{rs}V_s^{-1}y_s + \lambda'(MX - M_rV_{rs}V_s^{-1}X_s)\hat{\beta}_s$$

$$= \lambda'M_sy_s + \lambda'M_r[X_r\hat{\beta}_s + V_{rs}V_s^{-1}(y_s - X_s\hat{\beta}_s)]$$

minimises the mse (5.7.10), $\forall \lambda \in R^p$. Thus,

$$\hat{B}_{BLN} = R_*'y_s = M_sy_s + M_r[X_r\hat{\beta}_s + V_{rs}V_s^{-1}(y_s - X_s\hat{\beta}_s)]$$

is the minimum GMSE predictor of B_N in the class of all linear unbiased predictors. The next part follows as in the proof of Theorem 5.7.2.

EXAMPLE 5.7.1

Consider the superpopulation model $\psi(\beta, V)$ where $X = I_N, V = (1 - \rho)I_N + \rho 1_N 1'_N$; here $B_N = \bar{y}, \hat{\beta}_s = \bar{y}_s$. It can be shown that the best linear unbiased estimator $\hat{\beta}_{BLN} = \bar{y}_s$.

THEOREM 5.7.4 Under the GM - set up of Theorem 5.7.3, the best linear unbiased estimator of B is

$$\hat{B}_{BLN} = \hat{\beta}_s$$

Furthermore,

$$\text{GMSE } [\hat{\beta}_s] = \lambda'[(X'_s V_s X_s)^{-1} - (X'V^{-1}X)^{-1}]\lambda \qquad (5.7.13)$$

Proof

$$\hat{B}_{BLN} - \hat{\beta}_s = M_s y_s + M_r[X_r \hat{\beta}_s + V_{rs} V_s^{-1}(y_s - X_s \hat{\beta}_s)]$$
$$- MX\hat{\beta}_s \text{ (using } MX = I_p)$$
$$= (M_s + Q_r V_{rs} V_s^{-1})(y_s - X_s \hat{\beta}_s)) = 0$$

which proves the first part. The second part follows easily.

The results of Theorems 5.7.3 and 5.7.4 state that $\hat{\beta}_s$ is BUP of B_N under the model $\psi(\beta, V)$ (including the assumption of normality) both when V is diagonal or not. However, under the GM model (i.e. model ψ, but without the assumption of normality) $\hat{\beta}_s$ is BLN -unbiased predictor of B_N both when V is diagonal and not.

5.7.1   BAYES PREDICTION OF A FINITE POPU- LATION REGRESSION COEFFICIENT

Bolfarine and Zacks (1991) considered Bayes estimation of B_N. Consider model $\psi(\beta, V)$ with a normal prior for β

$$\beta \sim N(\nu, R) \qquad (5.7.14)$$

The model $\psi(\beta, V)$ together with the prior (5.8.1) of β will be denoted as ψ_R. For a predictor $\hat{\beta}_N$ of β, its generalised prediction risk under model ψ_R is given by

$$RG_{\psi_R}(\hat{\beta}_N, B_N) = E_{\psi_R}[\lambda'(\hat{\beta}_N - B_N)(\hat{\beta}_N - B_N)'\lambda] \qquad (5.7.15)$$

for some vector λ.

A Bayes predictor of B_N with respect to prediction risk (5.7.15) is given by

$$\hat{\beta}_{BN} = E_{\psi_R}[B_N \mid y_s] \qquad (5.7.16)$$

The corresponding ψ_R-generalised Bayes prediction risk is given by

$$RG_{\psi_R}(\hat{\beta}_{BN}, B_N) = \lambda' E_{\psi_R}[Var_{\psi_R}(B_n \mid y_s)\lambda \qquad (5.7.17)$$

As noted in section 5.7, case (a), when $V_{rs} = 0, B_N = A_s\hat{\beta}_s + A_r\beta_r$. Hence, in this case,

$$\begin{aligned}\hat{\beta}_{BN} &= A_s\hat{\beta}_s + A_r E_{\psi_R}(\beta_r \mid y_s) \\ &= A_s\hat{\beta}_s + A_r\hat{\beta}_R\end{aligned} \qquad (5.7.18)$$

where

$$\hat{\beta}_R = (X_s V_s^{-1} X_s + R^{-1})^{-1}(X_s' V_s^{-1} y_s + R^{-1}\nu) \qquad (5.7.19)$$

as given in (3.3.5). The corresponding ψ_R- generalised Bayes prediction risk is given by

$$RG_{\psi_R}(\hat{\beta}_{BN}, B_N) = \lambda' E_{\psi_R}[Var_{\psi_B}(A_s\hat{\beta}_s + A_r\beta_r \mid y_s)\lambda$$

$$= \lambda'(X'V^{-1}X)^{-1}[X_r' V_r^{-1}\Sigma_r V_r^{-1} X_r](X'V^{-1}X)^{-1}\lambda \qquad (5.7.20)$$

where $\Sigma_r = Var[y_r \mid y_s]$.

Some other works on B_N are due to Konign (1962), Hartley and Silken (1975), Shah et al (1977), Sarndal (1982), Hung (1990). Rodrigues and Elian (1995), Rai and Srivastava (1998).

5.8 Exercises

1. Let

$$z = (y, x) = (z_0, \dots, z_k)$$

be a $1 \times (k+1)$ vector of $(k+1)$ variables, $y, x = (x_1, \dots, x_k)$, where x are auxiliary variables and y is the main variable. Let z_{ih} be the value of the variable z_i on unit $h = 1, \dots, N$ (Clearly, $z_{ih} = x_{i-1,h}$). Let

$$m = (\bar{y}, \bar{x}_1, \dots, \bar{x}_k)'$$

$$\Sigma = ((\sigma_{ij}))$$

denote the finite population mean and covariance matrix of z where it is assumed that

$$\sigma_{ij} = \begin{cases} \sigma_i^2 & i = j \\ \rho\sigma_i\sigma_j & (i \neq j) \end{cases} \qquad (i)$$

and $-1/k < \rho < 1$. Assume also that z possesses the same moments as a $1 \times (k+1)$ multivariate normal variable up to eighth order and $\sigma_i^2 (i = 1, \ldots, k)$ are known while ρ and σ_0^2 are unknown. Let

$$s_{ij} = \frac{1}{n-1} \sum_{h=1}^{n} (z_{ih} - \bar{z}_i)(z_{jh} - \bar{z}_j), \quad i, j = 0, \ldots, k$$

Consider a multivariate ratio estimator of σ_y^2

$$\hat{\sigma}_{yRm}^2 = \sum_{i=1}^{k} W_i \hat{r}_i \sigma_i^2$$

where $\hat{r}_i = \frac{s_{ii}}{s_{oi}}, i = 1, \ldots, k$ and W_i are suitable weights, $0 < W_i < 1, \sum_{i=1}^{k} W_i = 1$. Show that

$$Cov(\hat{r}_i, \hat{r}_j) \simeq \frac{2}{n-1} \frac{\sigma_0^4}{\sigma_i^2 \sigma_j^2}$$

$$Var(\hat{r}_i) \simeq \frac{2}{n-1} \frac{\sigma_0^4}{\sigma_i^4}(1 - \rho^2)$$

by approximating $\hat{r}_i$, $\hat{r}_j$ by their first-order Taylor series approximation. Hence, show that minimisation of $Var(\hat{\sigma}_{yRm}^2)$ under condition $\sum_{i=1}^{k} W_i = 1$ yields $W_i = 1/k \ \forall \ i$. Show that the optimum value of the variance is $Var(\hat{\sigma}_{yRm}^2) \simeq \frac{1}{k(n-1)}[2(k+1)\sigma_0^4(1 - \rho^2)]$ to terms of order n^{-1}. Under assumptions that z possesses the same moments as the above,

$$Var(s_0^2) \simeq \frac{2\sigma_0^4}{n-1}$$

Therefore,

$$Var(\hat{\sigma}_{yRm}^2) \leq Var(s_0^2)$$

iff $|\rho| > \frac{1}{\sqrt{k+1}}$

(Isaki, 1983)

2. Consider the problem of predicting a finite population variance

$$V(\mathbf{y}) = \frac{1}{N} \sum_{i=1}^{N} (y_i - \bar{y})^2 = a_1 \sum_{i=1}^{N} y_i^2 - a_2 \sum_{i \neq i'=1}^{N} \sum y_i y_{i'}$$

where

$$a_1 = \frac{1}{N}\left(1 - \frac{1}{N}\right), \quad a_2 = \frac{1}{N^2}.$$

Assume the superpopulation model $\eta : y_1, \ldots, y_N$ are independently distributed such that the conditional distribution of y_k given x_k (known value of an auxiliary variable x) is normal with

$$\mathcal{E}(y_k \mid x_k) = 0$$

$$\mathcal{E}(y_k^2 \mid x_k) = v(x_k) = \sigma^2 w(x_k)$$

where σ^2 is a constant and $w(x_k)$ is a known function of x_k. Two measures of uncertainty of a strategy (p, e) are:

$$\mathcal{E}E[e - V]^2 = M_{1p} \text{ (say)}$$

$$E\mathcal{E}[e - \mathcal{E}(V)]^2 = M_{(2p)} \text{ (say)}$$

Consider the class $Q(V)$ of homogeneous quadratic (h.q.) predictors of $V(\mathbf{y})$:

$$e(s, \mathbf{y}) = \sum_{k \in s} b(s, kk) y_k^2 + \sum_{k \neq k' \in s} \sum b(s, kk') y_k y_{k'}$$

Denote the class of p-unbiased, η-unbiased, $p\eta$-unbiased h.q.-predictors of $V(\mathbf{y})$ as $Q_p(V), Q_\eta(V), Q_{p\eta}(V)$ respectively.

Show that

(i) If $e \in Q_\eta(V)$,

$$M_{1p}(e) = EV(e) + V(V) + 2[(\mathcal{E}(V))^2 - \mathcal{E}(V E(e))]$$

$$M_{2p}(e) = EV(e)$$

(ii) If $e \in Q_p(V)$,

$$M_{1p}(e) = \mathcal{E}V(e),$$

$$M_{2p}(e) = \mathcal{E}V(e) + V(V)$$

(iii) If $e \in Q_{p\eta}(V)$,

$$M_{1p}(e) = \mathcal{E}V(e) + \mathcal{E}[B(e)]^2$$

$$M_{2p}(e) = EV(e) + E[\mathcal{B}(e)]^2$$

where $B(e) = E(e) - V$, the p-bias in predicting $V, \mathcal{B}(e) = \mathcal{E}(e - V)$, the η-bias in predicting V.

Let p be any non-informative design $\in \rho_n$. Prove the following results.

(a) For any predictor $e \in Q_\eta(V)$,

$$M_{1p}(e_1^*) < M_{1p}(e)$$

$$M_{2p}(e_1^{*\prime}) < M_{2p}(e)$$

where

$$e_1^* = \frac{a_1}{n} \sum_{k \in s} \{n + \frac{\sum_{l \in \bar{s}} w(x_l)}{w(x_k)}\} y_k^2 - 2a_2 \sum \sum_{k < k' \in s} y_k y_{k'}$$

$$e_1^{*\prime} = \frac{a_1}{n} W(x) \sum_{k \in s} \frac{y_k^2}{w(x_k)}$$

Hence an optimum sampling design to use e_1^* is p^* where p^* selects the sample with the largest value of $\sum_{k \in s} w(x_k)$ with probability one. However, any $p \in \rho_n$ is optimum for $e_1^{*\prime}$.

(b) Let $e \in Q_p(V)$ such that

$$\sum_{s \ni (k,k')} b(s, kk) b(s, kk') p(s) \geq \frac{a_1^2 \pi_{kk'}}{\pi_k \pi_{k'}} \ \forall \ k \neq k' = 1, \ldots, N$$

The optimal predictor for any given p, both in the M_{1p}, M_{2p} sense is

$$e_2^* = a_1 \sum_{k \in s} \frac{y_k^2}{\pi_k} - 2a_2 \sum \sum_{k < k' \in s} \frac{y_k y_{k'}}{\pi_{kk'}}$$

If, further, we assume that

$$\pi_k \propto w(x_k) \ \forall = 1, \ldots, N, \tag{i}$$

then the optimal $s.d.$ to use e_2^* is one for which

$$\sum \sum_{k < k'=1}^{N} \frac{w(x_k) w(x_{k'})}{\pi_{kk'}}$$

is minimum.

(c) Let $e \in Q_{p\eta}(V)$. Then

$$M_{1p}(e_3^*,) < M_{1p}(e)$$

$$M_{2p}(e_3^{*\prime}) < M_{2p}(e)$$

where

$$e_3^* = e_3^{*\prime} - 2a_3 \sum \sum_{k < k' \in s} y_k y_{k'}$$

$$e_3^{*\prime} = a_1 \sum_{k \in s} \frac{y_k^2}{\pi_k}$$

If (i) is assumed, then the optimal design to use e_3^* is one for which

$$\sum_{k<k'=1}^{N} w(x_k)w(x_{k'})\pi_{kk'}$$

is maximum. Under assumption (i), any $s.d\ p \in \rho_n$ is optimum for $e_3^{*\prime}$.

(Mukhopadhyay, 1982)

3. Consider the following model: $y_1, \ldots, y_N$ are independent random variables such that the conditional distribution of y_k given x_k is normal with

$$\mathcal{E}(y_k \mid x_k) = \beta x_k$$

$$\mathcal{V}(y_k \mid x_k) = \sigma^2 x_k^2, \ k = 1, \ldots, N$$

where β, σ^2 are constants. Show that the following reslts hold under this model.

(i) The estimator

$$e_1^* = \frac{a_1}{n}\sum_{k=1}^{N} x_k^2 \sum_{jins} \frac{y_j^2}{x_j^2} - \frac{a_2}{n(n-1)}\sum \sum_{k\neq k'=1}^{N} x_k x_{k'} \sum \sum_{j\neq j'\in s} \frac{y_j y_{j'}}{x_j x_{j'}}$$

is optimal for V in the minimum M_{2p} sense in Q_{pm} for every fixed $p \in \rho_n$. Again, any $p \in \rho_n$ is optimal for using e_1^*.

(ii) When $x_k \neq \alpha$ (a constant) $\forall\ i$, there does not exist any p-unbiased optimal quadratic estimator for V in $Q_p, p \in \rho_n$.

When x_k is a constant for all k, the strategy (p_0, s'^2) is optimal for V, where

$$\pi_k = \frac{n}{N}, \quad \pi_{kk'} = \frac{n(n-1)}{N(N-1)}$$

(iii) Consider the class of necessarily non-negative unbiased ($nnnu$) estimators of V:

$$\Psi = -\frac{1}{2}\sum \sum_{k\neq k'\in s} b(s, k\ k')(y_k - y_{k'})^2$$

Show that Ψ is pm-unbiased iff

$$\sum_{s\ni(k,\ k')} b(s, k\ k')p(s) = -\frac{1}{N^2}\ \forall\ k \neq k' = 1, \ldots, N$$

Denote this class of estimators as Ψ_{pm}. Show that the estimator

$$\Psi^* = \frac{1}{2N^2} \sum \sum_{k \neq k' \in s} \frac{(y_k - y_{k'})^2}{\pi_{kk'}}$$

is optimal nnue for V in Ψ_{pm} provided $p \in \rho_n$ satisfies the following conditions.

$$p(s) > 0 \Rightarrow \{\sum_s x_j = \mu_1, \ \sum_s \frac{1}{x_j} = \mu_2\},$$

μ_1, μ_2 being constants and

$$\pi_{kk'} = n(n-1)x_k x_{k'} / (\sum \sum_{k \neq k'=1}^{N} x_k x_{k'})$$

In all the cases the optimality is to be understood in the minimum M_{2p} sense, as defined in Exercise 2.

(Mukhopadhyay and Bhattacharyya, 1989)

4. Consider the superpopulation model $\xi : y_1, \ldots, y_N$ are independent with $\mathcal{E}(y_i) = 0 \forall \ i = 1, \ldots, N$ The Bayes risk of a p-unbiased predictor $e(s, \mathbf{y})$ of $V = \frac{1}{N} \sum_{i=1}^{N} (y_i - \bar{y})^2$ under prior ξ is

$$\begin{aligned} r_\xi(e) &= \mathcal{E}[V(e(s, \mathbf{y})] \\ &= E\mathcal{E}(e^2) - \mathcal{E}(V^2) \\ &= E[(\mathcal{E}(e))^2] + E\mathcal{V}(e) - \mathcal{E}(V^2) \end{aligned} \qquad (i)$$

Now,

$$\begin{aligned} E[\mathcal{E}(e(s, \mathbf{y}) - \mathcal{E}(V)]^2 \\ = E\mathcal{E}^2(e(s, \mathbf{y})) - \mathcal{E}^2(V) \ \geq 0 \end{aligned} \qquad (ii)$$

Therefore, by (i),

$$r_\xi(e) \geq E\mathcal{V}(e) - \mathcal{V}(V) \qquad (iii)$$

Let

$$\begin{aligned} e(s, \mathbf{y}) &= e_L(s, \mathbf{y}) + [e(s, \mathbf{y}) - e_L(s, \mathbf{y})] \\ &= e_L(s, \mathbf{y}) + k(s, \mathbf{y}) \end{aligned} \qquad (iv)$$

Clearly, $E[k(s, \mathbf{y})] = 0$. Hence,

$$\sum_{s \ni i} k(s, \mathbf{y})p(s) = - \sum_{s:i \notin s} k(s, \mathbf{y})p(s)$$

$$\sum_{s \ni (i,j)} k(s, \mathbf{y})p(s) = - \sum_{s:(i,j) \notin s} k(s, \mathbf{y})p(s) - \sum_{s:i \in s, j \notin s} k(s, \mathbf{y})p(s) - \sum_{s:i \notin s, j \in s} k(s, \mathbf{y})p(s)$$

Show that
$$EC(e_L(s, \mathbf{y}), k(s, \mathbf{y}) = 0$$

Hence, from (iv), show that

$$E\mathcal{V}[e(s, \mathbf{y})] \geq E\mathcal{V}[e_L(s, \mathbf{y})]$$

Therefore, it follows from (iii) that

$$r_\xi(e) \geq E\mathcal{V}[e_L(s, \mathbf{y})] - \mathcal{V}(V) \tag{v}$$

The expression (v) gives a lower bound to the Bayes risk of any p-unbiased predictor $e(s, \mathbf{y})$.

Show that for any p-unbiased predictor e of V, the following relation holds:

$$\mathcal{E}V(e) = E\mathcal{V}(e) + E\{\mathcal{B}(e)\}^2 - \mathcal{V}(V)$$

where $\mathcal{B}(e) = \mathcal{E}(e - V)$, the ξ-bias of the predictor e.

Consider now designs $p \in \rho_n$ and assume that

$$\mathcal{E}(y_i^2) = \frac{N^2}{N-1}[\frac{\mathcal{E}(V)}{n}]\pi_i \; \forall \; i = 1, \ldots, N \tag{vi}$$

Show that under (vi), $\mathcal{B}(e_L) = 0$. Thus,

$$E\mathcal{E}V(e_L) = E\mathcal{V}[e_L] - \mathcal{V}(V) = r_\xi(e_L)$$

Therefore, for $p \in \rho_n$ and under condition (vi),

$$r_\xi(e) \geq r_\xi(e_L)$$

(Liu, 1974)

5. Consider the following superpopulation model. $\mathbf{y}$ follows a N-variate normal distribution with

$$\mathcal{E}(y_k \mid x_k) = \mu_k$$

$$\mathcal{V}(y_k \mid x_k) = \nu_k$$

$$\mathcal{C}(y_k, y_{k'} \mid x_k, x_{k'}) = \rho\sqrt{\nu_k \nu_{k'}}, \; 1/(N-1) \leq \rho \leq 1$$

We shall denote by ξ_1, the model with $\mu_k = \beta$ and $\nu_k = \sigma^2 \; \forall \; k$. Similarly, ξ_2 will denote the model with $\mu_k = \beta_0 + \beta x_k$ and $\nu_k = \sigma^2 \forall \; k$. Prove the following results.

(a) Under ξ_1, the statistic $e_1^* = s'^2$ is uniformly minimum variance model-unbiased predictor (UMVUP) of V in the class of all unbiased predictors of V. Also, any $p \in \rho_n$ is optimal for using e_1^*.

(b) Under model ξ_2, UMVUP of V is

$$e_2^* = \frac{N-1}{N}\delta_y^2 + V_x(\hat{\beta}^2 - \frac{\delta_y^2}{s_x^2})$$

where

$$\delta_y^2 = \frac{1}{n-2}\sum_{k\in s}(y_i - \hat{\beta}_0 - \hat{\beta}_1 x_i)^2$$

$$\hat{\beta}_0 = \bar{y}_s - \hat{\beta}_1 \bar{x}_s$$

$$\hat{\beta}_1 = \frac{\sum_s(x_i - \bar{x}_s)y_i}{\sum_s(x_i - \bar{x}_s)^2}$$

(c) Study the robustness of e_1^*, e_2^* under ξ_2, ξ_1, respectively.

(Mukhopadhyay and Bhattacharyya, 1991)

6. Let p be any non-census design with $\pi_{ij} > 0 \ \forall \ i \neq j = 1, \ldots, N$. Prove the following results.

(a) If there exists a uniformly minimum variance quadratic unbiased estimator of V, it must be Liu's estimator e_L.

(b) Consider a unicluster design p with $p(s_1) > 0, p(s_2) > 0, s_1 \neq s_2$. Suppose $i \in s_1 \cup s_2$, and $j \in s_1 \cup s_2^c$. Define an estimator e_0 as

$$e_0(s, \mathbf{y}) = \begin{cases} e_L - p(s_2)y_i^2 & \text{for } s = s_1 \\ e_L + p(s_1)y_i^2 & \text{for } s = s_2 \\ e_L & \text{for } s \neq s_1, s_2 \end{cases}$$

Show that e_0 is unbiased for V. Also, show that

$$\text{Var}\,(e_0) - \text{Var}\,(e_L) = y_i^2 p(s_1)p(s_2)[y_i^2\{p(s_1) + p(s_2)\} - 2a_1\frac{y_j^2}{\pi_j} + 4a_2\frac{y_iy_j}{\pi_{ij}}]$$

Hence, observe that $\text{Var}\,(e_0)$ can be made smaller than $\text{Var}\,(e_L)$ for certain values of y_i and y_j. Hence, show that for any given p with $\pi_{ij} > 0 \ \forall \ i \neq j$ there does not exist any UMVQUE of V.

(c) There does not exist any UMV estimator in the class of all unbiased estimators of V.

(Bhattacharyya, 1997)

7. Consider the model (5.6.1). Show that for a given design, the estimator

$$\hat{\beta}_s = \frac{1}{Mp(s)} \frac{\sum_{i \in s}(x_i - \bar{x}_s)y_i}{\sum_{i \in s}(x_i - \bar{x}_s)^2}$$

where M denotes the number of samples in $\mathcal{S}$, is pm-unbiased. Furthermore,

$$\text{Var}\,(\hat{\beta}_s) = \sigma^2 \sum_{s \in \mathcal{S}}\{p(s)M^2 \sum_{i \in s}(x_i - \bar{x}_s)^2\}^{-1} + \beta^2\{\sum_{s \in \mathcal{S}}(p(s)M^2)^{-1} - 1\}.$$

Show that for the population with $x_1 = 1, x_2 = 2, x_3 = 3$, and the design with $p(s_1) = 2/3, p(s_2) = 1/3, s_1 = (1,2), s_2 = (1,3)$, $\text{Var}\,(\hat{\beta}_s^*) = 1.5\sigma^2$ and $\text{Var}\,(\hat{\beta}_s) = (9\sigma^2 + \beta_0^2)/8$. Hence, $\text{Var}\,(\hat{\beta}_s^*)$ is not uniformly smaller than $\text{Var}\,(\hat{\beta}_s)$.

(Thomsen 1978)

8. *Two-Stage Sampling* The population is divided into H clusters of size $N_1, \ldots, N_H$. In the first stage a sample s of n clusters is selected. In the second stage a sample s_j of size m_j is selected from cluster j, $\forall j \in s$. Considering the model $\mathbf{y} = X\beta$ where

$$\mathbf{y} = (y_1', \ldots, y_H')' \ X = (1_{N_1}', \ldots, 1_{N_H}')',$$

$$V = \text{Diag}\,(V_1, \ldots, V_H), \ V_h = \sigma_h^2 I_{N_h} + \sigma_v^2 J_{N_h}$$

show that

$$B_N = \sum_{h=1}^{H} \frac{N_h \bar{y}_h}{\sigma_h^2 + N_h \sigma_v^2} \Big/ \sum_{h=1}^{H} \frac{N_h}{\sigma_h^2 + N_h \sigma_v^2}$$

Show also that

$$\hat{B}_{BLN} = \hat{\beta}_s = \frac{\sum_{h \in s} w_h \bar{y}_{s_h}}{\sum_{h \in s} w_h}$$

where

$$\bar{y}_{s_h} = \sum_{j \in s_h} y_{h_j}/m_h \text{ and } w_h = \frac{m_h \sigma_v^2}{\sigma_h^2 + m_h \sigma_v^2}$$

(Bolfarine, et al, 1994)

9. Consider the model

$$y_i = \sum_{j=0}^{p} \beta_j x_{ij} + e_i, i = 1, \ldots, N$$

$$Y_i = y_i + u_i, \ i \in s$$

$$u_i \underset{\sim}{iid}(0, \sigma_u^2), \; e_i \underset{\sim}{iid}(0, \sigma_e^2), \; e_i \underset{\sim}{ind} u_i$$

where the true value pf y, y_i cannot be observed, but a different value Y_i mixed with measurement error u_i is observed. Following Theorem 5.5.1 show that $\sqrt{n}(b - B)$ converges in law to a normal distribution, where $b = (X_s' X_s)^{-1} X_s' Y_s$

10. Show that $\hat{\beta}_s$ is minimax for B_N with respect to ψ_R-generalised prediction risk (5.7.17) under model ψ_R.

Chapter 6

Estimation of a Finite Population Distribution Function

6.1 INTRODUCTION

Estimation of a finite population distribution function has attracted considerable attention of survey statisticians over the last two decades. Our problem here is to estimate the finite population distribution function (*d.f.*)

$$F_N(t) = \frac{1}{N} \sum_{i=1}^{N} \Delta(t - y_i) \qquad (6.1.1)$$

where $\Delta(z)$ is a step function with

$$\Delta(z) = 1(0) \text{ if } z \geq 0 \text{ (otherwise) },$$

on the basis of a sample s selected according to a sampling design p with selection probability $p(s)$ and observations of the data.

The distribution function $F_N(t) = F(t)$ denotes the proportion of units in $\mathcal{P}$ for which the value of y does not exceed t. Such functions are of considerable interest in estimating functions like *Lorenz Ratio* where y is the income and the units are individuals or in the establishments survey where y may be the value added by manufacture and the units are the factories.

As usual, under the superpopulation model approach, we will consider $\mathbf{y} = (y_1, \ldots, y_N)$ as the realisation of a random vector $\mathbf{Y} = (Y_1, \ldots, Y_N)$ (y_i being a realised value of the random variable Y_i, having a joint distribution ξ). In this case, our problem will be to predict $F_N(t)$ i.e. to estimate $\mathcal{E}(F_N(t))$ on the basis of the observed data and the assumed superpopulation model ξ, estimating any unknown parameter involved in the model in the process. We shall, as before, for simplicity, use the same symbol y_i to denote the random variable Y_i as well as its realised value, the actual meaning will be clear from the context.

The following superpopulation models will often be used. Assume that x_i is the value of an auxiliary variable x, closely related to the main variable y and the values $x_i (i = 1, \ldots, N)$ are known.

(a)
$$y_i = \beta x_i + U_i v(x_i) \qquad (6.1.2)$$

where U_i are *iid* random variables with mean 0 and variance σ^2 and $v_i = v(x_i)$ is a known positive function of x_i and β is a unknown constant.

(b)
$$y_i = \alpha + \beta x_i + \epsilon_i \qquad (6.1.3)$$

where α, β are unknown constants and ϵ_i are *iid* random variables with mean zero.

6.2   DESIGN-BASED ESTIMATORS

A general class of design-based estimators of $F(t)$ is

$$\hat{F}(t) = \frac{H(t)}{\hat{N}} = \frac{\sum_{j=1}^{N} d_{js}\Delta(t - y_j)}{\sum_{j=1}^{N} d_{js}}$$

where the weight d_{js} may be a function of (j, s) but is independent of y ($d_{js} = 0$ if $j \notin s$). The weights should satisfy the unbiasedness condition

$$\sum_{s \ni j} d_{js} p(s) = 1, \ j = 1, \ldots, N$$

Taking $d_{js} = 1/\pi_j$, one gets the conventional design-based estimator of $F(t)$ which is the Haj'ek-type estimator

$$\hat{F}_0(t) = \sum_{j \in s} \frac{\Delta(t - y_j)}{\pi_j} / \sum_{j \in s} \frac{1}{\pi_j} \qquad (6.2.1)$$

Under *srswor* of size n, $\hat{F}_0(t)$ reduces to the sample empirical distribution function

$$\hat{F}_{s_n}(t) = \frac{1}{n} \sum_{j \in s} \Delta(t - y_j) \quad (6.2.2)$$

For small sample sizes it is always advantageous to smooth out (6.2.1). The design-based ratio estimator of $F(t)$ is obtained by treating $\Delta(t-y)$ as the main variable and $\Delta(t - \hat{R}x_i)$ as the auxiliary variable, where

$$\hat{R} = \frac{\sum_s d_{js} y_j}{\sum_s d_{js} x_j}$$

In particular taking

$$\hat{R} = \hat{R} = \frac{(\sum_s y_k/\pi_k)}{(\sum_s x_k/\pi_k)} \quad (6.2.3)$$

one gets a ratio predictor of $F(t)$ corresponding to $\hat{F}_0(t)$ as

$$\hat{F}_r(t) = \frac{1}{N}[\sum_{j \in s} \frac{\Delta(t - y_j)}{\pi_j} / \sum_{j \in s} \frac{\Delta(t - \hat{R}x_j)}{\pi_j}] \sum_{i=1}^{N} \Delta(t - \hat{R}x_i) \quad (6.2.4)$$

When $y \propto x$, $\hat{F}_r(t)$ reduces to $F(t)$. This property will be called *ratio estimator* property of $\hat{F}_r(t)$. This suggests that $\hat{F}_r(t)$ would be expectedly more efficient than $\hat{F}_0(t)$ when y is approximately proportional to x.

A design-based difference estimator of $F(t)$ is

$$\hat{F}_d(t) = \frac{1}{N}[\sum_{j \in s} \frac{\Delta(t - y_j)}{\pi_j} + d\{\sum_{i=1}^{N} \Delta(t - \hat{R}x_i) - \sum_{j \in s} \frac{\Delta(t - \hat{R}x_j)}{\pi_j}\}] \quad (6.2.5)$$

where d is a known constant. Clearly, $\hat{F}_d(t)$ is design-unbiased for $F(t)$. The optimum value of d is obtained by minimising the variance of $\hat{F}_d(t)$ with respect to d and is given by , for *srswor*,

$$d^* = \frac{\rho_\Delta S_{\Delta y}}{S_{\Delta x}}, \quad (6.2.6)$$

where

$$S_{\Delta y} = \frac{1}{N-1} \sum_{i=1}^{N} \{\Delta(t - y_i) - F(t)\}^2$$

$$S_{\Delta x} = \frac{1}{N-1} \sum_{i=1}^{N} \{\Delta(t - \hat{R}x_i) - \frac{1}{N} \sum_{i=1}^{N} \Delta(t - \hat{R}x_i)\}^2$$

and ρ_Δ is the finite population correlation coefficient between $\Delta(t-y)$ and $\Delta(t-\hat{R}x)$.

In general, the correlation between $\Delta(t-y)$ and $\Delta(t-Rx)$ is likely to be weaker than the correlation between y and x where $R = Y/X$. Consequently, the gain in efficiency of $\hat{F}_r(t), \hat{F}_d(t)$ over $\hat{F}_0(t)$ is likely to be smaller than those achieved by the customary ratio and regression estimator of population mean $\bar{y}$. The estimators $\hat{F}_0(t), \hat{F}_r(t), \hat{F}_d(t)$ are design-consistent under approporiate regularity conditions.

If there are p auxiliary variables $x_1, \ldots, x_p$ with known values x_{ji} on unit i $(j = 1, \ldots, p; i = 1, \ldots, N)$ the multivariate design-based ratio estimator is

$$\hat{F}'_r(t) = \frac{1}{N} \sum_{k=1}^{p} w_k [\sum_{j \in s} \frac{\Delta(t-y_j)}{\pi_j} / \sum_{j \in s} \frac{\Delta(t-\hat{R}_k x_j)}{\pi_j}] \sum_{i=1}^{N} \Delta(t - \hat{R}_k x_i) \quad (6.2.7)$$

where

$$\hat{R}_k = \frac{\sum\limits_{j \in s} y_j / \pi_j}{\sum\limits_{j \in s} x_{kj} / \pi_k}, \quad R_k = \frac{Y}{X_k}, \quad X_k = \sum_{i=1}^{N} x_{ki}$$

and $w_k (> 0)$ $(\sum w_k = 1)$ are constants to be suitably determined.

Similarly, the multivariate design-based difference estimator is

$$\hat{F}'_d(t) = \frac{1}{N} [\sum_{j \in s} \frac{\Delta(t-y_j)}{\pi_j} + \sum_{k=1}^{p} d_k \{\sum_{i=1}^{N} \Delta(t - \hat{R}_k x_{ki}) -$$

$$\sum_{j \in s} \frac{\Delta(t - \hat{R}_k x_{kj})}{\pi_j} \}] \quad (6.2.8)$$

where the constants d_k's are to be optimally chosen. The estimators (6.2.4), (6.2.5) are asymptotically design-unbiased but not m-unbiased under model (6.1.2). This is because

$$\mathcal{E}[\Delta(t - y_i)] \neq \Delta(t - \beta x_i)$$

Similar results hold for $\hat{F}'_r, \hat{F}'_d$ under the multiple regression models.

Silva and Skinner (1995) defined the following post-stratified estimator corresponding to $\hat{F}_0(t)$. Let L be the number of post-strata $\mathcal{P}_1, \ldots, \mathcal{P}_L$

$(\bigcup\limits_{g=1}^{G} \mathcal{P}_g = \mathcal{P})$. A unit $i \in \mathcal{P}_g$ if $x_{(g-1)} < x_i < x_{(g)}$ where $x_{(0)} = -\infty < x_{(1)} <$

$\ldots < x_{(L)} = \infty$. Let $s_1, \ldots, s_L$ be the corresponding partitioning of s so that

$$s_g = s \bigcap \mathcal{P}_g$$

Let N_g be the size of $\mathcal{P}_g$ and let

$$\hat{N}_g = \sum_{j \in s_g} \frac{1}{\pi_j}, \quad g = 1, \ldots, L \qquad (6.2.9)$$

The post-stratified estimator is

$$\hat{F}_{ps}(t) = \frac{1}{N} \sum_{g=1}^{L} \frac{N_g}{\hat{N}_g} \sum_{j \in s_g} \frac{\Delta(t - y_j)}{\pi_j}$$

$$= \frac{1}{N} \sum_{g=1}^{L} N_g \hat{F}_{0g}(t) \text{ (say)} \qquad (6.2.10)$$

It is desirable to define the post-strata such that the probability that s_g is empty is very small. In practice, any post-strata with $\hat{N}_g = 0$ are pooled with adjacent post-strata until all $\hat{N}_g$ are positive.

The predictor $\hat{F}_{ps}$ is exactly m-unbiased under a model for which y_i has a common mean within each post-stratum. It may, however, be m-biased under model (6.1.2).

Kuk (1988) considered homogeneous linear unbiased estimators of $F(t)$,

$$\hat{F}_L(t) = \frac{1}{N} \hat{H}(t) = \frac{1}{N} \sum_{i=1}^{N} d_{is} \Delta(t - y_i) \qquad (6.2.11)$$

where d_{is} has been defined earlier in this section. For any arbitrary sampling design, the choice $d_{is} = \frac{1}{\pi_i}$ (0) if $i \in s$ (otherwise) gives the HT-estimator

$$\hat{F}_{HT}(t) = \frac{1}{N} \sum_{i \in s} \frac{\Delta(t - y_i)}{\pi_i} \qquad (6.2.12)$$

For probabilitty proportional to aggregrate sample size $(ppas)s.d.$,

$$d_{is} = \frac{1}{\sum_{j \in s} p_j} (0) \quad \text{if } i \in s \text{ (otherwise)}$$

where $p_j = x_j/X$. Define the complementary function of $F(t)$ as

$$S(t) = 1 - F(t) = \frac{1}{N} \sum_{i=1}^{N} \Delta(y_i - t) \qquad (6.2.13)$$

Its estimator is, following (6.2.11),

$$\hat{S}(t) = \frac{1}{N} \sum_{i=1}^{N} d_{is}\Delta(y_i - t) \qquad (6.2.14)$$

An estimator of $F(t)$ is, therefore,

$$\hat{F}_R(t) = 1 - \hat{S}(t) = 1 - \frac{1}{N} \sum_{i=1}^{N} d_{is} + \hat{F}_L(t) \qquad (6.2.15)$$

Again, neither $\hat{F}_L(t)$, nor $\hat{F}_R(t)$ is a distribution function since their maximum values are not equal to one. A natural remedy is to divide $\hat{F}_L(t)$ or $\hat{F}_R(t)$ by its maximum value. The normalised version of $\hat{F}_L(t)$

$$\hat{F}_\nu(t) = \frac{\sum\limits_{i \in s} d_{is}\Delta(t - y_i)}{\sum\limits_{i \in s} d_{is}} \qquad (6.2.16)$$

is, however, not unbiased. The mse of $\hat{F}_L(t)$,

$$MSE(\hat{F}_L(t)) = \frac{1}{N^2} E(\sum_{i=1}^{N}(d_{is} - 1)\Delta(t - y_i))^2$$

$$= \frac{1}{N^2} \sum_{i=1}^{N} \sum_{j=1}^{N} \Delta(t - y_i)\Delta(t - y_j)a_{ij} \qquad (6.2.17)$$

where

$$a_{ij} = Cov(d_{is}, d_{js})$$

Similarly,

$$MSE(\hat{F}_R(t)) = \frac{1}{N^2} \sum_i \sum_j \Delta(y_i - t)\Delta(y_j - t)a_{ij} \qquad (6.2.18)$$

Now,

$$MSE(\hat{F}_R(t)) \le MSE(\hat{F}_L(t)) \Rightarrow \sum_{i=1}^{N} b_i \ge 2\sum_{i=1}^{N} \Delta(y_i - t)b_i \qquad (6.2.19)$$

where

$$b_i = \sum_{j=1}^{N} a_{ij}$$

Since $\hat{F}_L(t), \hat{F}_R(t), \hat{F}_\nu(t)$ are all step functions we need to compare them at $t = y_1, \ldots, y_N$ and at a value t such that $F(t) = 0$. Assume no ties and let

$$y_{(1)} \leq y_{(2)} \leq \cdots \leq y_{(N)} \qquad (6.2.20)$$

be the ordered y-values. Let y_0 be a value less than $y_{(1)}$. Let $D(i)$ be the anti-ranks so that

$$y_{(i)} = y_{D(i)}, \quad i = 1, \ldots, N \qquad (6.2.21)$$

The condition (6.2.19) imples that

$$MSE(\hat{F}_R(y_{(l)})) \leq MSE(\hat{F}_L(y_l))$$

if

$$\sum_{i=1}^{N} b_i \geq 2 \sum_{i=1}^{N} \Delta(y_{(i)} - y_{(l)}) b_{(i)}$$

$$= 2 \sum_{i=l+1}^{N} b_{D(i)} \qquad (6.2.22)$$

Since $\hat{F}_\nu(t)$ is a ratio estimator, its mse is approximately its variance and is given by

$$MSE(\hat{F}_\nu(t)) \simeq E\left(\frac{\sum_{i=1}^{N} d_{is} \Delta(t - y_i)}{\sum_{i=1}^{N} d_{is}} - F(t) \right)^2$$

$$\simeq \frac{E(\sum_{i=1}^{N} d_{is} \Delta(t - y_i) - \sum_{i=1}^{N} d_{is} F(t))^2}{E(\sum_{i=1}^{N} d_{is})^2}$$

$$= \frac{1}{N^2} \sum_{i \neq i'=1}^{N} \sum (\Delta(t - y_i) - F(t))(\Delta(t - y_{i'}) - F(t)) E(d_{is} d_{i's})$$

$$= \frac{1}{N^2} \sum_{i \neq i'=1}^{N} (\Delta(t - y_i) - F(t))(\Delta(t - y_{i'}) - F(t))a_{ii'} \qquad (6.2.23)$$

From (6.2.17) and (6.2.23), denoting asymptotic mse as AMSE,

$$\text{MSE} \ (\hat{F}_L(t)) - \text{AMSE} \ (\hat{F}_\nu(t))$$

$$= \frac{F(t)}{N} [2 \sum_i \Delta(t - y_i)b_i - F(t) \sum_i b_i] \qquad (6.2.24)$$

$$= l[2 \sum_{i=1}^{l} b_{D(i)} - l\bar{b}] \quad \text{for } t = y_{(l)}$$

$$\geq \quad 0$$

if

$$\sum_{i=1}^{l} b_{D(i)} \geq \frac{1}{2} l\bar{b} \qquad (6.2.25)$$

where $\bar{b} = \sum_{i=1}^{N} b_i/N$. From (6.2.18) and (6.2.23) we conclude that

$$\text{MSE} \ (\hat{F}_R(y_{(l)}) \ \leq \ \text{AMSE} \ (\hat{F}_\nu(t))$$

if

$$\sum_{i=l+1}^{N} b_{D(i)} \leq \frac{1}{2}(N - l)\bar{b} \qquad (6.2.26)$$

Conditions (6.2.19), (6.2.22), (6.2.25) are not useful in practice since $D(1), \ldots$, $D(N)$ are not known. Assume that the ordering of y-values agree with that of the x-values so that

$$x_{D(1)} \ \leq \ x_{D(2)} \ \leq \cdots \ \leq x_{D(N)}. \qquad (6.2.27)$$

In this case, we can compute $b_{D(i)}(i = 1, \ldots, N)$ and hence check the conditions (6.2.19), (6.2.22) and (6.2.25). The condition (6.2.27) implies

$$b_{D(1)} \ \geq b_{D(2)} \ \geq \cdots \ \geq b_{D(N)} \qquad (6.2.28)$$

for a number of sampling designs including Poisson, modified Poisson (Ogus and Clark, 1971), Collocated sampling (Brewer et al, 1972) and *ppas* sampling designs using x as the size measures.

If (6.2.28) holds, (6.2.25) holds in general implying $\hat{F}_\nu(t)$ is preferable to $\hat{F}_L(t)$. From (6.2.22) and (6.2.28) it follows that there is an $l_1 \in [1, Q]$ where

$Q = [\frac{N+1}{2}]$ such that $\text{MSE}(\hat{F}_R(y_l))) \leq \text{MSE} (\hat{F}_L(y_{(l)})) \; \forall \; l \geq l_1$. Generally, l_1 is sufficiently smaller than Q so that for estimating the population median ξ, $\hat{F}_R \succeq \hat{F}_L(t)$.

It follows from (6.2.22), (6.2.25) and (6.2.26) that $\hat{F}_L(t)$ is inferior to both $\hat{F}_\nu(t)$ and $\hat{F}_R(t)$. From (6.2.26) and (6.2.28) it follows that there is an l_2 such that $\text{MSE}(\hat{F}_R(t)) \leq \text{AMSE} (\hat{F}_\nu(t))$ for $t \geq y_{(l_2)}$.

The empirical studies considered by Kuk (1988) using $n = 30$ for three populations, – Dwellings (N=270; Kish, 1965, p.624), Villages (N=250; Murthy 1967, p.127), Metropolitan (x=1970 population, y = 1980 population for 250 metropolitan statistical areas in US) – confirmed the above findings.

Kuk and Mak (1989) considered the following cross-classified estimator. For any value of t, let $F_1(t)$ denote the proportion among those units in the sample with x values $\leq M_x$ (population median of x), that have y values $\leq t$. Similarly, let $F_2(t)$ denote the proportion among those units with x values $> M_x$. Let N_x denote the number of units in the population with x values $\leq M_x$. Then $F(t)$ can be estimated as

$$\hat{F}_{KM} = \frac{1}{N}[N_x F_1(t) + (N - N_x)F_2(t)]$$

$$\simeq \frac{1}{2}(F_1(t) + F_2(t)) \tag{6.2.29}$$

Mukhopadhyay (2000 c) considered calibration estimation of finite population d.f. under multiple regression model.

6.3 MODEL-BASED PREDICTORS

Following Royall (1970), Royall and Herson (1973), Rodrigues et al (1985), we consider in this section model-based optimal predictors of $F(t)$. After a sample has been selected, we may write,

$$F_N(t) = \theta_s + \theta_{sr} \tag{6.3.1}$$

where

$$\theta_s = \theta(y_s) = \frac{n}{N}\hat{F}_{s_n}(t)$$

and

$$\theta_{sr} = \theta(y_s, y_r) = \frac{N - n}{N}F_r(t)$$

where

$$F_r(t) = \frac{1}{N - n}\sum_{i \in r}\Delta(t - y_i), \tag{6.3.2}$$

$F_N(t) = F(t)$ and the other symbols have usual meanings. Hence a predictor of $F(t)$ is of the form

$$\hat{F}(t) = \theta_s + \hat{\theta}_{sr} \qquad (6.3.3)$$

where $\hat{\theta}_{sr}$ is a predictor of θ_{sr}.

DEFINITION 6.3.1 A predictor $\hat{F}(t)$ is model (m) -unbiased predictor of $F(t)$ with respect to the model (6.1.2) if

$$E_\psi[\hat{F}(t) - F(t)] = 0 \ \forall \ \psi = (\beta, \sigma^2) \in \Psi \text{ and } \forall s : p(s) \ge 0 \qquad (6.3.4)$$

where Ψ is the parameter space.

Chambers and Dunstan (CD)(1986), therefore, suggested a m-unbiased predictor of $F(t)$,

$$\hat{F}(t) = \frac{1}{N}[\sum_{j \in s} \Delta(t - y_j) + \hat{V}_s] \qquad (6.3.5)$$

where $\hat{V}_s$ is a m-unbiased predictor of $\sum_{i \in r} \Delta(t - y_i)$ i.e.

$$E(\hat{V}_s) = E(\sum_{i \in r} \Delta(t - y_i))$$

Now

$$E(\Delta(t - y_i)) = G(\frac{t - \beta x_i}{v(x_i)}) \qquad (6.3.6)$$

where $G(z) = P(U_i \le z)$ is the distribution function of U_i. An empirical estimator of $G(\frac{t - \beta x_i}{v(x_i)})$ is, therefore,

$$G_n(\frac{t - \beta x_i}{v(x_i)}) = \frac{1}{n}\sum_{j \in s} \Delta(\frac{t - b_n x_i}{v(x_i)} - \hat{U}_j) \qquad (6.3.7)$$

where

$$\hat{U}_j = U_{nj} = \frac{y_j - b_n x_j}{v(x_j)}$$

$$b_n = \hat{\beta} = \sum_{j \in s} \frac{x_j y_j}{v(x_j)} / \sum_{j \in s} \frac{x_j^2}{v(x_j)} \qquad (6.3.8)$$

Hence, an approximately m-unbiased predictor of $F(t)$ is

$$\hat{F}_{CD}(t) = \frac{1}{N}[\sum_{j \in s} \Delta(t - y_j) + \frac{1}{n}\sum_{i \in r}\sum_{j \in s} \Delta(\frac{t - b_n x_i}{v(x_i)} - U_{nj})] \qquad (6.3.9)$$

However, $\hat{F}_{CD}$ is not design-unbiased under repeated sampling. For small sample sizes it may be desirable to replace U_{nj} by its studentised equivalent under (6.1.2). Also, one could replace $G_n(t)$ in (6.3.7) by a smoother estimator of G, e.g., a kernel estimator of this function, obtained by integrating a kernel density estimator (Hill, 1985). Dorfman (1993) extended CD-estimator to multiple regression model.

Dunstan and Chambers (1989) extended CD-estimator to the case where only summary information is available for the auxiliary size variable x. We assume that only the histogram-type information on x is available enabling the population to be split up into H strata, defined by the end-points $x_{hL}, x_{hU}(h = 1, \ldots, L)$. Also, strata sizes N_h and strata means $\bar{x}_h$ are known. In this case, the double summation in (6.3.9) can be written as

$$\frac{1}{n} \sum_h \sum_{j \in s_h} \sum_{i \in r_h} \Delta\left(\frac{t - b_n x_{hi}}{v(x_{hi})} - U_{nj}\right) \tag{6.3.10}$$

Assuming x_{hi} to be an independent realisation of a random variable X_h with distribution function C_h,

$$\mathcal{E}\left\{\Delta\left(\frac{t - b_n x_h}{v(x_h)}\right) - z)\right\} = 1 - \Gamma_{ht}(z)$$

($\mathcal{E}$ denoting expectation with respect to *d.f.* C_h) where Γ_{ht} is the distribution function of the transformed variable $(t - b_n x_h)/v(x_h)$. Therefore, expectation of expression in (6.3.10) is

$$\sum_h (N_h - n_h)[1 - n^{-1} \sum_{j \in s_h} \Gamma_{ht}(U_{nj})]$$

The actual form of Γ_{ht} will depend on C_h and the form of the variance function $v(x)$. For example, when $v(x) = \sqrt{x}$, assuming $b_n > 0, t > 0$,

$$\Gamma_{ht}(u) = C_h\left[\frac{u^2 + 2b_n t - u\sqrt{u^2 + 4b_n t}}{2b_n^2}\right]$$

If approximation $\hat{C}_h$ to C_h and hence $\hat{\Gamma}_{ht}$ to Γ_{ht} are available from survey data, a limited information estimator corresponding to $\hat{F}_{CD}(t)$ is

$$\hat{F}_{CD}^{(L)}(t) = \frac{1}{N}\left[\sum_{j \in s} \Delta(t - y_j) + \sum_h (N_h - n_h)\{1 - n^{-1} \sum_{j \in s_h} \hat{\Gamma}_{ht}(U_{nj})\}\right] \tag{6.3.11}$$

The authors derive estimator of asymptotic prediction variance of $\hat{F}_{CD}^{(L)}(t)$ by obtaining the limited information approximation as above to the asymptotic variance of $\hat{F}_{CD}(t)$ derived in Theorem 6.5.1.

Model-dependent strategies can perform poorly in large samples under model-misspecification (Hansen, Madow, and Tepping, 1983). Rao, Kover and Mantel (RKM) (1990) noticed a similar poor performance of the model-dependent estimator $\hat{F}_{CD}(t)$ under model-misspecification and they, therefore, considered model-assisted approach. In this approach one considers design-consistent estimators, $\hat{F}_{dm}(t)$ (say) that are also model-unbiased (at least asymptotically) under the assumed model. Estimators of model-variance $\mathcal{V}(\hat{F}_{dm} - F)$ that are design-consistent and at the same time model-unbiased (at least asymptotically) can be obtained following Sarndal et al (1989), Kott (1990). The resulting pivot $[\hat{F}_{dm}(t) - F(t))]/\sqrt{\mathcal{V}(\hat{F}_{dm}(t) - F(t))}$ provides valid inference under the assumed model and at the same time protects against model mis-specifications in the sense of providing valid design-based inference under model-failures.

Rao, Kover and Mantel (RKM) (1990) considered the model (6.1.2) with $v(x) = \sqrt{x}$. Considering

$$G_i = \frac{1}{N} \sum_{j=1}^{N} \Delta \left(\frac{t - Rx_i}{\sqrt{x_i}} - V_{nj} \right) \tag{6.3.12}$$

where

$$V_{nj} = \frac{y_j - Rx_j}{\sqrt{x_j}}, \quad R = \frac{Y}{X},$$

as the value of an auxiliary variable, they defined a difference estimator

$$\tilde{F}_{RKM}(t) = \frac{1}{N} \left[\sum_{j \in s} \frac{\Delta(t - y_j)}{\pi_j} + \left\{ \sum_{i=1}^{N} G_i - \sum_{j \in s} \frac{G_j}{\pi_j} \right\} \right] \tag{6.3.13}$$

This estimator is both design-unbiased and asymptotically m-unbiased. Now, in G_i of $\sum_{i=1}^{N} G_i, V_{nj}$ will not be known for all j. Thus G_i requires to be estimated. A design-based estimator of G_i in $\sum_{i=1}^{N} G_i$ is

$$\hat{G}_i = \sum_{j \in s} \frac{\Delta}{\pi_j} \left(\frac{t - \tilde{R}x_i}{\sqrt{x_i}} - \hat{V}_{nj} \right) / \sum_{j \in s} \frac{1}{\pi_j} \tag{6.3.14}$$

where

$$\hat{V}_{nj} = \frac{y_j - \tilde{R}x_j}{\sqrt{x_j}}$$

$$\tilde{R} = (\sum_s \frac{x_i y_i}{v(x_i)^2 \pi_i}) / (\sum_s \frac{x_i^2}{v(x_i)^2 \pi_i})$$

$$= (\sum_s \frac{y_i}{\pi_i}) / (\sum_s x_i/\pi_i) = \hat{R} \text{ (here)} \tag{6.3.15}$$

Similarly, G_j in $\sum_s \frac{G_i}{\pi_j}$ requires to be estimated. G_j is estimated by

$$\hat{G}_{jc} = \frac{1}{\sum\limits_{k \in s} \pi_j/\pi_{jk}} \sum_{k \in s} \frac{\pi_j}{\pi_{jk}} \Delta(\frac{t - \hat{R}x_j}{\sqrt{x_j}} - \hat{V}_{nk}) \tag{6.3.16}$$

The estimator $\hat{G}_i$ is asymptotically design-unbiased for G_i while $\hat{G}_{jc}$ is asymptotically conditionally design-unbiased for G_j given $j \in s$.

The alternative model-assisted estimator is, therefore,

$$\hat{F}_{RKM} = N^{-1}[\sum_{j \in s} \frac{\Delta(t - y_j)}{\pi_j} + (\sum_{i=1}^{N} \hat{G}_i - \sum_{j \in s} \frac{\hat{G}_{jc}}{\pi_j})] \tag{6.3.17}$$

which is aymptotically both design-unbiased and model-unbiased.

Godambe (1990) derived (6.3.17) with slight modifications on the basis of optimal estimating functions.

Under *srswor* and $v_i = \sqrt{x_i}$,

$$\hat{G}_i = n^{-1} \sum_{j \in s} \Delta(\frac{t - \hat{R}x_j}{\sqrt{x_j}} - \hat{V}_{nj}) = \tilde{G}_i \text{ (say)} \tag{6.3.18}$$

and $\hat{G}_{jc} = \tilde{G}_j$.

Dorfman (1993), therefore, proposed a model-based generalisation of $\hat{F}_{RKM}(t)$ as

$$\hat{F}_{RKM*}(t) = N^{-1}[\sum_{j \in s} \frac{\Delta(t - y_j)}{\pi_j} + \sum_{i=1}^{N} \tilde{G}_i - \sum_{j \in s} \tilde{G}_j/\pi_j]$$

$$= N^{-1}[\sum_{j \in s} \frac{\Delta(t - y_j) - \tilde{G}_j}{\pi_j} + \sum_{i=1}^{N} \tilde{G}_i] \tag{6.3.19}$$

regarding π_j as the reflective of the proportion of sampled units near the data point x_j, not necessarily the inclusion probabilities. $\hat{F}_{RKM*}(t)$ is, therefore, free from the second order inclusion-probabilities which may be difficult to estimate. Godambe's (1989) estimator also shares this property.

Rao and Liu (1992) proposed a model-assisted estimator for the general weights d_{is} satisfying the design-unbiasedness condition . Assume first that G_i is known for all i. A model-assisted estimator is given by

$$\tilde{F}_{RL}(t) = \frac{1}{N}[\sum_{j \in s} d_{js}\Delta(t - y_j) + \{\sum_{i=1}^{N} G_i - \sum_{j \in s} d_{js}G_j\}] \qquad (6.3.20)$$

Now, replace G_i in $\sum_{i=1}^{N} G_i$ by

$$\tilde{G}_i = \{\sum_{j \in s} d_{js}\}^{-1} \sum_{j \in s} d_{js}\Delta(\frac{t - \hat{\tilde{R}}x_i}{\sqrt{x_i}} - \hat{\tilde{V}}_{nj})$$

where

$$\hat{\tilde{V}}_{nj} = \frac{y_j - \hat{\tilde{R}}x_j}{x_j}$$

Similarly G_j in $\sum_{j \in s} d_{js}G_j$ is replaced by

$$\tilde{G}_{jc} = \{\sum_{k \in s} d_{ks|j}\}^{-1}[\sum_{k \in s} d_{ks|j}\Delta(\frac{t - \hat{\tilde{R}}x_j}{\sqrt{x_j}} - \hat{\tilde{V}}_{nk})]$$

when the weights $d_{ks|j}$ satisfy $\sum_{s \ni (j,k)} d_{ks|j}p(s) = \pi_j$. The final model-assisted estimator of Rao and Liu (1992) is

$$\hat{F}_{RL}(t) = \frac{1}{N}[\sum_{j \in s} d_{js}\Delta(t - y_j) + \{\sum_{i=1}^{n} \tilde{G}_i - \sum_{j \in s} d_{js}\tilde{G}_{jc}\}] \qquad (6.3.21)$$

which is asymptotically both design-unbiased and model-unbiased.

Godambe's (1990) estimator based on estimating function theory is

$$\hat{F}_G(t) = \frac{1}{n}[\sum_{j \in s} d_{js}\Delta(t - y_j) + \{\sum_{i=1}^{N} \tilde{G}_i - \sum_{j \in s} d_{js}\tilde{G}_j\}] \qquad (6.3.22)$$

Wang and Dorfman (1996) combined CD-estimator and RKM estimator based on the model (6.1.3). The CD estimator is

$$\hat{F}'_{CD}(t) = \frac{1}{N}[\sum_{j \in s} \Delta(t - y_j) + \sum_{i \in r} \hat{H}(t - \hat{\alpha} - \hat{\beta}x_i)] \qquad (6.3.23)$$

where $\hat{H}(z) = \frac{1}{n}\sum_{j \in s}\Delta(z - \hat{\epsilon}_i)$ is an estimate of $H(z) = $ Prob. $(\epsilon \le z)$ and $\hat{\epsilon}_i = y_i - \hat{\alpha} - \hat{\beta}x_i$, $\hat{\alpha}, \hat{\beta}$ being least squared estimates of α, β respectively.

Rao et al (1990) estimator for *srswor* corresponding to the model (6.1.3) is

$$\hat{F}'_{RKM}(t) = \frac{1}{n}\sum_{j\in s}\Delta(t-y_j) + \frac{1}{N}\sum_{i=1}^{N}\hat{H}(t-\hat{\alpha}-\hat{\beta}x_i) - \frac{1}{n}\sum_{i\in s}\hat{H}(t-\hat{\alpha}-\hat{\beta}x_i)$$

(6.3.24)

Noting that both $\hat{F}'_{CD}$ and $\hat{F}'_{RKM}$ have desirable properties and deficiencies in certain situations, Wang and Dorfman (1996) considered a new estimator which is their convex combination,

$$\hat{F}_{WD}(t) = w\hat{F}'_{CD}(t) + (1-w)\hat{F}'_{RKM}(t)$$

$$= \frac{1}{N}\sum_{j\in s}\Delta(t-y_j) + (1-w)(\frac{1}{n}-\frac{1}{N})\sum_{j\in s}\{\Delta(t-y_j)-\hat{H}(t-\hat{\alpha}-\hat{\beta}x_i)\}+$$

$$\frac{1}{N}\sum_{i\in r}\hat{H}(t-\hat{\alpha}-\hat{\beta}x_i)$$

(6.3.25)

where $0 < w < 1$ depends on t and is optimally estimated by minimizing MSE$\{\hat{F}_{WD}(t)-F(t)\}$ under the assumption that both n and N increase to infinity such that $n/N \to f \in (0,1)$ and the sample and non-sample design points have a common asymptotic density.

Mukhopadhyay (1998 d) considered the design-model unbiased optimal prediction of finite population distribution function of a random vector following simple location model and linear regression model with one auxilary variable under measurement errors. This will be considered in the next chapter.

6.4 CONDITIONAL APPROACH

Consider the estimator $\tilde{F}_{LR}(t)$ in (6.3.20). Under *srs* $\tilde{F}_{RL}$ reduces to

$$\tilde{F}'_{RL}(t) = \bar{h}(t) + (\bar{G} - \bar{g})$$

(6.4.1)

where $\bar{h}(t) = \sum_{i\in s}h(t,y_i)/n = \hat{F}_{s_n}(t), h(t,y_i) = \Delta(t-y_i)$ and $\bar{G} = \sum_{i=1}^{N}G_i/N, \bar{g} \sum_{i=1}^{n}g_i/n$. The asymptotic conditional bias of $\tilde{F}'_{RL}(t)$ is

$$E(\tilde{F}'_{RL}(t) \mid \bar{x}) - F(t) = B^*(\bar{x}-\bar{X}) = 0_p(n^{-1/2})$$

(6.4.2)

where

$$B^* = \{Cov(\bar{h},\bar{x}) - Cov(\bar{g},\bar{x})\}/V(\bar{x})$$

$$= \frac{S_{xh} - S_{xG}}{S_x^2} \tag{6.4.3}$$

where S_{xh} and S_{xG} are, respectively, the population covariances between x and h and between x and G. A bias-adjusted estimator is, therefore, given by

$$\tilde{F}_{RLa}(t) = \tilde{F}'_{RL}(t) + s_x^{-2}(s_{xh} - s_{xG})(\bar{X} - \bar{x}) \tag{6.4.4}$$

where s_{xh} and s_{xG} are the sample covariances. The conditional bias of $\tilde{F}_{RLa}(t)$ is $0_p(n^{-1})$ and consequently $\tilde{F}_{RLa}(t)$ provides conditionally valid inference under large sample. $\tilde{F}'_{RLa}(t)$ is also model-unbiased since $E(B^*) = 0$ under (6.1.2).

In practice, one replaces G_i by $\tilde{G}_i$ to get

$$\tilde{F}'_{RLa}(t) = \tilde{F}'_{RL}(t) + s_x^{-2}(s_{xh} - s_{x\tilde{G}})(\bar{X} - \bar{x}) \tag{6.4.5}$$

where $s_{x\tilde{G}}$ is the sample covariance between x_i and $\tilde{G}_i$. If only the population mean $\bar{X}$ is known, $\bar{x}$, an estimate of $\bar{X}$ is an approximate ancillary statistic. The estimator $\hat{F}_G(t)$ in (6.3.22) or $\tilde{F}'_{RLa}(t)$ cannot be used in this case since they require the knowledge of $x_j (j = 1, \ldots, N)$. We, therefore, find the conditional bias of $\bar{h} = \hat{F}_{s_n}(t)$ to obtain a bias-adjusted estimator. The conditional asymptotic bias of $\hat{F}_{s_n}(t)$ is

$$E[\hat{F}_{s_n}(t) \mid \bar{x}] - F(t) = \tilde{B}(\bar{x} - \bar{X}) = 0_p(n^{-1/2}) \tag{6.4.6}$$

where $\tilde{B} = Cov(\bar{h}, \bar{x})/V(\bar{x}) = S_{xh}/S_x^2$. A bias-adjusted estimator is, therefore, given by

$$\hat{F}_a(t) = \hat{F}_{s_n}(t) + (s_{xh}/s_x^2)(\bar{X} - \bar{x}) \tag{6.4.7}$$

The conditional bias of $\hat{F}_a(t)$ is $0_p(n^{-1})$ and as such, $\hat{F}_a(t)$ provides conditionally valid inferences in large samples. However, $\hat{F}_a(t)$ is model-biased under model (6.1.2).

Quin and Chen (1991) used the empirical likelihood method to obtain a maximum likelihood estimator of $F(t)$ which has the same asymptotic variance as $\hat{F}_a(t)$.

6.5   Asymptotic Properties of the Estimators

We first recall a result due to Randles (1982).

Consider random variables which would have been U-statistics were it not for the fact that they contain an estimator. Let $X_1, \ldots, X_n$ be a random sample from some population. Let $h(x_1, \ldots, x_r, \gamma)$ be a symmetric kernel of order r with expected value

$$E_\lambda[h(X_1, \ldots, X_r; \gamma) = \theta(\gamma)] \qquad (6.5.1)$$

where λ denotes a $p \times 1$ vector. Here γ is a mathematical symbol whose one particular value may be $\hat{\lambda}$, a consistent estimator of λ. Both the kernel and its expected value may depend on γ.

The U-statistic corresponding to (6.5.1) is

$$U_n(\gamma) = \frac{1}{\binom{N}{n}} \sum_{a \in A^*} h(X_{a1}, \ldots, X_{ar}; \gamma) \qquad (6.5.2)$$

where A^* denotes the collection of all subsets of size r from the integers $\{1, \ldots, n\}$.

LEMMA 6.5.1 Under certain regularity conditions,

$$\sqrt{n}[U_n(\hat{\lambda}) - \theta(\lambda)] \underset{\rightarrow}{L} N(0, \tau^2) \qquad (6.5.3)$$

provided $\tau^2 > 0$ where τ^2 is given by either
(a)
$$\tau^2 = D'\Sigma D,$$

$$D' = (1, \frac{\partial \theta(.)}{\partial \gamma_1}, \ldots, \frac{\partial \theta(.)}{\partial \gamma_p}), \ \gamma = (\gamma_1, \ldots, \gamma_p),' \qquad (6.5.4)$$

Σ is the covariance-matrix of

$$\sqrt{n}[U_n(\lambda) - \theta(\lambda), \ (\hat{\lambda} - \lambda)']_{1 \times (p+1)}$$

or

$$\tau^2 = \text{var}[E\{h(X_1, \ldots, X_r; \lambda) \mid X_1\}] \qquad (6.5.5)$$

THEOREM 6.5.1 Assume the following regularity conditions:

- (1) As both N and n increase, the sampling fraction $n/N \to f \in (0, 1)$.

- (2) The *d.f.* $G(t)$ of the random variable $u_i = \frac{y_i - \beta x_i}{v(x_i)}$ is differentiable with derivative $g(t) > 0$.

- (3) The quantities x_i and $v(x_i)$ are bounded.

- (4) For arbitrary b define

$$S_j(t, b) = \frac{1}{N - n} \sum_{i \in r} \Delta\left(\frac{t - bx_i}{v(x_i)} - \frac{y_j - bx_j}{v(x_j)}\right)$$

$$F_r^*(t, b) = \frac{1}{n} \sum_{j \in s} S_j(t, b) \qquad (6.5.6)$$

Assume that as both n, N increase the mean and variance of $F_r^*(t, b)$ tend to a limit in $(0,\ 1)$.

- (5) The estimator b_n (defined in (6.3.8)) is asymptotically normlly distributed under model (6.1.2).

Let

$$D_r(t, \beta) = (1,\ h_s)'$$

where

$$h_s = \frac{1}{n} \sum_{j \in s} \frac{1}{N - n} \sum_{i \in r} [\{\frac{x_j}{v(x_j)} - \frac{x_i}{v(x_i)}\}]g(\frac{t - \beta x_i}{v(x_i)}) \qquad (6.5.7)$$

$$V_r^*(t, \beta) = \text{Cov. matrix of } (F_r^*(t, \beta) - E\{F_r^*(t, \beta)\},\ b_n - \beta) \qquad (6.5.8)$$

Define

$$W_r^*(t, \beta) = D_r(t, \beta)'V_r^*(t, \beta)D_r(t, \beta) \qquad (6.5.9)$$

$$W_r(t, \beta) = \frac{1}{(N - n)^2} \sum_{i \in r} G\{\frac{(t - \beta x_i)}{v(x_i)}\}[1 - G\{(t - \beta x_i)/v(x_i)\}] \qquad (6.5.10)$$

Then, as both N and n increase,

$$\{\hat{F}_{CD}(t) - F(t)\}/[(1 - \frac{n}{N})^2\{W_r^*(t, \beta) + W_r(t, \beta)\}^{1/2}] \underset{\rightarrow}{L} N(0,\ 1) \qquad (6.5.11)$$

Proof. When $b = \beta$, $F_r^*(t, \beta)$ is a U-statistic. Hence, by Randle's theorem

$$\sqrt{n}[F_r^*(t,\ b_n) - E\{F_r^*(t, \beta)\}] \underset{\rightarrow}{L} N(0,\ W_r^*(t, \beta))$$

where

$$D_r(t,\ \beta) = \left(1,\ \frac{\partial E\{F_r^*(t, b)\}}{\partial b}\Big]_{b=\beta}\right)$$

Now,

$$E\{F_r^*(t,\ b)\} = \frac{1}{N - n} \sum_{i \in r} G(\frac{t - \beta x_i}{v(x_i)})$$

$$= E_r(t, \; \beta) \; \text{(say)}$$

Hence,

$$\frac{\partial}{\partial b} E\{F_r^*(t, b)\}\bigg|_{b=\beta} = h_s$$

Therefore, for large $n, N, F_r^*(t, \; b_n) \sim AN(E_r(t, \; \beta), \; W_r^*(t, \; \beta))$ where we write AN to denote asymptotically normal. Now,

$$\hat{F}_{CD}(t) - F(t) = \frac{N-n}{N}[F_s^*(t, b_n) - F_r(t)]$$

where $F_r(t)$ is as defined in (6.3.2). Also $F_r(t)$ is independent of $F_r^*(t, \; b_n)$. Again, Var $[F_r(t)] = W_r(t, \beta)$. Hence the result (6.5.11).

Note 6.5.1

Suppose (6.1.2) holds but with variance function $a(x) \neq v(x)$. It can be shown that $\hat{F}_{CD}(t) - F(t)$ is still asymptotically normally distributed but with mean given by

$$(N-n)^{-1} \sum_{i \in s}([n^{-1} \sum_{j \in r} G\{h_{ij}(\frac{t - \beta x_i}{a(x_i)})\}] - G\{\frac{t - \beta x_i}{a(x_i)}\}])$$

where

$$h_{ij} = \frac{v(x_j)a(x_i)}{a(x_j)v(x_i)}$$

The asymptotic bias is approximately zero if the sample is such that $h_{ij} \simeq 1 \; \forall \; i$.

Using Lemma 6.5.1 and denoting the variance and variance-estimator of $\hat{T} = \sum_s y_i/\pi_i$ by $\tilde{V}(y_i)$ and $\tilde{v}(y_i)$, respectively, Rao et al (1990) showed that

$$V\{\hat{F}_d(t)\} \simeq N^{-2}\tilde{V}\{\Delta(t - y_i) - \Delta(t - Rx_i)\}$$

$$v\{\hat{F}_d(t)\} \simeq N^{-2}\tilde{v}\{\Delta(t - y_i) - \Delta(t - \hat{R}x_i)\}$$

$$V\{\hat{F}_r(t)\} \simeq N^{-2}\tilde{V}\{\Delta(t - y_i) - \frac{F(t)}{F_x(t/R)}\Delta(t - Rx_i)\} \tag{6.5.12}$$

$$v\{\hat{F}_r(t)\} \simeq N^{-2}\tilde{v}\{\Delta(t - y_i) - \frac{\hat{F}_0(t)}{\hat{F}_{0x}(t/\hat{R})}\Delta(t - \hat{R}x_i)\}$$

when $\hat{F}_{0x}(t)$ is the customary design-based estimators of $F_x(t)$ defined similarly as (6.2.1) and V, v denote, respectively, the design-variance and estimator of design-variance.

The predictor $\hat{F}_{RKM}(t)$ as well as $\tilde{F}_{RKM}(t)$ is asymptotically model-unbiased with respect to (6.1.2). The asymptotic design-variance of $\hat{F}'_{RKM}(t)$ which is the same as that of $\hat{F}_{RKM}(t)$ is given by

$$V\{\hat{F}_{RKM}(t)\} \simeq N^{-2}\tilde{V}\{\Delta(t - y_i) - G_i\} \qquad (6.5.13)$$

Similarly,

$$V(\hat{F}_{ps}(t)) \simeq N^{-2}\tilde{V}(\Delta(t - y_i) - F_{h(i)}(t)) \qquad (6.5.14)$$

where $h(i)$ is the post-stratum to which i belongs and

$$F_h(t) = \frac{1}{N_h} \sum_{i \in \mathcal{P}_h} \Delta(t - y_i) \qquad (6.5.15)$$

$$v(\hat{F}_{ps}(t)) \simeq N^{-1}\tilde{v}(\Delta(t - y_i) - F_{h(i)}(t))$$

A variance estimator with possibly superior conditional properties is obtained following Rao (1985) and Sarndal et al (1989) by replacing $\Delta(t - y_i) - F_{h(i)}(t)$ by $N_{h(i)}\{\Delta(t - y_i) - F_{h(i)}(t)\}/\hat{N}_{h(i)}$.

Chambers et al (1992) examined the consistency and asymptotic mse of $\hat{F}_{CD}(t)$ and $\hat{F}_{RKM}(t)$ based on the model (6.1.3) under the assumption that the sampling is by *srswor* and assumptions that (i) $n, N \to f \in (0, 1)$, (ii) non-sampled design points have a common asymptotic density d i.e.

$$\frac{1}{n} \sum_{i \in s} \Delta(x_i - x) \to \int_{-\infty}^{x} d(y)dy$$

$$\frac{1}{N - n} \sum_{i \in s} \Delta(x_i - x) \to \int_{-\infty}^{x} d(y)dy \qquad (6.5.15)$$

We shall call these assumptions as assumptions A. It then follows that model-bias of both $\hat{F}_{CD}(t)$ and $\hat{F}_{RKM}(t)$ are of order $0(\frac{1}{n})$ and the s.e. is of order $0(\frac{1}{\sqrt{n}})$ so that mse is approximately equal to the variance of the estimator. It is found that

$$ASV\{\hat{F}_{s_n}(t) - F(t)\} = ASV\{\hat{F}_{RKM}(t) - F(t)\}$$

where $\hat{F}_{s_n}(t)$ has been defined in (6.2.2) and ASV denotes asymptotic variance. The $ASV\{\hat{F}_{CD}(t)\}$ is found to be lower than that of $\hat{F}_{RKM}(t)$ in generel when the model (6.1.3) holds. However, this result does not hold under certain situations. The authors simulated conditions under which $ASV(\hat{F}_{CD})$ would be greater than that of $\hat{F}_{RKM}$ or even $\hat{F}_{s_n}(t)$ even when (6.1.3) holds. Two artificial populations each of size N=550 and with $\alpha = \beta = 1$ were employed. In the first population, the ϵ_k's were generated from a standard exponential distribution and the x_k's according to a double exponential, truncated on the left and shifted to the right to give positively skewed values. For the second population, ϵ_k and x_k were shifted from a mean-centred standard gamma distribution with slope parameter 0.1. In addition, a small bump was put in the extreme right to widen the gap between the mean and mode. For each population 500 simple random samples of size n= 100 were taken and $\hat{F}_{CD}(t)$, $\hat{F}_{RKM}(t)$ and $\hat{F}_{s_n}(t)$ were calculated for certain values of $t(= t_0)$ and population medians. For the first population and t_0 all the estimators were found to be approximately unbiased (average error approximately zero), $\hat{F}_{CD}$ having minimum variance among the three, being followed by $\hat{F}_{RKM}$. For the second population and t_0, $\hat{F}_{CD}$ performed worst both with respect to average error and average standard error. However, for $t =$ population median, the poor performance of $\hat{F}_{CD}$ was not reflected.

Wang and Dorfman (1996) found the asymptotic variance of $\hat{F}_{WD}$ and its estimator under the assumptions A. Kuk (1993) proved the pointwise consistency of $\hat{F}_K(t)$ (defined in (6.6.2)) under the assumption (i) of A and that the finite population values $(x_i, y_i)(i = 1, \ldots, N)$ are realisations of N independent random vectors having a continuous bivariate distribution.

6.6 Non-parametric Kernel Estimators

The last two estimators to be considered are the *nonparametric kernel estimators* proposed by Kuo (1988) and Kuk (1993), given, respectively, by

$$\hat{F}_{KO}(t) = N^{-1}[\sum_{i \in s} \Delta(t - y_i) + \sum_{j \in r} \sum_{i \in s} w_{ij} \Delta(t - y_i)] \qquad (6.6.1)$$

$$\hat{F}_K(t) = N^{-1} \sum_{i=1}^{N} \tilde{R}_j \qquad (6.6.2)$$

where

$$w_{ij} = \frac{K[(x_j - x_i)/b]}{\sum_{i \in s} K[(x_j - x_i)/b]} \qquad (6.6.3)$$

are weights for Kuo's estimator, $K(z) = e^{-z^2/2}$ is a standard normal density (kernel),

$$\tilde{R}_j = \frac{\sum\limits_{i \in s} u_{ji}/\pi_i}{\sum\limits_{i \in s} v_{ji}/\pi_i} \tag{6.6.4}$$

where

$$u_{ji} = w[(x_j - x_i)/b]W[(t - y_i)/b]$$

$$v_{ji} = w[(x_j - x_i)/b] \tag{6.6.5}$$

and $W(z) = \frac{e^z}{1+e^z}$ is the standard logistic distribution function with density $w(z) = \frac{e^z}{(1+e^z)^2}$ and b is the bandwidth parameter used to control the amount of smoothing.

$$V(\hat{F}_K) = \frac{1}{N^2}[\sum_{i=1}^{N} V(\tilde{R}_i) + 2\sum_{i<j=1}^{N} \text{Cov } (\tilde{R}_i, \tilde{R}_j)] \tag{6.6.6}$$

Since $\tilde{R}_j$ is a ratio estimator one can estimate $V(\tilde{R}_j)$ and Cov $(\tilde{R}_i, \tilde{R}_j)$ by standard methods. Under *srswor*,

$$V(\tilde{R}_j)]_{est} = \frac{N-n}{Nn(n-1)\bar{v}_j^2} \sum_{i \in s}(u_{ji} - \hat{R}_j v_{ji})^2$$

and

$$\text{Cov } (\tilde{R}_j, \tilde{R}_k) = \frac{N-n}{Nn(n-1)\bar{v}_j\bar{v}_k} \sum_{i \in s}(U_{ji} - \hat{R}_j v_{ji})(u_{ki} - \hat{R}_k v_{ki}) \tag{6.6.7}$$

$$\bar{v}j = \frac{1}{n}\sum_{i \in s} v_{ji}, \bar{v}_k = \frac{1}{n}\sum_{i \in s} v_{ki}$$

Substituting these in (6.6.6) one gets an unbiased estimator of $V(\hat{F}_k)$. The estimator $v(\hat{F}_K)$ is almost always non-negative and the corresponding confidence interval has good coverage property.

The estimator F_{KO} has been improved upon by Chambers et al (1992). Chambers, Dorfman and Wehrly (1993) considered nonparametric calibration estimator of $F(t)$.

6.7 DESIRABLE PROPERTIES OF AN ESTIMATOR

Kuk (1993), Silva and Skinner (1995) listed the following as desirable properties of an estimator of $F(t)$.

- (i) $\hat{F}(t)$ should have the properties of a distribution function, i.e. it should be monotonically increasing with $\hat{F}(-\infty) = 0$ and $\hat{F}(\infty) = 1$. This property holds for $\hat{F}_0(t), \hat{F}_{ps}(t), \hat{F}_{CD}(t), \hat{F}_{KO}(t)$ and $\hat{F}_K(t)$, However, as noted by Kuk (1993), none of the estimators $\hat{F}_r(t), \hat{F}_d(t), \hat{F}_{RKM}(t)$ is monotonically increasing in general.

- (ii) It is desirable that as y approaches x, value of an auxiliary variable $\hat{F}(t)$ should approach $F(t)$. In particular, if $y = x$, $\hat{F}(t)$ should equal $F(t)$. This property does not hold for $\hat{F}_0(t)$ as it makes no use of x-values. This property holds for each of $\hat{F}_{CD}, \hat{F}_r, \hat{F}_d, \hat{F}_{RKM}$ but not in general for $\hat{F}_K, \hat{F}_{KO}$. If $y_i = x_i \forall i$, then $\hat{F}_{ps}(t) = F(t)$ for $t = x_h, h = 1, \ldots, L$. For other values of t, equality will not hold in general.

- (iii) The estimator should make efficient and flexible use of the auxiliary information. Often the value of x on all the units in the population are not available, but some summary information of these values, eg. in the case of a continuous vriable like age, the number of persons N_g in an age-group in lieu of age of each individual. $\hat{F}_{CD}(t)$ was suggested with this aim in view. $\hat{F}_{ps}(t)$ can also be used with this limited information. The other estimators require individual x-values and cannot be used in these cases. As noted before, each of $\hat{F}_{CD}, \hat{F}_d, \hat{F}_r, \hat{F}_{RKM}$ can be extended to multiple regression model. The extensions of $\hat{F}_{KO}$ and $\hat{F}_K$ do not seem evident.

- (iv) *Simplicity of estimators* Computations of estimators are particularly simple if

$$\hat{F}(t) = \sum_{i \in s} w_i \Delta(t - y_i)$$

where the weights w_i depend only on the label i. This is particularly suitable for surveys with multiple characteristics. $\hat{F}_0, \hat{F}_{ps}, \hat{F}_{KO}$ possess this property. The estimators $\hat{F}_{CD}, \hat{F}_{RKM}, \hat{F}_K$ require intensive calclations.

- (v) *Uniqueness in the definition* The expressions for $\hat{F}_0, \hat{F}_d(d = 1), \hat{F}_r$ are unque. $\hat{F}_{ps}$ depends on choice of the strata. $\hat{F}_{CD}, \hat{F}_{RKM}$

depend on the choice of the model. $\hat{F}_{KO}, \hat{F}_K$ require the specification of the bandwidth b, $\hat{F}_K$ also requiring appropriate scaling of the response variable.

- (vi) *Availability of the variance-estimators* All the above estimators possess variance-estimators.

- (vii) The estimators should have good conditional properties. In particular, it should remain approximately unbiased over variations in values of auxiliary variable x.

6.8   EMPIRICAL STUDIES

We first consider two populations employed by different authors.

(i) Chamber and Dunstan (1980) (CD) population: The population consisted of 33﹡ sugarcane farms covered in the survey of Queensland sugarcane industry, Australia, 1982. The main variables were: $y_{(1)}$ (total cane harvested); $y_{(2)}$ (gross value of cane); $y_{(3)}$ (total farm expenditure). The auxiliary variable was x (area assigned for cane planting). This population with $y = y_{(2)}$ obey model (6.1.2) fairly well.

(ii) Beef cattle population of Chambers et al (1993): The population consisted of 430 farms with 50 or more beef cattle covered in the Australian Agriculture and Grazing Industries Surveys conducted by the Australian Bureau of Agriculture and Resource Economics in 1988 with y as the income from beef and x as the number of beef cattle in each farm. The true model for this population is a quadratic mean function with $v(x) \propto (x+20)^{3/4}$ in (6.1.2).

We denote by $\theta(\alpha)$, the αth population quantile,

$$\theta(\alpha) = \inf \{t : F(t) \geq \alpha\} \qquad (6.8.1)$$

Sometimes estimators are calculated for different quantiles θ_α, $\alpha = 1, \ldots m$. Also, let there be A samples and $\hat{F}^s$ denote the value of $\hat{F}$ on the sample s. Different measures of performance of $\hat{F}(t)$ are:

- (i) Relative Mean Error (RME)

$$= \frac{1}{A} \sum_{s=1}^{A} \mid \hat{F}^s(t) - F(t) \mid / F(t)$$

- (ii) Relative Root Mean Square Error (RRMSE)

$$= \sqrt{\frac{1}{A}\sum_{s=1}^{A}(\hat{F}^s(t) - F(t))^2/F(t)^2}$$

- (iii) Average Absolute Bias (AAB)

$$= \frac{1}{m}\sum_{\alpha=1}^{m} \text{Bias} \{\hat{F}(\theta_\alpha)\}$$

where

$$\text{Bias} \{\hat{F}(t)\} = \frac{1}{A}\sum_{s=1}^{A} | \hat{F}^s(t) - F(t) |$$

- (iv) Average Root Mean Square Error (ARMSE)

$$= \sqrt{\frac{1}{m}\sum_{\alpha=1}^{m}\{ \text{RMSE} (\hat{F}(\theta_\alpha)\}^2}$$

where

$$\text{RMSE} (\hat{F}(t)) = \sqrt{\frac{1}{A}\sum_{s=1}^{A}(\hat{F}^s(t) - F(t))^2}$$

- (v) Maximum Absolute Deviation of $\hat{F}(t)$ for a given sample s (MAD(s))

$$= \max_{\alpha} | \hat{F}^s(\theta_\alpha) - \alpha |$$

One should consider an estimator $\hat{F}'$ optimal if

$$\text{MAD(s)} (\hat{F}') = \min_{\hat{F}} \text{MAD(s)} (\hat{F}),$$

$\hat{F}, \hat{F}' \in \hat{\mathcal{F}}$, a class of estimators.

- (vi) Average Minimum Absolute Error (AMAE)

$$= \frac{1}{m}\sum_{\alpha=1}^{m}\max_{s} | \hat{F}^s(\theta_\alpha) - \alpha |$$

We shall write $t_1 \succ t_2$ with respect to a to denote that the estimator t_1 is better than t_2 with respect to the property a.

Chambers and Dunstan (1986) compared the performances of $\hat{F}_{CD}$ and $\hat{F}_0$ through a simulation study based on samples drawn from the CD-population using the following sampling designs: (a) *srs* (b) stratified random sampling with two strata and proportional allocation (c) same as in (b) but with optimum allocation. Strata boundaries were such as to make the strata sizes (total of x-values) constant over strata. For each sampling design, 1000 samples each of size $n = 30$ were drawn. From each sample, estimates of $F(t)$ for the quantiles $t = \theta_N(1/4), \theta_N(1/2), \theta_N(3/4)$ (where $\theta_N(\alpha) = \inf\{t : F(t) \geq \alpha\}$) were calculated for each of the study variables. $\hat{F}_{CD}(t)$ was found to be better than $\hat{F}_0(t)$ in terms of RRMSE. However, $\hat{F}_{CD}$ was slightly more biased than $\hat{F}_0$.

The 1000 samples were ordered by their x-sample means, split into 20 groups of size 50 each and RME of the estimators were calculated for each group. The RME of $\hat{F}_{CD}$ remained approximately unaffected over variations in $\bar{x}$ whereas $\hat{F}_0$ showed a linear decreasing trend.

Rao et al (1990) compared RME and RMSE of $\hat{F}_0, \hat{F}_r, \hat{F}_{RKM}$ and $\hat{F}_{CD}$ based on (6.1.2) with $v(x_i) = \sqrt{x_i}$ on the basis of CD-population with y as $y_{(2)}$. Sampling designs, number of samples, sample sizes and quantiles $\theta(\alpha)$ were same as in CD-simulation study. It was found that

$$\hat{F}_d, \hat{F}_{RKM} \succ \hat{F}_r \text{ (specially, for small d) } \succ \hat{F}_{CD}$$

with respect to RME. Also,

$$\hat{F}_{RKM} \succ \hat{F}_d \succ \hat{F}_{s_n} \text{ for srs}$$

$$\hat{F}_d \succ \hat{F}_r$$

with respect to RMSE. The model-based estimator $\hat{F}_{CD}$ was found to be significantly more efficient than the design-based $\hat{F}_0(= \hat{F}_{s_n}$ for *srs*), $\hat{F}_d, \hat{F}_r, \hat{F}_{RKM}$, since the data seemed to obey the model (6.1.2).The conditional performance of the estimators was studied as in Chambers and Dunstan (1986). It was found that the RME of $\hat{F}_d, \hat{F}_{CD}, \hat{F}_{RKM}, \hat{F}_r$ remained more or less stable over variations in $\bar{x}($ RME $(\hat{F}_{RKM}) \in [-.03, .03]$, RME $(\hat{F}_r)$, RME $(\hat{F}_d)$ varied over $[-.09, .05]$, RME $(\hat{F}_{CD}) \approx -.04)$, while that of $\hat{F}_0$ showed a linear trend (RME $(\hat{F}_{s_n}) \in [-.2, .2]$). The authors also studied the performance of these estimators with respect to Hansen, Madow and Tepping (1983) population.

Silva and Skinner (1995) considered Monte Carlo comparison of $\hat{F}_{CD}, \hat{F}_d, \hat{F}_r, \hat{F}_{RKM}, \hat{F}_K, \hat{F}_{KK}, \hat{F}_{ps}$ by selecting 1000 samples of sizes 30 and 50 by *srswor* from each of CD- population with y as income and Chambers et al (1993)-beef population. Three alternative schemes of post-stratification were used:

(a)The choice $x_{(1)} < \ldots < x_{(L-1)}$ such that $N_h = N \ \forall \ h = 1, \ldots, L$

(b) The choice for which

$$\sum_{i \in \mathcal{P}_h} \sqrt{x_i} = \sum_{i=1}^{N} \sqrt{x_i}/L \ \forall \ h$$

(c) The choice for which

$$\sum_{i \in \mathcal{P}_h} x_i = \sum_{i=1}^{N} x_i/L \ \forall \ h$$

For each sample estimates were calculated for 11 different quantiles $\theta(\alpha), \alpha = \frac{1}{12}, \ldots, \frac{11}{12}$. The numerical study indicated there was considerable gain in efficiency for $\hat{F}_{ps}$ over $\hat{F}_0$. For these populations four seemed to be the optimum number of strata. Also AAB($\hat{F}_{ps}$) was small. $\hat{F}_{ps}$ was found to be better than $\hat{F}_r, \hat{F}_d, \hat{F}_K$ and worse than $\hat{F}_{CD}, \hat{F}_{RKM}$ from the point of view of ARMSE.

Kuk (1993) compared $\hat{F}_0, \hat{F}_{RKM}, \hat{F}_{CD}$ (both based on model (6.1.2) with $v(x) = \sqrt{x}$) and $\hat{F}_K$ on the basis of CD-population with $y = y_{(2)}$ and beef population . Samples of size n = 30 were drawn using *srs*, stratified random sampling with x-stratification and proportional allocation, and *ppswr* from CD-data. For each sample $\hat{F}(\theta_\alpha)$ was computed for $\alpha = \frac{1}{12}, \ldots, \frac{11}{12}$. The criteria of comparison were average bias, ARMSE and AMAE . It followed that $\hat{F}_K \succ \hat{F}_0, \hat{F}_{RKM}$ with respect to ARMSE and AMAE. However, $\hat{F}_K$ was not as efficient as $\hat{F}_{CD}$. The conditional behaviour of the estimators was studied by splitting the 200 ordered (according to $\bar{x}$-values) samples into 10 groups of equal size and then calculating the conditional bias for each group. The conditional relative bias of $\hat{F}_K$ was found to be small and invariant while $\hat{F}_0$ exhibited trend over the variation of $\bar{x}$. Since the data were highly skewed, transformed variables $x' = x^{1/4}$ and $y' = (y/100)^{1/4}$ were used. Samples, 200 in number, each of sizes 30, 60 and 90 were drawn by *pps* and $\hat{F}_{CD}, \hat{F}_{RKM}$ (both based on (6.1.2) with $v(x) = x$) were calculated for $\alpha = \frac{1}{2}, \frac{3}{4}$. It was found that for $\alpha = \frac{3}{4}, \hat{F}_{CD} \succ \hat{F}_K \succ \hat{F}_{RKM}$ with respect to RMSE. For $\alpha = \frac{1}{2}, \hat{F}_{CD}$ was inferior to $\hat{F}_K$ and $\hat{F}_{RKM}$. For $\alpha = \frac{1}{4}$, the relative bias of $\hat{F}_{CD}$ was close to 20% with root mse much larger than that of $\hat{F}_K$ which was best. The bias of $\hat{F}_{CD}$ remained constant over changes in sample sizes meaning $\hat{F}_{CD}$ was not asymptotically m-unbiased with respect to the data. In terms of MAE, $\hat{F}_K$ was comparable to $\hat{F}_{CD}$ for n = 30 and was better than $\hat{F}_0, \hat{F}_{CD}, \hat{F}_{RKM}$ for n = 90. With respect to conditional relative bias $\hat{F}_K$ was better than each of $\hat{F}_{CD}, \hat{F}_{RKM}$ and $\hat{F}_0$.

Dorfman (1993) made empirical comparison of $\hat{M}_{CD}, \hat{M}_{RKM*}$ (based on a quadratic model, $y_i = \alpha + \beta x_i + \gamma x_i^2 + \sqrt{x_i}\epsilon_i, \mathcal{E}(\epsilon_i) = 0, \mathcal{V}(\epsilon_i) = \sigma^2$), $\hat{M}_{RKM}$ and $\hat{M}_0$ where $M(t) = \frac{1}{N-n} \sum_{i \in s} \Delta(t - y_i)$ and $\hat{M}_{..}$ is obtained from $\hat{F}_{..}$ (for example, $\hat{M}_{RKM*}$ is obtained from $\hat{F}_{RKM*}$ 'in (6.3.19)). Simple random samples, 1000 in number, each of size $n = N/10$ were selected from each of five populations based on data collected in beef cattle population and the above estimators were calculated for each sample at each quartile of the population. It was found that $\hat{M}_0$ was invariably less efficient than $\hat{M}_{RKM}$ or $\hat{M}_{RKM*}$, which were very close, indicating that the last two estimators were stable under changes of the model. $\hat{M}_{CD}$ was far better than the other estimators in three populations while in two other populations $\hat{H}_{RKM}$ fared better than $\hat{M}_{CD}$ particularly for $\alpha = 1/4$. The same trend was found for $n = N/5$.

Dunstan and Chambers (1989) compared the performance of $\hat{F}_{CD}(t)$, $\hat{F}_{CD}^{(L)}(t)$ and $\hat{F}_0(t)$ as well as their variance estimators $v(\hat{F}_{CD}), v(\hat{F}_{CD}^{(L)})$, $v(\hat{F}_o)$ (with $v(x) = \sqrt{x}$ in (6.1.2) everywhere) on the basis of 1000 samples each of size 30 drawn independently from CD-population using the following sampling dsigns: (a) *srswor* with post-stratification into (i) six post-strata, suitably defined (ii) three post-strata formed by collapsing six post-strata into pairs (b) stratified random sampling with proportinal allocation using the six post-strata in (a i) above as strata. The estimates for each of $y_{(1)}, y_{(2)}, y_{(3)}$ were calculated for each sample for all the quantiles along with their variance estimates. The criterion was repeated sampling average, $\bar{\hat{F}}(t) = \frac{1}{1000} \sum_{s=1}^{1000} \hat{F}^s(t)$ and RMSE. It was found that $\hat{F}_{CD}$ and $\hat{F}_{CD}^{(L)}$ had very similar performance, both having slight repeated sampling bias and having RMSE smaller than $\hat{F}_0$. The performance of $v(\hat{F})$ was assessed by comparing $\frac{1}{1000} \sum_{s=1}^{1000} \sqrt{v^s(\hat{F})}$ with RMSE($\hat{F}$) where $v(\hat{F}^s)$ is the value of $v(\hat{F})$ for the sth sample and by checking the coverage property of the confidence intervals generated by these strategies. The coverage was closest to its nominal value for all the three estimators at $\alpha = 1/2$.

6.9   BEST UNBIASED PREDICTION (BUP) UNDER GAUSSIAN SUPERPOPULATION MODEL

Bolfarine and Sandoval (1993) considered best unbiased prediction (BUP) of $F(t)$ under multiple regression model with errors having a Gaussian distribution. Consider the model

$$\mathbf{y} = X\beta + e, \quad e \sim N_N(0, \sigma^2 W) \tag{6.9.1}$$

where $\mathbf{y} = (y_1, \ldots, y_N)', e = (e_1, \ldots, e_N)', X = ((x_{ij}, i = 1, \ldots, N; j = 1, \ldots, p))_{N \times p}, x_{ij}$ being the value of variable x_j on unit i, $\beta = (\beta_1, \ldots, \beta_p)'$, a vector of p regression coefficients, σ^2 a known constant and W a $N \times N$ known diagonal matrix. We shall denote by y_s, X_s, e_s, W_s the parts of $\mathbf{y}, X, e, W$, respectively, corresponding to sample s, the same symbols with r in place of s will denote the parts corresponding to non-sampled units. If the sample is drawn by *srs*, then the model (6.9.1) holds for the sampled elements as well.

Rodrigues et al (1985) considered the following definition and proved Theorem 6.9.1 in the case of survey sampling.

DEFINITION 6.9.1 *Complete and Totally Sufficient Statistics* A statistic $S = S(y_s)$ is said to be totally sufficient for the family $\{\xi_\theta, \theta \in \Theta\}$ where ξ_θ is the *pdf* of y_s depending on some unknown parameter θ, if (i) the conditional distribution of y_s given S is independent of θ (ii) y_s, y_r are conditionally independent given S.

A totally sufficient statistic S is said to be complete if the induced family ξ^* of sampling distributions of S is complete.

Condition (ii) means that S contains all information contained in y_s about y_r. In case y_s and y_r are independent (ii) always holds.

THEOREM 6.9.1 Let $S = S(y_s)$ be a complete and sufficient statistic for θ which is also totally sufficient. If $\hat{\theta}(y_s) = \theta(y_s) + \hat{\theta}_{rs}(S)$ is ξ-unbiased for $\theta(\mathbf{y}) = \theta(y_s) + \theta_{rs}(\mathbf{y})$ then $\hat{\theta}(y_s)$ is the unique best unbiased predictor (BUP) (in the sense of having minimum ξ-variance among all ξ-unbiased predictors) for $\theta(\mathbf{y})$.

Clearly, if we have an m-unbiased predictor of $F(t)$ which depends on y_s

only through S, then it is the BUP of $F(t)$. Under model (6.9.1),

$$\hat{\beta}_s = (X_s' W_s^{-1} X_s)^{-1} X_s' W_s^{-1} y_s \qquad (6.9.2)$$

is complete and sufficient for β and since W is diagonal it is also totally sufficient and complete for β. The BUP of $F(t)$ is then obtained using Theorem 6.9.1 and some results about estimation of $F(t)$ in infinite population due to Olkin and Ghurye (1969).

THEOREM 6.9.2 Under model (6.9.1) the BUP of $F(t)$ is given by

$$\hat{F}_{BU}(t) = \frac{n}{N} F_{s_n}(t) + \frac{1}{N} \sum_{i \in r} \Phi \left(\frac{t - x_i' \hat{\beta}_s}{\sigma \sqrt{w_i} \sqrt{1 - w_i^{-1} x_i' (X_s' W_s^{-1} X_s)^{-1} x_i}} \right)$$

$$(6.9.3)$$

provided

$$1 - x_i' (X_s' W_s^{-1} X_s)^{-1} x_i / w_i > 0 \ \forall i \in r$$

where Φ is the distribution function of the standard normal deviate and $W = \text{Diag.} (w_1, \ldots, w_N)$ and $x_i = (x_{i1}, \ldots, x_{ip})'$.

PROOF. When $W = I_N$, according to Olkin and Ghurye (1969),

$$E \left[\Phi \left(\frac{t - \mathbf{x}_i' \hat{\beta}_s}{\sigma \sqrt{1 - \mathbf{x}_i' (X_s' X_s)^{-1} x_i}} \right) \right] = E[\Delta(t - y_i)]$$

which shows that if $x_i' (X_s' X_s)^{-1} x_i < 1 \ \forall \ i \in r$, $\hat{F}_{BU}(t)$ is an unbiased estimator of $F(t)$ by (6.3.1). Since $\hat{F}_{BU}(t)$ is a function of sufficient statistic the result follows by Theorem 6.9.1. The result for $W = \text{Diag.} (w_1, \ldots, w_N)$ follows from the case $W = I_N$ by making the transformation $y_i^* = y_i / \sqrt{w_i}$, $x_i^* = x_i / \sqrt{w_i}$, $e_i^* = e_i / \sqrt{v_i}$.

EXAMPLE 6.9.1

Suppose $X = 1_N, W = I_N$ in model (6.9.1). The complete and totally sufficient statistic is $\hat{\beta}_s = \bar{y}_s$. in this case

$$\hat{F}_{BU}(t) = \frac{n}{N} \hat{F}_{s_n}(t) + (1 - \frac{n}{N}) \Phi (\sqrt{\frac{n}{n-1}} (\frac{t - \bar{y}_s}{\sigma})) \qquad (6.9.4)$$

$$= \hat{F}^*(t) \quad (\text{say})$$

NOTE 6.9.1

Under model (6.9.1) with $X = 1_N, W = I_N, \hat{F}_{s_n}(t)$ is a m-unbiased predictor of $F(t)$. If ξ is a family of continuous distributions (not necessarily normal), $\hat{F}_{s_n}(t)$ being a symmetric function of order statistic $y_{(s)}$ (order statistic corresponding to y_s) is a totally sufficient statistic and hence, by Theorem 6.9.1 is BUP for $F(t)$.

Considering the model of example 6.9.1 under assumptions that the finite populatiuon sequence $\mathcal{P}_\nu$ of size N_ν is an increasing sequence such that as $\nu \to \infty, N_\nu - n_\nu \to \infty$ with $n_\nu/N_\nu \to f, f \in [0, 1]$, and applying Lindeberg-Levy CLT,

$$\sqrt{n_\nu}(\hat{F}^*(t) - F_{N\nu}(t)) \underset{\to}{L} N(0, \ (1-f)^2[\phi^2(\frac{t-\beta}{\sigma})$$

$$+\frac{f}{1-f}\Phi(\frac{t-\beta}{\sigma})][1 - \Phi(\frac{t-\beta}{\sigma})] \qquad (6.9.5)$$

and

$$\sqrt{n_\nu}(\hat{F}_{s_\nu}(t) - F_{N_\nu}(t)) \underset{\to}{L} N(0, (1-f)\Phi(\frac{t-\beta}{\sigma})[1 - \Phi(\frac{t-\beta}{\sigma})] \qquad (6.9.6)$$

where $\phi(.)$ is the density function of a standard Normal distribution, $\hat{F}^*(t)$ is obtainable from (6.9.4) and $F_{N_\nu}(t), \hat{F}_{s_\nu}(t)$ are, respectively, population d.f. and sample d.f. for $\mathcal{P}_\nu$. From (6.9.5) and (6.9.6), asymptotic relative efficiency of $\hat{F}_{s_\nu}(t)$, with respect to $\hat{F}_{BU}(t)$ is

$$\text{ARE } (F_{s_\nu}(t) : \hat{F}_{BU}(t)) = \frac{APV(\hat{F}_{BU}(t))}{APV(\hat{F}_{s_\nu}(t))}$$

$$= f + \frac{(1-f)\phi^2((t-\beta)/\sigma)}{\Phi(\frac{t-\beta}{\sigma})(1 - \Phi(\frac{t-\beta}{\sigma}))} \qquad (6.9.7)$$

where APV denotes asymptotic prediction variance. For the case $t = \beta$ this reduces to

$$ARE = f + .637(1 - f)$$

Again, ARE is a decreasing function of $| t - \beta |$ and as $| t - \beta | \to \infty, ARE \to f$.

6.9.1 EMPIRICAL STUDY

Considering model (6.9.1) with $X = (x_1, \ldots, x_N)', W = \text{Diag.}(w_1, \ldots, w_N)$ and a Gaussian distribution of error, CD-type estimator, obtained by using Royall's approach is

$$\hat{F}''_{CD}(t) = \frac{n}{N} F_{s_n}(t) + \frac{1}{N} \sum_{i \in r} \Phi(\frac{t - x'_i \hat{\beta}_s}{\sigma \sqrt{v_i}})$$

which is closely related to BUP $\hat{F}_{BU}(t)$, especially if n is large.

Bolfarine and Sandoval compared $\hat{F}_{BUP}, \hat{F}''_{CD}, \hat{F}_{s_n}$ and $\hat{F}_r$ on the basis of 1000 *srs* each of size n = 10 drawn from a population of size N = 1000 generated according to the model

$$y_i = 3x_i + e_i \ (i = 1, \dots, 100)$$

$e_i \sim N(0, \ 8^2 x_i)$, the x_i's being generated according to Uniform (10, 200). For each sample estimates of quartiles $\hat{F}(t)$ were calculated for $t = \theta(1/4), t = \theta(1/2), t = \theta(3/4)$. The estimates were compared with respect to repeated sampling mse. It was found that

$$\hat{F}_{BUP}, \hat{F}''_{CD} \succ\succ \hat{F}_{s_n} \succ \hat{F}_r$$

$$\hat{F}_d, \hat{F}_{RKM} \succ \hat{F}_{s_n}, \hat{F}_r$$

The estimator $\hat{F}_{CD}$ performed closely with $\hat{F}_{BUP}$; performance of $\hat{F}_{CD}$ was poor for $\alpha = 1/4$. One may, therefore, conclude that under normal super-population models, the model-based predictors provide improvement over design-based predictors, specially, for small values of α.

As in Chambers and Dunstan (1986), 1000 samples were ordered according to $\bar{x}_s$ values and divided into 20 groups of 50 samples each. The average bias

$$\frac{1}{50} \sum_{s=1}^{50} (\hat{F}^s(t) - F(t))$$

was plotted against the $\bar{x}_s$ values. It was found that $\hat{F}_r$ was more affected by variation in $\bar{x}_s$ -values than were $\hat{F}_{BUP}$ and $\hat{F}_{CD}$.

Similar studies with large sample prediction variance (as in (6.9.7)) showed that variance decreased as $\bar{x}_s$ increased. The optimum sampling design is, therefore, to choose a sample with the largest $\bar{x}_s$ values with probability one.

6.10   ESTIMATION OF MEDIAN

Since many real life populations are highly skewed, the estimation of median is often of interest. Kuk and Mak (1989) suggested the following method for estimating the finite population median $M_y = M = \theta(1/2)$. In the absence of auxiliary information x, a natural estimator of M is sample median,

$$\hat{M}_{s_n} = m_y \qquad (6.10.1)$$

When the values of the auxiliary variable x are available, the ratio estimator of M_y is

$$\hat{M}_r = \frac{m_y}{m_x} M_x \qquad (6.10.2)$$

Let $y_{(1)} \leq \cdots \leq y_{(n)}$ be the ordered values of y in s. Let i_0 be an integer such that

$$y_{(i_0)} \leq M_y \leq y_{(i_0+1)}$$

and $p = i_0/n$. Thus M_y is approximated by the pth sample quantile Z_p. Since M_y is unknown p is unobservable. If $\hat{p}$ is a guessed value of p, an estimate of M is

$$\hat{M}_{(p)} = Z_{\hat{p}}$$

Let n_x be the number of units in the sample with x values $\leq M_x$. Let P_{11} be the proportion of population values with y-values $\leq M_y, x-$ values $\leq M_x$; P_{12} the same with y-values $\leq M_y, x$ values $> M_x$; P_{21} the same with y-values $> M_y$ and x-values $\leq M_x$ and $P_{22} = 1 - P_{11} - P_{12} - P_{21}$. If P_{ij}'s are known, an estimate of p is

$$\hat{p}_0 = \frac{1}{n}\{\frac{n_x P_{11}}{P_{01}} + \frac{(n - n_x)P_{12}}{P_{02}}\}$$

$$\approx \frac{2}{n}[n_x P_{11} + (n - n_x)(\frac{1}{2} - P_{11})] \qquad (6.10.3)$$

since $P_{01} \approx \frac{1}{2}, P_{10} \approx \frac{1}{2}$. In practice, the P_{ij}'s are usually unknown and are estimated by the sample proportion p_{ij} obtained by similar cross-classification of the values in the sample against the sample median $m_y = m$ and m_x.

Therefore, from (6.10.3), a sample-based estimate of p is

$$\hat{p}_1 \simeq \frac{2}{n}\{n_x p_{11} + (n - n_x)(1/2 - p_{11})\}$$

and an estimator of M_y is

$$\hat{M}_P = \hat{M}_{yp} = Z_{\hat{p}_1} \qquad (6.10.4)$$

and is referred to as the 'position estimator'.

Another estimator of M_y is derived from Kuk and Mack (1989) estimator of $d.f.$ as

$$\hat{M}_{(KM)} = \inf\{t : \hat{F}_{KM}(t) \geq 1/2\}.$$

We consider now asymptotic properties of the estimates. Assume that as $N \to \infty, n/N \to f \in [0, 1]$ and the distribution of (X, Y) approaches a bivariate continuous distribution with marginal densities $f_X(x)$ and $f_Y(y)$, respectively, and that $f_X(M_x) > 0, f_Y(M_y) > 0$. Under these conditions, the sample median m_y is consistent and asymptotically normal with mean M_y and variance

$$\frac{1-f}{4n} \frac{1}{\{f_Y(M_y)\}^2} = \sigma_y^2$$

(Gross, 1980). It follows that the asymptotic distribution of $(m_x - M_x, m_y - M_y)$ is bivariate normal with mean zeroes and variances σ_y^2, σ_x^2 (defined similarly) and covariance

$$\sigma_{xy} = \frac{(1-f)(P_{11} - \frac{1}{4})}{n f_x(M_x) f_y(M_y)} \qquad (6.10.5)$$

Now,

$$\hat{M}_r - M_y = \frac{M_x m_y - M_y m_x}{m_x}$$

Since, $m_x/M_x \to 1, \hat{M}_r - M_y$ has the same distribution as

$$\frac{M_x m_y - M_y m_x}{M_x} = m_y - M_y - \frac{M_y}{M_x}(m_x - M_x)$$

Thus, $\hat{M}_r - M_y$ is asymptotically normal with mean 0 and variance

$$\sigma_y^2 + (\frac{M_y}{M_x})^2 \sigma_x^2 - 2(\frac{M_y}{M_x})\sigma_{xy}$$

Consequently, $\hat{M}_r$ is asymptotically more efficient than m_y if

$$\rho_c > \frac{1}{2}[\{f_x(M_x)\}^{-1}M_x^{-1}]/[\{f_y(M_y)\}^{-1}M_y^{-1}]$$

where $\rho_c = 4(P_{11} - \frac{1}{4}) \in [-1, 1]$ as $P_{11} \in [0, \frac{1}{2}]$. The quantity P_{11} can be regarded as a measure of concordance between x and y. Similarly, the authors considered asymptotic distribution of $\hat{M}_{(p)}$ and $\hat{M}_{KM}$ both of which are found to be more efficient than $\hat{M}_{s_n}$.

In an empirical study the authors show that for populations showing a strong linear relationship between x and $y, \hat{M}_r, \hat{M}_P, \hat{M}_{(KM)}$ perform considerably better than m_y. However, if the correlation coefficient between x and y is week (P_{11} small), $\hat{M}_r$ performs very poorly while $\hat{M}_P, \hat{M}_{(KM)}$ retain their superiority relative to m_y.

Two estimators of ξ due to Kuk (1988) are

$$\hat{\xi}_L = \hat{F}_L^{-1}(1/2), \ \hat{\xi}_R = \hat{F}_R^{-1}(1/2)$$

Again,

$$\hat{F}_\lambda = \lambda \hat{F}_L(t) + (1 - \lambda)\hat{F}_R(t), \ 0 < \lambda < 1 \qquad (6.10.6)$$

is also an estimator of $F(t)$. An estimator of ξ is, therefore,

$$\hat{\xi}_\lambda = \hat{F}_\lambda^{-1}(1/2)$$

Behaviour of $\hat{\xi}_\lambda$ depends largely on the behaviour of $\hat{F}_\lambda$ near ξ. Now,

$$V\{\hat{F}_\lambda(\xi)\} = \lambda^2 V\{\hat{F}_L(\xi)\}^2 + (1 - \lambda)^2 V\{\hat{F}_R(\xi)\}$$

$$+2\lambda(1 - \lambda) \ \text{Cov} \ \{\hat{F}_L(\xi), \hat{F}_R(\xi)\}$$

The optimal value of λ is

$$\lambda^* = \sum_{i=1}^{N} b_i \Delta(\xi - y_i) / \sum_{i=1}^{N} b_i \qquad (6.10.7)$$

(b_i has been defined in (6.2.19)). Assuming that the ordering of y-values agrees with that of x, an estimate of λ^* is

$$\hat{\lambda}^* = \sum_{i=1}^{N} b_i \Delta(\eta - x_i) / \sum_{i=1}^{N} b_i \qquad (6.10.8)$$

where η is the median of x. Therefore,

$$\hat{\xi}_{\hat{\lambda}^*} = \hat{F}_{\hat{\lambda}^*}^{-1/2}(1/2) \qquad (6.10.9)$$

If $a_L(t)$ and $a_R(t)$ denotes the mse's of $\hat{F}_L(t)$ and $\hat{F}_R(t)$, respectively, then it can be shown from that

$$\frac{a_L(\xi) - a_R(\xi)}{a_L(\infty)} = 1 - 2\lambda^*$$

Hence,

$$\frac{b_L(\xi) - b_R(\xi)}{b_L(\infty)} = 1 - 2\hat{\lambda}^*$$

where $b_L(t), b_R(t)$ denote the mse's of $\hat{G}_L(t)$ and $\hat{G}_R(t)$, respectively, $\hat{G}_L(t) = \hat{F}_L(t)\Big]_{y_i = x_i}$ and similarly for $\hat{G}_R(t)$.

Empirical studies reported that $\hat{\xi}_R$ is considerably better than $\hat{\xi}_L$ and $\hat{\xi}_\nu(t)$ (in conformity with the result $\hat{F}_R$ is always better than $\hat{F}_L$ and $\hat{F}_\nu$, defined in (6.2.11)). The performance of $\hat{\xi}_\lambda$ and $\hat{\xi}_{\hat{\lambda}}$ are usually at least as good as that of $\hat{\xi}_R$.

Since $\hat{F}_{CD}(t)$ is a monotonically non-increasing function of t, Chambers and Dunstan (1986) obtained estimation of $\theta_N(\alpha)$ as

$$\hat{\theta}_{N;CD}(\alpha) = inf\{t; \hat{F}_{CD}(t) \geq \alpha\} \tag{6.10.10}$$

Since, $\hat{F}_{CD}(t)$ is asymptotically unbiased under (6.1.2), $\hat{\theta}_{N;CD}$ is also so. From Serfling (1980, Theorem 2.5.1) one can note the Bahadur representation of $\theta_N(\alpha)$ as

$$\theta_N(\alpha) = \theta(\alpha) + [\alpha - F_N\{\theta(\alpha)\}]/e_N\{\theta(\alpha)\} + o_p(N^{1/2})$$

where $\theta(\alpha)$ is defined by $E[F_N(\theta(\alpha))] = \alpha$ and $e_N(t) = \frac{d}{dt}E\{F_N(t)\}$. Assuming a similar representation for $\hat{\theta}_{N;CD}(\alpha)$ for N, n large,

$$\hat{\theta}_{N;CD}(\alpha) = \theta(\alpha) + [\alpha - \hat{F}_{CD}\{\theta(\alpha)\}]/e_N\{\theta(\alpha)\} + o_p(n^{-1/2})$$

asymptotic variance of $\hat{\theta}_{N;CD}(\alpha) - \theta(\alpha)$, following Theorem 6.5.1, is

$$(1 - n/N)^2 [W_r^*\{\theta(\alpha)\} + W_r\{\theta(\alpha)\}]/e_N^2\{\theta(\alpha)\} \tag{6.10.11}$$

Rao et al (1990) obtained ratio estimator of $\theta_N(\alpha)$ as

$$\hat{\theta}_{N;r}(\alpha) = \frac{\hat{\theta}_{N(y)}(\alpha)}{\hat{\theta}_{N(x)}(\alpha)}\theta_N(x)(\alpha) \tag{6.10.12}$$

where

$$\hat{\theta}_{N(y)}(\alpha) = inf\{t; \hat{F}_y(t) \geq \alpha\}$$

$$\hat{\theta}_{N(x)}(\alpha) = inf\{t; \hat{F}_x(t) \geq \alpha\},$$

and $\theta_x(\alpha) = inf\{t; F_x(t) \geq \alpha\}$ is the known finite population α-quantile for x. Similarly, a difference estimator for $\theta(\alpha)$ is

$$\hat{\theta}_d(\alpha) = \hat{\theta}_y(\alpha) + \hat{R}\{\theta_x - \hat{\theta}_x(\alpha)\} \tag{6.10.13}$$

where $\hat{R}$ is defined in (6.2.3). Both $\hat{\theta}_r(\alpha)$ and $\hat{\theta}_d(\alpha)$ have ratio estimation property.

Rao et al (1990) compared the RME and RRMSE of $\hat{\theta}_0(\alpha)$, $\hat{\theta}_d(\alpha), \hat{\theta}_r(\alpha)$ for $\alpha = 1/4, 1/2$ and 3/4 on the basis of samples drawn from

CD-population by (i) simple random sampling and (ii) stratified random sampling with x-stratification and proportional allocation. The relative bias of all the estimators was found to be small. For simple random sampling, $\hat{\theta}_r(\alpha)$ and $\hat{\theta}_d(\alpha)$ were found to be considerably more efficient than $\hat{\theta}_0(\alpha)$ with respect to RRMSE while their performance were almost identical for stratified random sampling as above. The conditional relative mean error of $\hat{\theta}_r$ and $\hat{\theta}_d$ remained more or less stable for variations in $\bar{x}$ while that of θ_0 showed linear trends for $\alpha = 1/2$. Rao et al also considered variance estimates of these estimates. Sedransk and Meyer (1978) considered confidence intervals for the quantiles of a finite population under simple random sampling and stratified random sampling. Some other references on estimation of quantiles are McCarthy (1965), Loynes (1966), Meyer (1972), Sedransk and Meyer (1978), David (1981), Sedransk and Smith (1983), Meeden (1985), Francisco and Fuller (1991) and Bessel et al (1994).

Chapter 7

Prediction in Finite Population under Measurement Error Models

7.1 INTRODUCTION

In practical sample survey situations the true values of the variables are rarely observed but only values mixed with measurement errors. Consider again a finite population $\mathcal{P}$ of a known number N of identifiable units labelled $1, \ldots, i \ldots, N$. Associated with i is a value y_i of a study variable 'y'. We assume that y_i cannot be observed correctly but a different value Y_i which is mixed with measurement errors is observed.

We also assume that the true value y_i in the finite population is actually a realisation of a random variable $\mathcal{Y}_i$, the vector $\mathcal{Y} = (\mathcal{Y}_1, \ldots, \mathcal{Y}_N)'$ having a joint distribution ξ. However, both y_i and $\mathcal{Y}_i$ are not observable and we cannot make any distinction between them. Our problem is to predict the population total T $(= \sum_{i=1}^{N} y_i)$ (population mean $\bar{y} = T/N$), the population variance $S_y^2 (= \sum_{i=1}^{N} (y_i - \bar{y})^2/N)$ or the population distribution function $F_N(t) = \frac{1}{N} \sum_{i=1}^{N} \Delta(t - y_i)$ by drawing a sample according to a sampling design $p(s)$, observing the data and using ξ. Note that in the previous chapters we used Y_i to denote the random variable corresponding to y_i. In this chapter we shall use Y_i to denote the observed value of y on unit i and

y_i will denote both the true value of y on unit i and the random variable corresponding to it.

A general treatment for inference under additive measurement error models has been considered in Fuller (1987, 1989) and the same for multiplicative measurement error models in Hwang (1986). In section 7.2 we review the prediction problems in finite population under additive measurement error models. The next section considers the problems under multiplicative measurement error models.

7.2   ADDITIVE MEASUREMENT ERROR MODELS

7.2.1   THE LOCATION MODEL WITH MEASUREMENT ERRORS

Consider the simple location model with measurement error:

$$y_i = \mu + e_i, \ e_i \sim (0, \ \sigma_{ee}),$$

$$Y_i = y_i + u_i, \ u_i \sim (0, \ \sigma_{uu}), \tag{7.2.1}$$

$$e_i \underset{\sim}{ind} u_j \ (i, j = 1, 2, \ldots)$$

where $\mu, \sigma_{ee}(> 0), \sigma_{uu}(> 0)$ are constants and $z_i \sim (0, A)$ denotes the random variables z_i are iid with mean zero and variance A. The model (7.2.1) was considered by Bolfarine (1991), Mukhopadhyay (1992, 1994 a), Bhattacharyya (1997), among others. Here , and subsequently, $E, V(E_p, V_p)$ will denote expectation, variance with respect to superpopulation model (design).

Consider the class of linear predictors

$$e(s, Y_s) = b_s + \sum_{k \in s} b_{ks} Y_k \tag{7.2.2}$$

where b_s, b_{ks} are constants not depending on Y-values and $Y_s = \{Y_i, \ i \in s\}$ denotes the set of observations on y on the units in s.

As noted in definition 2.2.2, a predictor g is said to be design-model (pm-) unbiased predictor of $\theta(\mathbf{y})$ or pm-unbiased estimator of $E(\theta(\mathbf{y}))$ where $\mathbf{y} = (y_1, \ldots, y_N)$ if

$$E_p E(g(s, Y_s)) = E(\theta(\mathbf{y})), \tag{7.2.3}$$

for all the possible values of the parameters involved in the model. Hence, $e(s, Y_s)$ will be pm-unbiased iff

$$E_p E(b_s + \sum_{k \in s} b_{ks} Y_k) = E(\bar{y}) = \mu$$

ie. iff

$$E_p(b_s) = 0 \tag{6.2.4.1}$$

$$E_p(\sum_{k \in s} b_{ks}) = 1 \tag{7.2.4.2}$$

Following the usual variance-minimisation criterion, a predictor g^* will be said to be optimal in a class of predictors G for predicting $\theta(\mathbf{y})$ for a fixed p, if

$$E_p E(g^* - \theta)^2 \leq E_p E(g - \theta)^2 \tag{7.2.5}$$

for all $g \in G$.

To find an optimal pm-unbiased predictor of $\bar{y}$, we consider the following theorem on UMVU-estimation (Rao, 1973). Let C denote a class of *pm*-unbiased estimators of τ and C_0 the corresponding class of *pm*-unbiased estimators of zero.

THEOREM 6.2.1 A predictor g^* in C is optimal for τ iff for any f in $C_0, E_p E(g^* f) = 0$.

From the above theorem, Theorem 7.2.2 readily follows.

THEOREM 6.2.2 Under model (7.2.1) optimal pm-unbiased predictor of $\bar{y}$ in the class of all linear pm-unbiased predictors, where $p \in \rho_n$, is given by $\bar{Y}_s (= \frac{1}{n} \sum_{k \in s} Y_k)$. Again, any $p \in \rho_n$ is optimal for using $\bar{Y}_s$.

Again, if $\mathcal{V}$ denotes variance operation with respect to joint operation of model (7.2.1) and s.d.p,

$$\mathcal{V}(\bar{Y}_s - \bar{y}) = E_p V(\bar{Y}_s - \bar{y}) + V_p E(\bar{Y}_s - \bar{y})$$

$$= (\frac{1}{n} - \frac{1}{N}) \sigma_{ee} + \frac{1}{n} = \frac{r_0}{N^2} \text{ (say)} \tag{7.2.6}$$

Theorem 7.2.2 states that any $FS(n-)$ design including a purposive sampling design is optimal for predicting $\bar{y}$. However, for purpose of robustness under model- failures (as shown in a different context by Godambe and Thompson (1977)) one should use a probability sampling design $p \in \rho_n$ along with $\bar{Y}_s$.

We now consider optimal prediction of S_y^2. For this we shall confine to the class of pm-unbiased quadratic predictors

$$e_q(s, Y_s) = b_s + \sum_{k \in s} b_{ks} Y_k^2 + \sum_{k \neq k' \in s} b_{kk's} Y_k Y_{k'}$$

where $b_s, b_{ks}, b_{kk's}$ are suitable constants that do not depend on Y_s.
By virtue of Theorem 7.2.1 the following result can easily be verified.

THEOREM 6.2.3 Under model (7.2.1) and assumptions $E(y_i^4) < \infty$, $E(Y_i^4 \mid y_i) < \infty$,

$$s_Y^2 = \frac{1}{n-1} \sum_{i \in s} (Y_i - \bar{Y})^2$$

is the UMV quadratic pm-unbiased predictor of S_y^2 for any $p \in \rho_n$. Again, any $p \in \rho_n$ is optimal for using s_Y^2.

In the next part we consider Bayes prediction of population total and variance under model (7.2.1).

Bayes Predictor of T

Assume that the variables e_i, u_i are normally distributed with variances $(\sigma_{ee}, \sigma_{uu})$, assumed to be known. As the distribution of a large number of socio-economic variables is (at least approximately) normal in large samples, we consider a normal prior for μ,

$$\mu \sim N(0, \theta^2) \tag{7.2.7}$$

The posterior distribution of μ is, therefore,

$$\mu \mid Y_s, \theta \sim N \left(\frac{n \bar{Y}_s \theta^2}{n\theta^2 + \sigma^2}, \frac{\sigma^2 \theta^2}{n\theta^2 + \sigma^2} \right) \tag{7.2.8}$$

where

$$\sigma^2 = \sigma_{ee} + \sigma_{uu} \tag{7.2.9}$$

Again, posterior distribution of y_i's, given (Y_s, μ) are independent:

$$y_i \mid (Y_s, \mu) \sim N \left(\frac{Y_i \sigma_{ee} + \mu \sigma_{uu}}{\sigma^2}, \sigma_0^2 \right) \quad (i \in s) \tag{7.2.10}$$

$$y_i \mid (Y_s, \mu) \sim N(\mu, \sigma_{ee}) \quad (\notin s) \tag{7.2.11}$$

where

$$\sigma_0^2 = \frac{\sigma_{uu}\sigma_{ee}}{\sigma^2} \qquad (7.2.12)$$

Therefore, under squared error loss function, Bayes predictor of T is

$$\hat{T}_B = E(\sum_{i=1}^{N} y_i \mid Y_s) = E\{E(\sum_{i=1}^{N} y_i \mid Y_s, \mu) \mid Y_s\}$$

$$= n\bar{Y}_s \left[\frac{\sigma_{ee}}{\sigma^2} + \frac{(N-n)\sigma_{ee} + N\sigma_{uu}}{\sigma^2(n\theta^2 + \sigma^2)}\theta^2 \right] \qquad (7.2.13)$$

Also,

$$V(T \mid Y_s) = E_\mu\{V(T \mid \mu, Y_s)\} + V_\mu\{E(T \mid \mu, Y_s)\}$$

$$= \frac{n\sigma_{ee}\sigma_{uu}}{\sigma^2} + (N-n)\sigma_{ee} + (N - \frac{n\sigma_{ee}}{\sigma^2})^2 \frac{\sigma^2\theta^2}{n\theta^2 + \sigma^2}$$

$$= r_\theta(\hat{T}_B) \text{ (say)} \qquad (7.2.14)$$

which, being independent of s, is also Bayes risk of T. Again,

$$lim_{\theta\to\infty}(\hat{T}_B) = \frac{N(N-n)}{n}\sigma^2 + \frac{N^2}{n}\sigma_{uu}$$

$$= r_0 \text{ (say)}$$

It is seen from (7.2.6) that the risk of the predictor $N\bar{Y}_s$ is given by r_0. Hence, by Theorem 3.2.2 $N\bar{Y}_s$ is a minimax predictor of T under the assumption of normality of the distribution of e_i's and u_i's as considered above. Again, since expression (7.2.6) was obtained without any assumption about the form of the distributions, it follows by Theorem 3.2.3 that the predictor $N\bar{Y}_s$ is minimax in the general class of distributions (not necessarily normal).

THEOREM 7.2.4 The predictor $N\bar{Y}_s$ is a minimax predictor of T under the general class of prior distributions of errors (e_i's and u_i's) which satisfy (7.2.1).

We now assume that $k = \sigma_{ee}/\sigma_{uu}$ is a known positive constant but $\tau = 1/\sigma_{ee}$ is unknown. Assuming a normal-gamma prior for (μ, τ) (eg. Broemling, 1985) with parameters (ν, α, β), the joint distribution of μ, τ is

$$P(\mu, \tau) \propto \tau^{\alpha-1/2} \exp\left\{-\frac{\tau}{2}[(\mu - \nu)^2 + 2\beta]\right\}$$

$$\mu, \nu \in R^1, \tau > 0, \alpha > 0, \beta > 0$$

The marginal posterior distribution of μ is a Student's t-distribution with $(n + 2\alpha)$ d.f. and posterior mean and variance given, respectively, by

$$E(\mu \mid Y_s) = \frac{\nu + nq\bar{Y}_s}{1 + nq}$$

$$V(\mu \mid Y_s) = \frac{2\beta + q\sum_{i \in s} Y_i^2 + \nu^2 - \frac{\nu + nq\bar{Y}_s}{(1+nq)}}{(1 + nq)(n + 2\alpha)}$$

where $q = k/(k+1)$. It is assumed that $n > 1$.

Marginal posterior distribution of τ is a gamma with parameters

$$\alpha^* = \frac{n + \alpha}{2},$$

$$\beta^* = \beta + \frac{\nu^2}{2} + \frac{q}{2}\sum_{i \in s} Y_i^2 - \frac{(\nu + nq\bar{Y}_s)^2}{2(1 + nq)}$$

Hence, Bayes predictor of T is

$$\hat{T}_B^{(1)} = E\{E(\sum_{i \in s} y_i + \sum_{i \, not \, in \, s} y_i \mid \mu, \tau, Y_s) \mid Y_s\}$$

$$= \frac{kn(N+1)}{k(n+1)+1}\bar{Y}_s + \frac{N + k(N-n)}{k(n+1)+1}\theta \qquad (7.2.15)$$

To use $\hat{T}_B^{(1)}$ one needs to know only the value of $k = \sigma_e^2/\sigma_u^2$. Note that when $n = N, \hat{T}_B^{(1)} \neq T$ and hence $\hat{T}_B^{(1)}$ is not a consistent estimator in Cochran's (1977) sense. Similarly, one can calculate Bayes predictor of T and S_y^2 under Jeffrey's non-informative prior (Exercise 1).

Bhattacharyya (1997) considered the Bayes prediction of S_y^2 under the model (7.2.1). Consider the identity

$$S_y^2 = \frac{n}{N}s_y^2 + \frac{N-n}{N}s_{yr}^2 + \frac{n(N-n)}{N^2}(\bar{y}_r - \bar{y}_s)^2$$

where

$$s_y^2 = \sum_{i \in s}(y_i - \bar{y}_s)^2/n, \ s_{yr}^2 = \sum_{i \in r}(y_i - \bar{y}_r)^2/(N-n)$$

$$\bar{y}_s = \sum_{i \in s} y_i/n, \ \bar{y}_r = \sum_{i \in r} y_i/(N-n), \ r = \bar{s} = \mathcal{P} - s$$

Now,

$$\frac{ns_y^2}{\sigma_0^2} \mid Y_s \sim \text{non-central } \chi^2((n-1), \lambda) \qquad (7.2.16)$$

where

$$\lambda = \frac{n\sigma_{ee}^2 s_Y^2}{\sigma^4 \sigma_0^2}, \quad \text{and} \quad s_Y^2 = \sum_{i \in s}(Y_i - \bar{Y}_s)^2/n \tag{7.2.17}$$

Also,

$$(N-n)s_{yr}^2/\sigma_{ee} \sim \chi_{(N-n-1)}^2 \tag{7.2.18}$$

Again,

$$(\bar{y}_r - \bar{y}_s \mid Y_s, \mu) \sim N(\mu_1, \sigma_1^2) \tag{7.2.19}$$

where

$$\mu_1 = \frac{\sigma_{ee}}{\sigma^2}(\mu - \bar{Y}_s)$$

$$\sigma_1^2 = \frac{\sigma_{ee}\sigma_{uu}}{n\sigma^2} + \frac{\sigma_{ee}}{N-n} = \frac{\sigma_{ee}}{n(N-n)\sigma^2}[N\sigma^2 - (N-n)\sigma_{ee}] \tag{7.2.20}$$

From (7.2.16) - (7.2.20), Bayes estimate of S_y^2 can easily be calculated.

Again, for a square error loss function, Bayes prediction risk of $\hat{S}_{yB}^2$ is given by $E[V(S_y^2 \mid Y_s)]$ where the expectation is taken with respect to prediction distribution of Y_s. The posterior variance of S_y^2 is

$$V(S_y^2 \mid Y_s) = E_\mu[V(S_y^2 \mid \mu, Y_s) \mid Y_s] + V_\mu[E(S_y^2 \mid \mu, Y_s) \mid Y_s]$$

$$= \frac{2}{N^2}\frac{\sigma_{ee}^2}{\sigma^4}\{(n-1)\sigma_{uu} + (N-n-1)\sigma^4\} + 4\frac{n}{N^2}$$

$$\frac{\sigma_{ee}^3\sigma_{uu}}{\sigma^6}s_y^2 + \frac{2n^2(N-n)^2}{N^4}\{\frac{\sigma_{ee}}{\sigma^2}\sigma^2(\frac{\sigma_{uu}}{n} + \frac{\sigma^2}{N-n}$$

$$+\frac{\sigma_{ee}\theta^2}{(n\theta^2 + \sigma^2)}\}^2 + \frac{4n^2(N-n)^2}{N^4}\frac{\sigma_{ee}^2\bar{Y}_s^2}{(n\theta^2 + \sigma^2)^2}$$

$$\{\frac{\sigma_{ee}}{\sigma^2}(\frac{\sigma_{uu}}{n} + \frac{\sigma^2}{N-n} + \frac{\sigma_{ee}\theta^2}{(n\theta^2 + \sigma^2)})\} \tag{7.2.21}$$

Again, the predictive distribution of $\bar{Y}_s$ under the above models is $N(0, \frac{\sigma^2}{n} + \theta^2)$ (Bolfarine and Zacks, 1991). Hence, taking expectation of (7.2.21) Bayes risk of $\hat{S}_{yB}^2$ is

$$E[V(S_y^2 \mid Y_s)] = \frac{2\sigma_{ee}^2}{N^2\sigma^4}\{(n-1)\sigma_{uu} + (N-n-1)\sigma^4 + 2(n-1)\sigma_{ee}\sigma_{uu}\}$$

$$+\frac{2n^2(N-n)^2}{N^4}\{\frac{\sigma_{ee}}{\sigma^2}(\frac{\sigma_{uu}}{n} + \frac{\sigma^2}{N-n} + \frac{\sigma_{ee}\theta^2}{(n\theta^2 + \sigma^2)})\}^2 +$$

$$4\frac{n^2(N-n)^2}{N^4}\frac{\sigma_{ee}^2}{(n\theta^2 + \sigma^2)}\{\frac{\sigma_{ee}}{\sigma^2}(\frac{\sigma_{uu}}{n} + \frac{\sigma^2}{N-n} + \frac{\sigma_{ee}\theta^2}{(n\theta^2 + \sigma^2)})\} \tag{7.2.22}$$

Allowing $\theta \to \infty$, the limiting value of risk of $\hat{S}^2_{yB}$ is

$$\frac{2\sigma^2_{ee}}{N^2\sigma^4}\{(N-1)\sigma^4 - (n-1)\sigma^2_{ee}\}$$

NOTE 7.2.1

Mukhopadhyay (1994 c) considered a variation of model (6.2.1). Consider the general class of superpopulation model distribution ξ of **y** such that for given μ, ξ_1 is a distribution in hyperplanes in R_N with $\bar{y}(=\sum_{i=1}^{N} y_i/N) = \mu$ and

$$E\sum_{i=1}^{N}(y_i - \mu)^2 \leq (N-1)\sigma_{ee}, \qquad (7.2.23)$$

σ_{ee} a constant and E denoting expectation with respect to superpopulation model ξ_1 (and other models relevant to the context). The distribution ξ_2 of $Y_s = \{Y_i, i \in s\}$ is considered to be a member of the class with the property that the conditional distribution of Y_i given y_i is independent and

$$E(Y_i \mid y_i) = y_i, \ V(Y_i \mid y_i) = \sigma_{uu} \qquad (7.2.24)$$

Let C denote the class of distributions $\{\xi = \xi_1 \times \xi_2\}$. Consider the subclass $C_o = \{\xi = \xi_{10} \times \xi_{20}\}$ of C where ξ_{10} is such that given μ, **y** is distributed as a N-variate singular normal distribution with mean vector $\mu 1$ and dispersion matrix Σ having constant values of

$$\sigma_{ii} = \frac{N-1}{N}\sigma_{ee} \ (\forall i) \text{ and } \sigma_{ij} = -\frac{\sigma_{ee}}{N}(\forall i \neq j)$$

ξ_{20} is a *pdf* on R^N such that the conditional distribution of Y_i given y_i is independent normal with mean and variances as stated in (7.2.24).

It is assumed that μ is distributed a priori normally with mean 0 and variance θ^2. Bayes predictor of $\bar{y}$ based on a random sample of size n for the class of distributions in C_0 is found to be

$$E(\mu \mid Y_s) = \frac{\bar{Y}_s}{1 + \frac{\sigma'^2}{n\theta^2}} = \delta_\theta \text{ (say)} \qquad (7.2.25)$$

where

$$\sigma'^2 = \sigma_{uu} + \frac{N-n}{N}\sigma_{ee} \qquad (7.2.26)$$

and posterior variance

$$V(\mu \mid Y_s) = \frac{\sigma'^2}{n + \frac{\sigma'^2}{\theta^2}}$$

An appeal to theorems 3.2.2 and 3.2.3 (using $\theta \to \infty$) showed that $\bar{Y}_s$ is minimax for the class of distributions in C. Bayes estimation of domain total was also considered.

The above model was extended to the case of stratified random sampling. It was found that for the loss function

$$L(F, \delta) = (\delta - F)^2 + \sum_{h=1}^{L} c_h n_h$$

for estimating F by δ, where c_h is the cost of sampling a unit and n_h the sample size in the hth stratum, minimax estimate of $\sum_{h=1}^{L} a_h \bar{y}_h (a_h$ a constant) is $\sum_{h=1}^{L} a_h \bar{Y}_{s_h}$ and a minimax choice of n_h is

$$n_h^* = \sqrt{\frac{a_h^2(\sigma_{uuh} + \sigma_{eeh})}{c_h}}$$

where $\sigma_{eeh}, \sigma_{uuh}$ have obvious interpretations. In particular, if $F = \bar{\bar{y}} = \sum_h W_h \bar{y}_h, W_h = \frac{N_h}{N}$,

$$n_h^* \propto \sqrt{N_h^2(\sigma_{uuh} + \sigma_{eeh})/c_h}$$

The model was also extended to the case of two-stage sampling (Mukhopadhyay, 1995 a).

7.2.2 LINEAR REGRESSION MODELS

We first consider regression model with one auxiliary variable. Assume that associated with each i there is a true value x_i of an auxiliary variable x closely related to the main variable y. The values x_i's ,however, cannot be observed without error and instead some other values X_i's are observed. It is assumed that $x_1, \ldots, x_N$ are unknown fixed quantities. Consider the model

$$y_i = \beta_0 + \beta x_i + e_i, \ e_i \sim (0, \sigma_{ee})$$

$$X_i = x_i + v_i, \ v_i \sim (0, \sigma_{vv}) \qquad (7.2.27)$$

$$Y_i = y_i + u_i, \ u_i \sim (0, \sigma_{uu})$$

where e_i, v_i, u_i are assumed to be mutually independent.

The following theorem easily follows from Theorem 7.2.1.

THEOREM 7.2.4 Under model (7.2.27) the best linear pm-unbiased predictor of $\bar{y}$ for any given $p \in \rho_n$ is

$$e_1^* = (\sum_{k=1}^{N} \frac{\pi_k}{Z_k})^{-1} \sum_{k \in s} \frac{Y_k}{Z_k} \qquad (7.2.28)$$

where $Z_k = \beta_0 + \beta x_k$. Again,

$$E_p E(e^* - \beta_0 - \beta\bar{x})^2$$

$$= \frac{1}{(\sum_{k=1}^{N} \frac{\pi}{Z_k})^2}[n^2 + (\beta^2\sigma_{vv} + \sigma_{uu} + \sigma_{ee})\sum_{k=1}^{N} \frac{\pi_k}{Z_k^2}] - \bar{Z}^2$$

$$= \delta_n \text{ (say)},$$

a constant dependent only on n. Hence, any $p \in \rho_n$ is an optimal *s.d.* for using e_1^*.

NOTE 7.2.2

In deriving e_1^* it has been assumed that β_0, β are known, which may not be the case. For estimating the parameters we assume

- (i) e_i, u_i, v_i are normally distributed with the parameters as stated in (6.2.17);

- (ii) $x_k \sim N(\mu_x, \sigma_{xx})$ and is independent of $e_i v_j$, $u_l(i, j, k, l = 1, 2, \ldots)$

We call these assumptions as assumptions A. Under these assumptions (X_i, Y_i) have a bivariate normal distribution with mean vector

$$\begin{bmatrix} \mu_Y \\ \mu_X \end{bmatrix} = \begin{bmatrix} \beta_0 + \beta\mu_x \\ \mu_x \end{bmatrix}$$

and dispersion matrix

$$\begin{bmatrix} \beta^2\sigma_{xx} + \sigma_{uu} + \sigma_{ee} & \beta\sigma_{xx} \\ \beta\sigma_{xx} & \sigma_{xx} + \sigma_{vv} \end{bmatrix}$$

We denote by m_{XX}, m_{YY}, m_{XY} the sample variances of X and Y and covariance of (X, Y) respectively, $m_{XY} = \frac{1}{n}\sum_s (X_i - \bar{X}_s)(Y_i - \bar{Y}_s)$, etc. The parameters are estimated in the following situations:

(i) the ratio $\sigma_{xx}/\sigma_{XX} = \sigma_{xx}/(\sigma_{xx} + \sigma_{vv}) = k_{xx}$ called the *reliability ratio*, is known. Here,

$$\hat{\beta} = \hat{\beta}_U \text{ (say)} = \frac{\hat{\beta}_{LS}}{k_{xx}}$$

where
$$\hat{\beta}_{LS} = m_{XY}/m_{XX}$$

is the ordinary least squares estimator of β. $\hat{\beta}_U$ is unbiased for β.

(ii) The measurement error variance σ_{vv} is known. Here

$$\hat{\beta} = \hat{\beta}_{F1} = \frac{m_{XY}}{m_{XX} - \sigma_{vv}}$$

(iii) The ratio $(\sigma_{uu} + \sigma_{ee})/\sigma_{vv} = \delta$ is known. Here

$$\hat{\beta} = \hat{\beta}_{F2} = \frac{1}{2m_{XX}}[m_{YY} - \delta m_{XX} + \sqrt{(m_{YY} - \delta m_{XX})^2 + 4\delta m_{XY}^2}]$$

In all the cases $\hat{\beta}_0 = \bar{Y}_s - \hat{\beta}\bar{X}_s$. The above derivations follow from Fuller (1987).

NOTE 7.2.3

In case, X_k's are known only for $k \in s, e_1^*$ in (7.2.28) may be replaced by

$$e_1^{*\prime} = \frac{\sum_{k \in s} \frac{Y_k}{Z_k}}{\sum_{k \in s} \frac{1}{Z_k}},$$

a Ha'jek(1959)- type predictor. The predictor $e_1^{*\prime}$ is pm-biased.

Following Royall (1970) and Rodrigues et al (1985), Bolfarine, Zacks and Sandoval (1996) obtained best linear m-unbiased predictor of T under model (7.2.27) where y_i's are measured without error. Assume that X_k's are available for $k = 1, \ldots, N$ and as in the case of note (7.2.2), $x_k \sim N(\mu_x, \sigma_{xx}), e_k \sim N(0, \sigma_{ee})$ and $v_k \sim N(0, v_{kk})$ and the variables are independent. A predictor of T is

$$\begin{aligned} \hat{T}_G &= \sum_s Y_k + \sum_{\bar{s}} \hat{Y}_k \\ &= N\bar{Y}_s + (N - n)(\bar{X}_r - \bar{X}_s)\hat{\beta}_G \end{aligned} \qquad (7.2.29)$$

where $\bar{X}_r = \sum_{k \notin s} X_k/(N - n)$ and $\hat{\beta}_G$ is a predictor of β. The predictors of T using predictors $\hat{\beta}_{LS}, \hat{\beta}_U, \hat{\beta}_{F1}, \hat{\beta}_{F2}$ of β will be denoted as $\hat{T}_{LS}, \hat{T}_U, \hat{T}_{F1}, \hat{T}_{F2}$, respectively. We recall below some of their important results.

THEOREM 7.2.5 Under models (7.2.27) and assumptions A (with $\sigma_{uu} = 0$)

$$\sqrt{n}(\hat{\beta}_{LS} - k_{xx}\beta) \to^L N(0, \sigma_{XX}^{-2}\{B - \beta^2\sigma_{vv}^2\})$$

$$\sqrt{n}(\hat{\beta}_U - \beta) \rightarrow^L N(0, \sigma_{XX}^{-2} k_{xx}^{-2} \{\sigma_{XX} B - \beta^2 \sigma_{vv}^2\}) \tag{7.2.30}$$

$$\sqrt{n}(\hat{\beta}_{F1} - \beta) \rightarrow^L N(0, \sigma_{xx}^{-2} \{\sigma_{XX} B + \beta^2 \sigma_{vv}^2\})$$

$$\sqrt{n}(\hat{\beta}_{F2} - \beta) \rightarrow^L N(0, \sigma_{xx}^{-2} \{B - \beta^2 \sigma_{vv}^2\})$$

as $n \rightarrow \infty, N \rightarrow \infty$ and $N - n \rightarrow \infty$, where

$$B = \beta^2 \sigma_{vv} + \sigma_{ee}$$

THEOREM 7.2.6 Under model of Theorem 5.2.8 and under the assumption that $\sqrt{n}(\hat{\beta}_G - \beta)$ converges to some proper distribution,

$$\sqrt{n}(\hat{T}_G - T)/N \rightarrow^L N(0, (1 - f)B) \tag{7.2.31}$$

as $n/N \rightarrow f$ and as $n, N \rightarrow \infty$.

Proof.

$$\hat{T}_G - T = (N - n)(\bar{X}_r - \bar{X}_s)(\hat{\beta}_G - \beta) +$$
$$(N - n)[\bar{Y}_s - \beta - \bar{X}_s\beta - (\bar{Y}_r - \beta_0 - \bar{X}_r\beta)]$$

$$\sqrt{n}(\hat{T}_G - T)/N = (1 - n/N)(\bar{X}_r - \bar{X}_s)\sqrt{n}(\hat{\beta}_G - \beta) +$$
$$(1 - n/N)n^{-1/2} \sum_{k \in s}(Y_k - \beta_0 - \beta X_k) +$$
$$\sqrt{n/N}\sqrt{1 - n/N}(N - n)^{-1/2} \sum_{k \in s}(Y_k - \beta_0 - \beta X_k)$$

Now, $\bar{X}_r - \bar{X}_s \rightarrow^P 0$. Again,

$$n^{-1/2}(\sum_{k \in s}Y_k - \beta_0 - \beta X_k) = n^{-1/2} \sum_{k \in s}(e_k - \beta v_k) \rightarrow^L N(0, B)$$

as $n \rightarrow \infty$, by the Central Limit Theorem (CLT). Similarly,

$$(N - n)^{-1/2} \sum_{k \in r}(Y_k - \beta_0 - \beta X_k) \rightarrow^L N(0, B)$$

as $N - n \rightarrow \infty$.

Theorems 7.2.5 and 7.2.6 imply that $\hat{T}_U, \hat{T}_{F1}, \hat{T}_{F2}$, when properly normalised, are each asymptotically distributed as (7.2.31).

THEOREM 7.2.7 Under model of Theorem 7.2.5,

$$\sqrt{n}(\hat{T}_{LS} - T)/N \rightarrow^L N(0, (1 - f)(\sigma_{ee} + \beta^2 k_{xx}\sigma_{vv})) \tag{7.2.32}$$

Defining asymptotic relative efficiency (ARE) of $\hat{T}_1$ with respect to $\hat{T}_2$ as

$$e_{12} = \frac{AV(\hat{T}_2)}{AV(\hat{T}_1)} \tag{7.2.33}$$

where AV means asymptotic variance, RE of $\hat{T}_{LS}$ with respect to $\hat{T}_U($ or $\hat{T}_{F1}$, $\hat{T}_{F2}$, since these estimators are asymptotically equivalent by Theorem 7.2.6) is

$$e_{12} = \frac{B}{\sigma_{ee} + k_{xx}\beta^2\sigma_{vv}} = \frac{\delta + \beta^2}{\delta + k_{xx}\beta^2} > 1 \tag{7.2.34}$$

where $\delta = \sigma_{ee}/\sigma_{vv}$. Unlike $\hat{T}_U, \hat{T}_{F1}$, and $\hat{T}_{F2}, \hat{T}_{LS}$ does not depend on any extra knowledge about the population. Also, e_{12} increases as k_{xx} decreases i.e. as measurement error becomes severe.

The authors performed simulation studies to investigate the behaviour of $\hat{T}_{LS}, \hat{T}_U, \hat{T}_{F1}, \hat{T}_{F2}$ and

$$\hat{T}_{F1}^* = N\bar{Y}_s + (N - n)(\bar{X}_r - \bar{X}S)\hat{\beta}_{F1}^*$$

where

$$\hat{\beta}_{F1}^* = \begin{cases} \hat{\beta}_{F1}, & \text{if } \hat{\lambda} \geq 1 + \frac{1}{n-1} \\ \frac{m_{XY}}{m_{XX} - \lambda\sigma_{vv} + \frac{2\sigma_{vv}}{n-1}}, & \text{if } \hat{\lambda} < 1 + \frac{1}{n-1} \end{cases}$$

$$\hat{\lambda} = \{m_{XY} - \frac{m_{XY}^2}{m_{YY}}\}/\sigma_{vv}^2 \tag{7.2.35}$$

as suggested by Fuller (1987) as a small sample improvement over $\hat{\beta}_{F1}$.

Mukhopadhyay(1994 b) obtained Bayes predictor of T under model (7.2.27) when the x_i values are measured without error. Assume further that e_i's, u_i's are independently normally distributed with known values of σ_{uu}, σ_{ee}. Suppose also that $x_k(k = 1, \ldots, N)$ are known positive quantities. Assume that prior distribution of $\beta = (\beta_0, \beta)'$ is bivariate normal with mean $b^0 = (b_0^0, b^0)'$ and precision matrix qS^0 $(q = \frac{k}{(k+1)}, k = \frac{\sigma_{ee}}{\sigma_{uu}})$ where S^0 is a 2×2 positive semidefinite matrix. The posterior distribution of β given Y_s, X_s where $X_s = [1, x_k; k \in s]_{n \times 2}$ is normal with

$$E(\beta \mid Y_s, X_s) = B^{00} = S^{00^{-1}}(Sb + S^0b^0) \tag{7.2.36}$$

and dispersion matrix

$$D(\beta \mid Y_s, X_s) = S^{00^{-1}} = \frac{1}{q\tau}(S^0 + S)^{-1}, \tag{7.2.37}$$

where $b = (\bar{Y}_s - b\bar{x}_s, \ b)', b = \sum_s(Y_k - \bar{Y}_s)(x_k - \bar{x}_s)/\sum_s(x_k - \bar{x}_s)^2 = m_{xY}/m_{xx}$,

$$S = \begin{bmatrix} n & \sum_s x_k \\ \sum_s x_k & \sum_s x_k^2 \end{bmatrix} \tag{7.2.38}$$

Bayes predictor of T is

$$\hat{T}_B' = (N - \frac{nk}{k+1})b_0^{00} + (\frac{n\bar{x}_s}{k+1} + (N-n)\bar{x}_r)b^{00} + \frac{nk\bar{Y}_s}{k+1}, \tag{7.2.40}$$

where $b^{00} = (b_0^{00}, b^{00})$.

In particuklar, if we assume a natural conjugate prior of β so that $b^0 = b, S^0 = S$, then $b^{00} = b$ and $\hat{T}_B'$ in (7.2.39) reduces to

$$\hat{T}_B'' = N[\bar{Y}_s + b(\bar{X} - \bar{x}_s)] \tag{7.2.40}$$

the linear regression predictor of T.

Bolfarine et al (1996) extended their work on simple regression model in Theorems 7.2.5 - 7.2.7 to multiple regression model. Consider the model

$$Y_k = \alpha + x_k'\beta + e_k$$

$$X_k = x_k + v_k, \quad k = 1, \ldots, N \tag{7.2.41}$$

where $\beta = (\beta_1, \ldots, \beta_r)', x_k = (x_{k1}, \ldots, x_{kp})', X_k = (X_{k1}, \ldots, X_{kr})', v_k = (v_{k1}, \ldots, v_{kp})', x_{kj}(X_{kj}), v_{kj}$ are the true (observed) value of the auxiliary variable x_j on k and the corresponding measurement error $(j = 1, \ldots, p)$. Assume that

$$\begin{bmatrix} e_k \\ x_k \\ v_k \end{bmatrix} \sim N_{2p+1} \begin{bmatrix} \begin{bmatrix} 0, \\ \mu_x \\ 0 \end{bmatrix} & \begin{bmatrix} \sigma_{ee} & 0 & 0 \\ 0 & \Sigma_{xx} & 0 \\ 0, & 0 & \Sigma_{vv} \end{bmatrix} \end{bmatrix} \tag{7.2.42}$$

where Σ_{xx}, Σ_{vv} are non-singular with Σ_{vv} known. Consider an unbiased predictor $\tilde{\beta}_U$ of β,

$$\tilde{\beta}_U = (M_{XX} - \Sigma_{vv})^{-1} M_{XY} \tag{7.2.43}$$

where

$$M_{XX} = \frac{1}{n} \sum_{k \in s} (X_{k-s})(X_k - \bar{X}_s)'$$

$$M_{XY} = \frac{1}{n} \sum_{k \in s} (X_k - \bar{X}_s)(Y_k - \bar{Y}_s)$$

$$\bar{X}_s = (\bar{X}_{s1} \ldots, \bar{X}_{sp})' \quad \bar{X}_{sj} = \sum_{k \in s} X_{kj}/n.$$

THEOREM 7.2.8 Under models (7.2.41), (7.2.42),

$$\sqrt{n}(\tilde{\beta}_U - \beta) \to N(0, G)$$

where

$$G = \Sigma_{xx}^{-1}\{(\beta'\Sigma_{vv}\beta + \sigma_{ee})(\Sigma_{XX} + \Sigma_{vv}\beta\beta'\Sigma_{vv})\}\Sigma_{xx}^{-1} \qquad (7.2.44)$$

and $\Sigma_{XX} = \Sigma_{xx} + \Sigma_{vv}$.

The proof follows from Theorem 2.2.1 of Fuller (1987). The corresponding predictor of T is

$$\tilde{T}_U = N\bar{Y}_s + (N - n)(\bar{X}_r - \bar{X}_s)'\tilde{\beta}_U \qquad (7.2.45)$$

The following theorem is an extention of Theorem 7.2.6.

THEOREM 7.2.9 Under models (7.2.41), (7.2.42),

$$\sqrt{n}(\tilde{T}_U - T)/N \to^L N(0, (1 - f)(\sigma_{ee} + \beta'\Sigma_{vv}\beta)) \qquad (7.2.46)$$

as $N \to \infty$. The least square predictor of β is

$$\tilde{\beta}_{LS} = M_{XX}^{-1}M_{XY}$$

The corresponding predictoopr of T is

$$\tilde{Y}_{LS} = N\bar{Y}_s + (N - n)(\bar{X}_r - \bar{X}_s)'\tilde{\beta}_{LS} \qquad (7.2.47)$$

THEOREM 7.2.10 Under model (7.2.41), (7.2.42)

$$\sqrt{n}(\tilde{T}_{LS} - T)/N \to^L N(0, (1 - f)(\sigma_{ee} + \beta'\Sigma_{xx}\Sigma_{XX}^{-1}\Sigma_{vv}\beta)) \qquad (7.2.48)$$

It follows that the asymptotic relative efficiency of $\tilde{T}_{LS}$ with respect to $\tilde{T}_U$ is

$$e'_{12} = \frac{\sigma_{ee} + \beta'\Sigma_{vv}\beta}{\sigma_{ee} + \beta'\Sigma_{XX}^{-1}\Sigma_{vv}\beta}, \qquad (7.2.49)$$

which is always greater than one since

$$\Sigma_{xx}\Sigma_{XX}^{-1}\Sigma_{vv} = (\Sigma_{vv}^{-1} + \Sigma_{xx}^{-1})^{-1} = \Sigma_{vv} - \Sigma_{XX}^{-1}\Sigma_{vv}.$$

Comparison between *matrix mse* of $\tilde{\beta}_{LS}$ and $\tilde{\beta}_U$ poses an interesting problem.

Mukhopadhyay (1995 b, 1998 b) considered multiple regression model with measurement errors on all the variables, His model is

$$y_i = \mathbf{x}_i'\beta + e_i$$

$$X_i = x_i + v_i \tag{7.2.50}$$

$$Y_i = y_i + u_i$$

Writing $\epsilon_i = (e_i, u_i, v_i')'$, $E(\epsilon_i) = 0$, and

$$V(\epsilon_i) = \begin{bmatrix} \sigma_{ee} & 0 & 0 \\ 0 & \sigma_{uu} & \Sigma_{uv} \\ 0 & \Sigma_{uv}' & \Sigma_{vv} \end{bmatrix} \tag{7.2.51}$$

Under the above models he considered best linear optimal pm-unbiased prediction of $\bar{y}$ for any given $p \in \rho_n$, Bayes prediction of $\bar{y}$ under normal theory set up and linear Bayes prediction of $\bar{y}$. Under a slightly different model he (1999 a) he derived a strongly consistent estimator for β (Exercise 3).

7.2.3   BAYES ESTIMATION UNDER TWO-STAGE SAMPLING ERROR-IN-VARIABLES MODELS

Consider as in Bolfarine (1991) the following two-stage sampling superpopulation models with error-in-variables. Superpopulation models for multistage sampling (without measurement errors) were earlier considered by Scott and Smith (1969), Royall (1976), Malec and Sedransk(1985), among others. Mukhopadhyay(1995 a) considered a different model in two-stage sampling accommodating measurement errors.

The finite population consists of K mutually exclusive and exhaustive subpopulations (clusters) of size M_h (Number of elementary units) , $h = 1, \ldots, K$. In the first stage a sample of n clusters is selected from the K available clusters. In the second stage a sample s_h of m_h elementary units is selected from each cluster h in the sample s. Let y_{hi} denote the true vlaue of the characteristic y on the ith elementary unit belonging to the cluster h. We assume that whenever (h, i) is in the sample $s_0 = \cup_{h \in s} s_h$, y_{hi} cannot be observed but a different value Y_{hi} , mixed with measurement error is observed.

The following model is assumed:

$$y_{hi} = \mu_h + e_{hi}$$

$$\mu_h = \mu + \nu_h \quad i = 1, \ldots, M_h$$

$$Y_{hi} = y_{hi} + u_{hi}, \quad h = 1, \ldots, K \tag{7.2.52}$$

where e_{hi}, ν_h, u_{hi} are all independent,

$$e_{hi} \underset{\sim}{iid} N(0, \sigma_{eeh})$$

$$\nu_h \underset{\sim}{iid} N(0, \delta^2)$$

$$u_{hi} \underset{\sim}{iid} N(0, \sigma_{uuh})$$

The above models correspond to the exchangeability between elements in a cluster. Let $Y_s = \{Y_{hi}, i \in s_h, \text{ and } h \in s\}$ denote the observed value corresponding to sample s_0. Now Posterior distribution of μ_h give Y_s is

$$\mu_h \mid Y_s \sim N(\mu_h^*, \frac{\delta^2 \sigma_{wwh}}{m_h \delta^2 + \sigma_{wwh}}), h \in s \tag{7.2.53}$$

$$\mu_h \mid Y_s \sim N(\mu, \delta^2), h \notin s$$

where

$$\mu_h^* = \frac{m_h \bar{Y}_{s_h} \delta^2 + \mu \sigma_{wwh}}{m_h \delta^2 + \sigma_{wwh}},$$

$$\sigma_{wwh} = \sigma_{eeh} + \sigma_{uuh}, \quad \bar{Y}_{s_h} = \frac{1}{m_h} \sum_{i \in s_h} Y_{hi}$$

Also, posterior distribution of y_{hi} given μ_h and Y_s is

$$y_{hi} \mid \mu_h \sim N \left(\frac{m_h \bar{Y}_{sh} \sigma_{eeh} + \mu_h \sigma_{uuh}}{m_h \sigma_{eeh} + \sigma_{uuh}}, \frac{\sigma_{eeh} \sigma_{uuh}}{m_h \sigma_{eeh} + \sigma_{uuh}} \right), i \in s_h, \ h \in s$$

$$y_{hi} \mid \mu_h \sim N(\mu_h, \sigma_{eeh}), (h, i) \notin s_0$$

After the sample s has been selected, we may write the population total T as

$$T = \sum_{h \in s} \sum_{i \in s_h} y_{hi} + \sum_{h \in s} \sum_{i \in \bar{s}_h} y_{hi} + \sum_{h \in \bar{s}} \sum_{i=1}^{M_h} y_{hi} \tag{7.2.54}$$

where $\bar{s}$ denotes the set of clusters not included in s and $\bar{s}_h$ is the set of elementary units in the hth cluster not included in the sample s_h. Hence, Bayes predictor of T is

$$\hat{T}^B = E[E\{T \mid Y_s, \mu_h\} \mid Y_s]$$

$$= \sum_{h \in s} \sum_{i \in s_h} \frac{m_h \bar{Y}_{s_h} \sigma_{eeh} + \mu_h^* \sigma_{uuh}}{m_h \sigma_{eeh} + \sigma_{uuh}} + \sum_{h \in s} \sum_{i \in \bar{s}_h} \mu_h^* + \sum_{h \in \bar{s}} \sum_{i=1}^{M_h} \mu_h \qquad (7.2.55)$$

Mukhopadhyay (1999 c) derived Bayes estimators of a finite population total in unistage and two-stage sampling under simple location model with measurement error under the Linex loss function due to Varian (1975) and Zellner (1986). Bayes estimator of a finite population variance under the same model and same loss function was also derived (exercise 6).

7.2.4 PREDICTION OF A FINITE POPULATION DISTRIBUTION FUNCTION

We first consider the model (6.2.1) along with the assumptions that e_i, u_i are normally distributed. Consider the class of predictors linear in $\Delta(t - Y_i)$ for $F_N(t)$ (defined in (6.1.1)),

$$\hat{F}_s(t) = b_s + \sum_{k \in s} b_{ks} \Delta(t - Y_k) \qquad (7.2.56)$$

where b_s, b_{ks} are constants not depending on y-values. Clearly, $\hat{F}_s(t)$ will be pm-unbiased for $F(t)$ iff

$$E_p E(\hat{F}_s(t)) = E(F(t)) = \Phi(\frac{t - \mu}{\sigma_e}) \qquad (7.2.57)$$

where $\sigma_{ee} = \sigma_e^2$ and $\Phi(z)$ denotes the area under a standard normal curve up to the ordinate at z. The condition (7.2.57) implies

$$E_p(b_s) = 0$$

$$E_p(\sum_{k \in s} b_{ks}) = \frac{\Phi(\frac{t-\mu}{\sigma_e})}{\Phi(\frac{t-\mu}{\sigma})} \qquad (7.2.58)$$

where $\sigma^2 = \sigma_{ee} + \sigma_{uu}$. The following theorem easily follows as Theorem 7.2.2.

THEOREM 7.2.12 Under model (7.2.1), optimal pm-unbiased predictor of $F(t)$ in the class of predictors (7.2.56), where $p \in \rho_n$ is given by

$$\hat{F}_s^*(t, \mu) = \frac{\Phi_0}{n \Phi_1} \sum_{k \in s} \Delta(t - Y_k) \qquad (7.2.59)$$

where

$$\begin{aligned} \Phi_0 &= \Phi_0(t, \mu) = \Phi(\frac{t-\mu}{\sigma_e}) \\ \Phi_1 &= \Phi_1(t, \mu) = \Phi(\frac{t-\mu}{\sigma}) \end{aligned} \qquad (7.2.60)$$

Again, any $p \in \rho_n$ is optimal for using $\hat{F}_s^*$.

Assume that μ is not known (σ_{ee}, σ_{uu} are, however, assumed known) and is estimated by $m(Y_s) = m$ (say). In this case, a predictor of $F(t)$ is

$$\hat{F}^*(t, m) = \frac{\hat{\Phi}_0}{n\hat{\Phi}_1} \sum_{k \in s} \Delta(t - Y_k) \qquad (7.2.61)$$

where

$$\hat{\Phi}_0 = \Phi_0(t, m) = \Phi(\frac{t - m}{\sigma_e})$$

$$\hat{\Phi}_1 = \Phi_1(t, m) = \Phi(\frac{t - m}{\sigma})$$

Note that $\hat{F}_s^*(t, m)$ is a U-statistic except for the estimate m of μ.

$$\hat{F}_s^*(t, m) = \frac{1}{n} \sum_{k \in s} \hat{F}_{ks}^*(t, m)$$

where

$$\hat{F}_{ks}^*(t, m) = \frac{\Phi_0(t, m)}{\Phi_1(t, m)} \Delta(t - Y_k)$$

is a symmetric kernel of degree one. Taking $m = \bar{Y}_s = \sum_{i \in s} Y_i/n$ it follows by Randles' (1982) Theorem (Lemma 6.5.1) that under some regularity conditions $\hat{F}_s^*(t, m)$ is asymptotically normally distributed with mean $\Phi_0(t, \mu)$ and asymptotic prediction variance

$$V(\hat{F}_s^*(t, m) - F_N(t)) = \frac{\tau^2}{n} + \frac{\Phi_0(2\Phi_1 - \Phi_0 - 1)}{N}$$

where

$$\tau^2 = D'\Sigma D$$

$$D' = (1, \ A(t, \mu))$$

$$A(t, \mu) = -\{\frac{1}{\sigma_e}\phi(\frac{t - \mu}{\sigma_e})\Phi_1(t, \mu) + \frac{1}{\sigma}\phi(\frac{t - \mu}{\sigma})\Phi_0(t, \mu)\}/\{\Phi_1(t, \mu)\}^2$$

$$\Sigma = D(\hat{F}_s^*(t, \mu), \bar{Y}_s)$$

$D(.)$ denoting the dispersion matrix (Mukhopadhyay, 1998 d).

We now consider Bayes prediction of $F(t)$. Assume that σ_{ee}, σ_{uu} are known. Also, consider a Normal prior $N(o, \theta)$ for μ. It then follows as in (7.2.13), Bayes predictor of $F(t)$ under a squared error loss function is

$$\hat{F}_B(t) = E(F(t) \mid Y_s) = E\{E(\frac{1}{N} \sum_{i=1}^{n} \Delta(t - y_i) \mid \mu, Y_s) \mid Y_s\}$$

$$= \frac{1}{N} E\{E(\sum_{i \in s} \Delta(t - y_i) + \sum_{i \in r} \Delta(t - y_i) \mid \mu, Y_s) \mid Y_s\} \qquad (7.2.62)$$

Now, for $i \in s$,

$$E\{E(\Delta(t - y_i) \mid \mu, Y_s) \mid Y_s\} = E(\Phi(\frac{t - Y_i^*}{\sigma_0}) \mid Y_s)$$

$$= H_i \text{ (say)}$$

where

$$Y_i^* = \frac{Y_i \sigma_{ee} + \mu \sigma_{uu}}{\sigma^2}, \ \sigma_0^2 = \frac{\sigma_{uu} \sigma_{ee}}{\sigma^2}$$

Similarly, for $i \in \bar{s}$,

$$E\{E(\Delta(t - y_i) \mid \mu, Y_s) \mid Y_s\} = G \text{ (say)}$$

Therefore, from (7.2.61),

$$\hat{F}_B(t) = \frac{1}{N}[\sum_{i \in s} H_i + (N - n)G] \qquad (7.2.63)$$

Consider now optimal prediction of $F(t)$ under model (7.2.27). Here

$$E_p E(F(t)) = \frac{1}{N} \sum_{i=1}^{N} \Phi(\frac{t - \beta_0 - \beta x_i}{\sigma_e}) = \Phi_3 \text{ (say)}$$

Hence, $\hat{F}_s(t)$ in (7.2.56) is pm-unbiased for $F(t)$ iff

$$E_p(b_s) = 0$$

$$\sum_{k=1}^{N} \Phi_{4k} \sum_{s \ni k} b_{ks} p(s) = \Phi_3$$

where

$$\Phi_{4k} = E(\Delta(t - Y_k)) = \Phi(\frac{t - \beta_0 - \beta x_k}{\sigma})$$

Hence, we have the following theorem.

THEOREM 7.2.13 Under model (7.2.27) the optimal pm-unbiased predictor of $F(t)$ in the class of predictors (7.2.56) with $b_s \geq 0 \ \forall \ s$, $b_{ks} \geq 0 \ \forall \ (k,s)$ and for any given $p \in \rho_n$ with $p(s) > 0 \ \forall \ s$ is

$$\hat{F}_{BU}^* = \frac{\Phi_3}{N} \sum_{k \in s} \frac{\Delta(t - Y_k)}{\phi_{4k} \pi_k} \qquad (7.2.64)$$

Again,

$$E_p E(\hat{F}^*_{BU}(t) - \Phi_3)^2$$

$$= \Phi_3^2 [\frac{1}{N^2} \{\sum_{k=1}^{N} \frac{1}{\pi_k \Phi_{4k}} + \sum_{k \neq k'=1}^{N} \frac{\pi_{kk'}}{\pi_k \pi_{k'}} \} - 1]$$

7.3 PREDICTION UNDER MULTIPLICATIVE ERROR-IN-VARIABLES MODEL

Hwang (1986) considered the multiplicative error-in-variables model as follows: Let

$$\begin{aligned} Y &= x\beta + e \\ X &= x * V \end{aligned}$$ (7.3.1)

where $x = ((x_{kj}, k = 1, \ldots, n; j = 1, \ldots, p))$ are not observed directly and the observed data are $Y = (Y_1, \ldots, Y_n)$ and

$$X = ((X_{kj})) = ((x_{kj} v_{kj}))$$ (7.3.2)

which is the *Hadamard Product* of the matrix x and v. We assume

$$e = (e_1, \ldots, e_n)' \sim (0, \sigma_{ee} I_n), \sigma_{ee} < \infty$$

$$V = \begin{bmatrix} v_1' \\ \cdot \\ \cdot \\ v_n' \end{bmatrix} \text{ where } v_i \sim iid(1_p, M)$$ (7.3.3)

$$x = \begin{bmatrix} x_1' \\ \cdot \\ \cdot \\ x_n' \end{bmatrix}$$

where x_i's are *iid* with mean μ_x and finite variance. Also, e_i, v_j, x_k are assumed uncorrelated $(i, j, k = 1, 2, \ldots)$. We define

$$x\%V = ((x_{kj}/v_{kj}))$$ (7.3.4)

The above model was used by Hwang in analysing the data collected by the US Department of energy concerning energy consumption and housing characteristics of households. To preserve confidentiality some of the predicting variables that might identify household owners were multiplied by v_{kj}, which were assumed to be truncated normal random variables with known variances and means all equal to one. Clearly, such a model can also be used in randomized response trials.

THEOREM 7.3.1 Assume that $E(x_1 x_1')$ exists and is non-singular and $M = Cov(v_1)$ exists and is known. Then

$$\hat{\beta}_c = \{(X'X)\%M\}^{-1}X'Y \tag{7.3.5}$$

is strongly consistent for β.

Proof. Consider the *lse*

$$\hat{\beta}_{LS} = (X'X)^{-1}XY$$

Now

$$X'X/n = X'x\beta/n + X'e/n$$

Again,

$$X'e/n = \sum_{i=1}^{n} e_i X_i/n \rightarrow^{as} E(e_1 X_1) = E(e_1)E(X_1) = 0$$

where *as* denotes 'almost surely'. Also,

$$X'x/n = \sum_{i=1}^{n} X_i x_i'/n \rightarrow^{as} E(X_i x_i') = E(X_1 x_1') = A \text{ (say)}$$

Hence,

$$X'Y/n \rightarrow^{as} A\beta$$

Again,

$$X'X/n = \sum_{i=1}^{n} X_i X_i'/n \rightarrow^{as} E(X_1 X_1') = A * M$$

Therefore,

$$\hat{\beta}_{LS} \rightarrow^{as} (A * M)^{-1}A\beta \tag{7.3.6}$$

provided $A * M$ is non-singular. This is so, because

$$A * M - A = A * (M - 1_p 1_p')$$

Again, A is positive definite, $M - 1_p 1_p' = E(v_1 v_1') - E(v_1)E(v_1') = V(v_1)$ is positive semidefinite. Therefore, matrix in the right hand side of (6.3.6) is non-negative definite (vide Theorem 3.1 of Styan, 1973). Therefore, $A * M$ is p.d. and hence nonsingular, since A is so. Hence

$$\hat{\beta} = A^{-1}(A * M)\hat{\beta}_{LS} = M(X'X)^{-1}XY$$

$$= \{(X'X)\%M\hat{-}1XY$$

is strongly consistent for β.

COROLLARY 7.3.1 For p=1,

$$\hat{\beta}_{LS} = \frac{M'_{XY}}{M'_{XX}} \to M^{-1}\beta \tag{7.3.7}$$

where $M'_{XX} = \sum_{i=1}^{n} X_i^2/n$, $M'_{XY} = \sum_{i=1}^{n} X_i Y_i/n$. Hence

$$\hat{\beta}_{LS} \to^{as} M^{-1}\beta$$

where

$$M = E(v_i^2) = 1 + \sigma_{vv} \tag{7.3.8}$$

Hence a consistent estimator of β is

$$\hat{\beta}_c = M\hat{\beta}_{LS} = M'_{XY}(M'_{XX}M^{-1})^{-1} \tag{7.3.9}$$

THEOREM 7.3.2 In addition to the assumptions of Theorem 6.3.1, suppose that the elements of x and V have finite fourth moments and that e_i has a finite second moment. Then $\sqrt{n}(\hat{\beta} - \beta)$ is asymptotically normal with mean zero.

Proof

$$\sqrt{n}(\hat{\beta} - \beta) = \sqrt{n}[\{(X'X)\%M\}^{-1}X'(x\beta + e) - \beta]$$
$$= \sqrt{n}[\{(X'X)\%M\}^{-1}\{(X'x - (X'X)\%M)\beta + X'e\}]$$
$$= \{n^{-1}(X'X)\%M\}^{-1}S/\sqrt{n} \tag{7.3.10}$$

where

$$S = \{X'x - (X'X)\%M\}\beta + X'e$$

Now

$$S/\sqrt{n} = \frac{1}{\sqrt{n}} \sum_{i=1}^{n} [\{X_i x_i' - (X_i X_i')\%M\}\beta + X_i e_i] \tag{6.3.11}$$

The random vectors within the braces form a sequence of iid random vectors, $i = 1 \ldots, n$, having finite covariance matrices and zero mean. Therefore, by Central Limit Theorem, $S/\sqrt{n}$ is asymptotically normally distributed. Again, $\{\frac{1}{n}(X'X)\%M\}^{-1} \to^{as} A^{-1}$. Hence the result.

Huang (1986) also obtained an asymptotic expression for $Cov\{\sqrt{n}(\hat{\beta} - \beta)\}$ and its estimate.

THEOREM 7.3.3 Under (7.3.1)-(7.3.3) with p=1,

$$\sqrt{n}(\hat{\tilde{\beta}} - M^{-1}\beta) \to^L N(0, \sigma_{LS}^2)$$

$$\sqrt{n}(\hat{\tilde{\beta}}_c - \beta) \to^L N(0, \sigma_c^2)$$

as $n \to \infty$ where

$$\sigma_{LS}^2 = \{E[x_1^2]M\}^{-2}\{E[x_1^4]E[v_1 - v_1^2/n]^2\beta^2 + E[x_1^2]\sigma_{ee}M\}$$

$$\sigma_c^2 = M^2\sigma_{LS}^2$$

with

$$E(x_1^2) = \mu_x + \sigma_{xx}, M = 1 + \sigma_{uu} \tag{7.3.12}$$

Gasco, Bolfarine and Sandoval (1997) considered estimation of finite population total under multiplicative measurement error superpopulation models. Consider the following variant of the model (7.2.41):

$$Y_i = \alpha + \beta'x_i + e_i$$

$$X_i = x_i * v_i \tag{7.3.13}$$

with

$$E(x_i, e_i, v_i) = (\mu_x, 0, 1_p), \ V(x_i, e_i, v_i) = \text{Diag}(\Sigma_{xx}, \sigma_{ee}, \Sigma_{vv})$$

where, as before, observed data are $(Y_i, X_i, i \in s)$ and v_i is a measurement error corresponding to the true value x_i of the auxiliary variable $x = (x_1, \ldots, x_p)'$ on unit i. It is assumed that $X_1, \ldots, X_N$ are all known. For $r \geq 1$, two predictors of T are:

$$\hat{\tilde{T}}_{LS} = N\bar{Y}_s + (N - n)(\bar{X}_r - \bar{X}_s)'\hat{\tilde{\beta}}_{LS}$$

$$\hat{\tilde{T}}_c = N\bar{Y}_s + (N - n)(X_r - X_s)'\hat{\tilde{\beta}}_c$$

THEOREM 7.3.4 Under model (7.3.14) with $p \geq 1, \hat{\tilde{\beta}}_{LS}$ is asymptotically normal about mean $B\beta$ and $\hat{\tilde{\beta}}_c$ is asymptotically normal about mean β where

$$B = (E[x_1x_1'] * M)^{-1}E[x_1x_1'].$$

THEOREM 7.3.5 Under moder (7.3.14), with $p \geq 1$,

$$\sqrt{n}(\hat{\tilde{T}}_{LS} - T)/N \to^L N(0, (1 - f)\sigma_{LS}^2)$$

$$\sqrt{n}(\hat{\hat{T}}_c - T)N \to^L N(0, (1-f)\sigma_c^2)$$

where

$$\sigma_{LS}^2 = \sigma_{ee} + \beta'\{(E[x_1 x_1'] * \Sigma_{vv})B - (I - B')E[x_1](E[x_1])'(I - B)\}\beta$$

$$\sigma_c^2 = \sigma_{ee} + \beta'(E[x_1 x_1'] * \Sigma_{vv})\beta.$$

As in the case of additive measurement error models, $pm-$ unbiased optimal predictor, Bayes and minimax predictor of T and S_y^2 can be found out in the present case (vide exercise 5).

7.4 EXERCISES

1. Consider model (7.2.1) with e_i, u_j's having independent normal distributions. Show that for Jeffrey's (1959) non-informative prior

$$P(\mu, \tau) \propto \frac{1}{\tau}, \mu \in R_1, \tau > 0$$

Bayes predictor of T is given by $N\bar{Y}_s$ with posterior variance of T,

$$V(T \mid Y_s) = \frac{N(k+1) - nk}{(k+1)^2} s_Y^2 [\frac{2k(n-1)}{n-3} + \frac{N(k+1) - nk}{n}]$$

$$\simeq \frac{N^2(k+1)^2 - n^2 k^2}{n(n+1)^2} s_Y^2 \text{ (assuming } \frac{n-1}{n-3} \simeq 1)$$

where $s_Y^2 = \frac{1}{n-1} \sum_{i \in s}(Y_i - \bar{Y})^2$. Also, Bayes predictor of S_y^2 is

$$\hat{S}_{yB}^2 = E(S_y^2 \mid \mathbf{Y}_s) = \frac{k^2}{(k+1)^2} \frac{n(N+n-2)}{N(n-1)} \bar{Y}_s^2 +$$

$$\frac{ks_Y^2}{(k+1)^2}[k + \frac{k(N-n)}{N(N-1)} + \frac{2(n-1)}{N(n-3)}((N-n)(k+1) + 1)]$$

(Mukhopadhyay, 1994 a)

2. Let a sample s of n clusters out of K clusters constituting a finite population and a sample s_h of m_h second stage units be selected from each cluster h (consisting of M_h ssu's) in s. Let y_{hj} be the quantity of interest associated with unit j in cluster h. Consider the following error-in-variables superpopulation model

$$y_{hj} = \mu_h + e_{hj}$$

$$\mu_h = \mu + u_h$$

$$Y_{hj} = y_{hj} + \nu_{hj}$$

where Y_{hj} is the observed value on (h, j) and $e_{hj} \sim N(0, \sigma_{eeh})$, $u_h \sim N(0, \delta^2)$, $\nu_{hj} \sim N(0, \sigma_{\nu\nu h})$, the variables are all independent and Y_s denotes the set of all observed y values on the sample. Show that

$$E(y_{hj} \mid Y_s, \mu_h) = \begin{cases} \mu_h + \frac{\sigma_{eeh}(Y_{hj} - \mu_h)}{\sigma_{eeh} + \sigma_{\nu\nu h}} & h \in s, j \in s_h \\ \mu_h & \text{otherwise} \end{cases}$$

Also,

$$E(\mu_h \mid Y_s) = \lambda_h \bar{Y}_{sh} + (1 - \lambda_h)\bar{Y}_s$$

where $\bar{Y}_{sh} = \sum_{j \in s_h} Y_{hj}/m_h$, $\bar{Y}_s = \sum_{h \in s} \lambda_h \bar{Y}_{sh}/\sum_{j \in s} \lambda_j$ and

$$\lambda_h = \frac{\delta^2}{\delta^2 + (\sigma_{eeh} + \sigma_{uuh}/m_h} \text{ for } h \in s(0 \text{ otherwise}) .$$

Hence find Bayes estimator of T.   (Bolfarine, 1991)

3. Consider the model

$$Y = x\beta + e$$

$$X = x + v$$

with

$$E(e) = \mathbf{0}, V(e) = \sigma_{ee}I_N, E(v) = 0, V(v) = M$$

where

$$x = [x_1, \ldots, x_n]', x_i = (x_{i1}, \ldots, x_{ip})', v = [v_1, \ldots v_n]'$$

$Y = (Y_1, \ldots, Y_n)', x_{ij}$ being the value of the auxiliary variable x_j on unit i. Assume that $E(x_1 x_1') = A$, a nonsingular matrix and $v_i, x_j, e_k(i, j, k = 1, \ldots, n)$ are mutually uncorrelated. Show that under the assumption that $(X'X/n - M)$ is nonsingular with probability one, the estimator $\hat{\beta} = (X'X/n - M)^{-1}X'Y/n$ is strongly consistent for β.

(Mukhopadhyay, 1999 a)

4. Consider the model and the assumptions of exercise 3. Show that

$$\sqrt{n}(\hat{\beta} - \beta) = (X'X/n - M)^{-1}\sqrt{n}S$$

where

$$S = \{X'x/n - (X'X/n - M)\}\beta + X'e/n$$

and $\hat{\beta}$ is defined in exercise 3. Show also that $\sqrt{n}S$ is asymptotically normally distributed with zero mean. Also, asymptotic covariancwe matrix of $\sqrt{n}(\hat{\beta} - \beta)$ is

$$\Sigma = A^{-1}Cov\{1/\sqrt{n}\sum_{i=1}^{n}(Z_{1i} + Z_{2i})\}A^{-1}$$

$$= A^{-1}Cov(Z_1 + Z_2)\mathbf{A}^{-1}$$

where

$$Z_{1i} = X_i e_{1i}, Z_{2i} = (X_i x_i' - X_i X_i' + M)\beta$$

$$Z_{1i} = Z_1, Z_{2i} = Z_2, (Z_{1i}, Z_{2i} \text{ are iid})$$

Also $Cov(Z_1, Z_2) = 0$. Hence, $\Sigma = A^{-1}(\Sigma_1 + \Sigma_2)\mathbf{A}^{-1}, \Sigma_1 = $ Cov $(Z_1) = \sigma_{ee}E(X_1 X_1')$, $\Sigma_2 = $ Cov (Z_2). Again, show that

$$\hat{\sigma}_{ee} = \frac{1}{n}[Y'Y - Y'X\hat{\beta}]$$

is strongly consistent for σ_{ee}. However, $\hat{\sigma}_{ee}$ may be negative. A positive part estimator of σee, is, therefore,

$$\hat{\hat{\sigma}}_{ee} = \text{max } (0, \sigma_{ee}).$$

5. Consider model $y_i = \mu + e_i, Y_i = y_i u_i, e_i \sim (0, \sigma_{ee}), u_i \sim (1, \sigma_{uu}), e_i \underset{\sim}{ind} u_j$ $(i, j = 1, 2, \ldots)$. Show that under this model $\bar{Y}_s$ is best linear pm-unbiased predictor of $\bar{y}$ for any given $p \in \rho_n$. Also,

$$e_q^* = \frac{1}{n}(1 + \sigma_{uu})\sum_{k \in s}Y_k^2 - \frac{1}{n(n-1)}\sum_{k \neq k' \in s}\sum Y_k Y_{k'}$$

is best quadratic pm-unbiased predictor of S_y^2 for any given $p \in \rho_n$. In both the cases any $p \in \rho_n$ is an optimal sampling design.

(Mukhopadhyay, 1999 a)

6. Consider the model (7.2.1) and a normal prior $\mu \sim N(0, \theta^2)$ for μ. Find Bayes estimates of population total and variance under the Linex loss function (3.4.1). Also, find Bayes estimate of T under the model (7.2.27) with respect to the same loss function.

(Mukhopadhyay, 1999e)

Chapter 8

Miscellaneous Topics

8.1 INTRODUCTION

In this chapter we consider three special topics - (1) calibration estimators (predictors), estimators having some calibration property, (2) post-stratification (iii) conditional inference in survey sampling. Calibration estimators, initiated by Deville and Sarndal (1992) considers estimators which satisfy certain properties with respect to some known auxiliary variables, in certain cases model-unbiasedness property. In post-stratification analysis one important consideration is whether one should consider conditional analysis based on the observed sample sizes in different strata or unconditional analysis.

When a superpopulation model is available and is exploited to build up suitable predictors of population characteristics (mean, total, etc.), the conditionality principle (Cox and Hinkley, 1974) states that the infrence should be based on the model for a given sample. In this chapter, we however, consider conditional inference in the design-based sense, that is, we restrict ourselves to a part of the sample space containing samples on which some statistics often based on x (auxiliary variables)-values have certain fixed values (properties) (eg. sample mean $\bar{x}_s = m_x$, a fixed value) and find conditionally unbiased estimators of population parameters (mean, total, etc.). Under certain circumstances, the conditionally weighted estimators have less variance than unconditional estimators. This principle of conditional inference was first invoked by Rao (1985). We review some of these results in this chapter.

8.2 CALIBRATION ESTIMATORS

Assume that with each unit k there is a vector $x_k = (x_{k1}, \ldots, x_{kp})'$ of values of auxiliary variables $x_1, \ldots, x_p$. It is known that the Horvitz Thompson estimator HTE, $\hat{T}_{y\pi} = \hat{T}_\pi = \sum_{k \in s} d_k y_k$ where $d_k = 1/\pi_k$ is unbiased for the population total T. It is desired to find calibration estimator $\hat{T}_{ycal} = \sum_{k \in s} w_k y_k$ where the weights w_k are as close to d_k as possible (in some sense) and which satisfy

$$\sum_{k \in s} w_k x_k = \sum x_k = T_x = (T_{x1}, \ldots, T_{x_p})', \qquad (8.2.1)$$

$T_{x_j} = \sum_{j=1}^{N} x_{kj}$, where it is assumed that the population totals in the right hand side are known. Consider a distance function G with argument $z = w/d$ and with the properties: (a) G is positive and strictly convex (b) $G(1) = G'(1) = 0$ (c) $G''(1) = 1$. Here $G(w_k/d_k)$ measures the distance between w_k and d_k and $\sum_{k \in s} d_k G(w_k/d_k)$ is a measure of distance between $\{w_k\}$ and $\{d_k\}$ for the whole sample. Our objective is to find w_k such that for a given s, $\sum_{k \in s} d_k G(w_k/d_k)$ is minimum subject to the condition (8.2.1). The minimising equation is, therefore,

$$\frac{\partial}{\partial w_k} [\sum_{k \in s} d_k G(w_k/d_k) - \lambda'(\sum_{k \in s} w_k x_k - \sum_{k=1}^{N} x_k)] = 0$$

or

$$g(w_k/d_k) = x_k'\lambda = 0 \qquad (8.2.2)$$

where $g(z) = dG(z)/dz$ and λ is a vector of constants. From (8.2.2), $w_k = d_k F(x_k'\lambda)$ where $F(u) = g^{-1}(u)$, the inverse function of $g(u)$. The above properties of G imply that $F(u)$ exists and $F(0) = 1, F'(u) = 1$. To calculate the new weights w_k the value of λ is determined from the calibration equation

$$\sum_{k \in s} d_k F(x_k'\lambda) x_k = T_x \qquad (8.2.3)$$

where λ is only unknown. The systems of equations (8.2.3) can be solved by numerical methods.

Two functions G and the corresponding F functions are:

(a) The linear method: $G(x) = \frac{1}{2}(x-1)^2, x \in \mathcal{R}; F(u) = u + 1$

(b) The multiplicative method: $G(x) = x \log x - x + 1; g(u) = \log u; u = e^{g(u)}; F(u) = e^u(> 0)$. Deville and Sarndal (1992), Deville, Sarndal and Sautory (1993) listed seven different distance functions and examined the statistical properties of the corresponding calibration estimators.

In the linear method, the calibrated weights are given by

$$w_k = d_k F(x'_k \lambda) = d_k (1 + x'_k \lambda) \qquad (8.2.4)$$

where λ is determined by the solution of the equation

$$\sum_{k \in s} d_k (1 + x'_k \lambda) x_k = T_x$$

or

$$\sum_{k \in s} d_k x_k x'_k \lambda = t_x - \hat{T}_{x\pi} \qquad (8.2.5)$$

In this case the estimator $\hat{T}_{ycal}$ is known as the *generalised regression estimator* (GREG) and can be written as

$$\hat{T}_{ygreg} = \sum_{k \in s} w_k y_k = \sum_{k \in s} d_k (1 + x'_k \lambda) y_k$$

$$= \sum_{k \in s} d_k y_k + \sum_{k \in s} d_k x'_k \lambda y_k$$

$$= \hat{T}_{y\pi} + (T_x - T_{x\pi})' \hat{B}_s$$

where $\hat{T}_{x\pi} = \sum_{k \in s} d_k x_k$ denotes the *HTE* for the x vector and $\hat{B}_s$ is obtained by solving the sample-based equations

$$(\sum_s d_k x_k x'_k) \hat{B}_s = \sum_s d_k x_k y_k \qquad (8.2.6)$$

It is well-known that the asymptotic variance (AV) of $\hat{T}_{ygreg}$ (Section 2.5) is

$$AV(\hat{T}_{ygreg}) = \sum \sum_{k \neq l = 1}^{N} \Delta_{kl} (d_k E_k)(d_l E_l)$$

$$= \sum \sum_{k \neq l = 1}^{N} (\pi_{kl} - \pi_k \pi_l)(\frac{y_k - x'_k B}{\pi_k})(\frac{y_l - x'_l B}{\pi_l}) \qquad (8.2.7)$$

where B is determined as the solution of the normal equations of the hypothetical census fit corresponding to (8.2.6). That is, B satisfies

$$\sum_{k=1}^{N} x_k x'_k B = \sum_{k=1}^{N} x_k y_k \qquad (8.2.8)$$

Deville and Sarndal (1992) advocated the use of the variance estimator

$$v(\hat{T}_{ygreg}) = \sum \sum_{k \neq k' \in s} (\frac{\Delta_{kl}}{\pi_{kl}})(w_k e_k)(w_l e_l) \qquad (8.2.9)$$

where $e_k = y_k - x'_k \hat{B}_s$.

Any member of the class (8.2.4) is asymptotically equivalent to $\hat{T}_{yreg}$. The calibration estimators generated by different F functions share the same large sample variance. Thus (8.2.9) can be used to estimate the variance of any estimator $\hat{T}_{ycal}$ in this class.

Tille' (1995) considered conditions under which an estimator can be design-unbiased as well as calibrated. Dupont (1995) considered calibration estimators of T when the auxiliary information is obtained at different phases of sampling in a two-phase sampling procedure. Mukhopadhyay (2000 b) derived calibration estimator of a finite population variance. His estimator is found to be a member of his class of generalised predictor of a finite population variance (Mukhopadhyay, 1990). He (2000 c) also considered calibration estimator of a finite population distribution function.

The'berge (1999) considered an extension of calibration estimators as follows. Let c be a $N \times 1$ vector of constants, $X = ((x_{kj}))$, a $N \times p$ matrix of auxiliary values x_{kj}, the value of x_j on unit $k(j = 1, \ldots, p; k = 1, \ldots, N)$. let

$$D = \text{Diag} (\frac{1}{\pi_1}, \ldots, \frac{1}{\pi_N}) \qquad (8.2.10)$$

A matrix or a vector with suffix s will denote the part of the same with elements corresponding to the units in s only. The problem is to estimate $y'_s c_s$ where w_s is a $n \times 1$ vector of calibration weights for the sampled units. The weights w_s are so chosen as to minimise the distance between w_s and $D_s c_s$ under the constraints

$$X'_s w_s = X'c \qquad (8.2.11)$$

In other words, the constraints state that application of calibration weights to each auxiliary variable reproduces the known calibration totals.

Let α be a $n \times 1$ vector and U be a $N \times N$ positive diagonal matrix. The squared norm of α is defined as $||\alpha||^2_{U_s} = \alpha' U_s \alpha$. Using the squared norm as a distance, the vector of weights w_s should minimise $||w_s - D_s c_s||^2_{U_s}$ while satisfying the calibration equation (8.2.11). The matrix U_s serves to weight the contribution of each unit in the distance measure between w_s and $D_s c_s$. Many of the commonly used estimators are obtained by taking $U = D^{-1}Q^{-1}$ where Q is a diagonal nonnegative matrix. For Sarndal's (1982) procedure,

$U = D^{-1}Q^{-1}$ where Q is a diagonal variance matrix under a model. In Brewer (1994), $U = (D - I)^{-1}Z$ (assuming $\pi_k \neq 0$), where Z is a diagonal matrix of some size measure.

Often, there does not exist any solution to the calibration problem. The author, therefore, proposes a generalised problem so that if a solution to the calibration problem exists this comes as a particular case of the general problem. Let T be a $p \times p$ positive diagonal matrix used to weight the contribution of each of p constraints in the distance measure between $X_s'w_s$ and $X'c$. When $T = I_p$, equal importance is given to each constraint. Consider the following problem. Among the set of weights w_s that minimise $||X_s'w_s - X'c||_T^2$, find the one that also minimises $||w_s - D_s c_s||_{U_s}^2$.

The solution is given as

$$w_s = D_s c_s + U_s^{-1} X_s T^{1/2} (T^{1/2} X_s' U_s^{-1} X_s T^{1/2})^+ T^{1/2} (X'c - X_s' D_s c_s) \quad (8.2.12)$$

where F^+ is the unique Moore-Penrose inverse of F. The estimator of $y'c$ then becomes

$$y_s' w_s = \hat{y}'c + (y - \hat{y})' D \Delta c \quad (8.2.13)$$

where

$$\hat{y} = X T^{1/2} (T^{1/2} X_s' U_s^1 X_s T^{1/2})^+ T^{1/2} X_s' U_s^{-1/2} y_s \quad (8.2.14),$$

Δ being the diagonal matrix of $\delta_k, \delta_k = 1(0)$ if $k \in (\notin)s$.

If the calibration equation is solvable, the minimum of $||X_s'w_s - X'c||_T^2$ is 0, regardless of T. In this case T may be set equal to I_p.

If U has the form $U = D^{-1}Q^{-1}$, where Q is a diagonal variance matrix under a model and if $T = I_p$ and X_s is of rank p, then there is a solution to the calibration equation (8.2.12) and (8.2.13) yields Sarndal's generalised regression estimator. However, in many situations the calibration equations can not be solved or X_s is of rank less than p. In all these cases The'berge's solution holds. His approach is thus very promising, though it needs some statistical justification.

8.3 POST-STRATIFICATION

Sometimes, after the sample has been selcted from the whole population, the units are classified according to variables like age, sex, occupation, education and similar other facors, information regarding which are not available before sampling and collection of data. The units are thus classified into H post-strata according to these classifying variables. The censuses provide information on all these variables at aggregrate levels and thus the post-strata sizes, henceforth called strata sizes N_h are generally known. These provide estimates of population total for each stratum and hence estimate of population total. This procedure is known as *post-stratification* or stratification after sampling.

The method is more flexible than stratification before sampling or *prior stratification*, because after sampling the stratification factors can be chosen in different ways for different sets of variables of study in order to maximise the gain in precision. The technique is particularly suitable for multipurpose surveys when stratification factors selected prior to sampling may be poorly correlated with large number of secondary variables.

Let y_{hi} denote the value of y on the ith unit belonging to the hth post-stratum ($i = 1, \ldots N_h$ in the population ; $i = 1, \ldots, n_h$(in the sample); $h = 1, \ldots, L$). Let Z be a stratifying variable (like age group, occupation class, etc). Clearly, the sample size n_h in stratum h is a random variable, $\sum_h n_h = n$. We assume that n is large enough or stratification is such that the probability that some $n_h = 0$ is negligibly small. If $n_h = 0$ for some strata, two or more strata can be combined to make the sample size non-zero for each stratum. Fuller (1966) gave an alternative procedure which is superior to that of collapsing strata (Exercise 6). Let

$$\bar{y}_h = \sum_{i=1}^{N_h} y_{hi}/N_h, \ \bar{y} = \sum_h N_h\bar{y}_h/N, \ S_h^2 = \sum_{i=1}^{N_h}(y_{hi} - \bar{y}_h)^2/(N_h - 1)$$

$$P_h = N_h/N, \ \bar{y}_{hs} = \sum_{i=1}^{n_h} y_{hi}/n_h, \ s_h^2 = \sum_{i=1}^{n_h}(y_{hi} - \bar{y}_{hs})^2/(n_h - 1) \quad (8.3.1)$$

The post stratified estimator of $\bar{y}$ is

$$\bar{y}_{post} = \sum_h P_h\bar{y}_{hs} \qquad (8.3.2)$$

Clearly,

$$E[\bar{y}_{post}] = EE[\bar{y}_{post} \mid \mathbf{n}] = E(\bar{y}) = \bar{y}$$

where $\mathbf{n} = (n_1, \ldots, n_H)'$. The unconditional variance is

$$V[\bar{y}_{post}] = \sum_h P_h^2 S_h^2 E(1/n_h - 1/N_h)$$

$$\simeq (1/n - 1/N) \sum_h P_h S_h^2 + \sum_h (1 - P_h)S_h^2/n^2 \qquad (8.3.3)$$

using Stephan's (1945) approximation to $E(1/n_h)$. The difference between the variance of the unbiased estimator of $\bar{y}$ under prior stratification with proportional allocation and (8.3.3) is of order n^{-2} and can be neglected in large samples. However, for large value of H the difference between these two quantities may not be negligible. The unconditional variance (8.3.3) is a measure of the long term performance of $\bar{y}_{post}$ and should be considered at the planning stage of a survey (Deming 1960, p.323) while the conditional variance measures the performance of the strategy in the present survey. Durbin (1969) considered $\mathbf{n}$ as an anciliary statistic and suggested that the inference should be made conditional on $\mathbf{n}$. Royall (1971), Cox and Hinkley (1974), Royall and Eberhardt (1975), Kalton [in the discusson of Smith's (1976) paper], Oh and Scheuren (1983) also advocated the use of conditional inference under such circumstances.

If we apply the self-weighted estimator

$$\bar{y}_s = \sum_h n_h \bar{y}_h / n \tag{8.3.4}$$

for the post-stratified sampling procedure,

$$E(\bar{y}_s \mid \mathbf{n}) = \bar{y} - \sum_h \bar{y}_h (P_h - n_h/n)$$

$$MSE(\bar{y}_s \mid \mathbf{n}) = \sum_h (n_h/n)^2 (1 - f_h) S_h^2 / n_h + \{\sum_h \bar{y}_h (P_h - n_h/n)^2\} \tag{8.3.5}$$

The conditional bias of $\bar{y}_s$ is zero when either $n_h/n = P_h$ or $\bar{y}_h = $ constant $\forall h$. Also

$$MSE(\bar{y}_s \mid \mathbf{n}) - V(\bar{y}_{post} \mid \mathbf{n})$$

$$= \{\sum_h \bar{y}_h (P_h - n_h/n)\}^2 + \sum_h \{(n_h/n)^2 - P_h\}(1 - f_h) S_h^2 / n_h \tag{8.3.6}$$

The quantity (8.3.6) may be positive or negative. Thus no definite conclusion can be drawn from the comparison between these strategies. Numerical studies by Holt and Smith (1979) show that $\bar{y}_{post}$ is superior to $\bar{y}$ unless the sample size is small and the ratio of between stratum to within stratum variance is small.

Similarly, considering the usual variance estimator s^2/n for $\bar{y}_s$ where $s^2 = \sum_{h=1}^{H} \sum_{i=1}^{n_h} (y_{hi} - \bar{y}_s)^2 / (n - 1)$ it is seen that

$$(n - 1)E(s^2 \mid \mathbf{n}) = \sum_h n_h (\bar{y}_h - \bar{y}_s)^2 + \sum_h n_h S_h^2 (1 - n^{-1}) \tag{8.3.7}$$

Holt and Smith (1979) numerically studied the variations in the values of (8.3.5) and $E(s^2 \mid \mathbf{n})$ over values of $\mathbf{n}$ for fixed n and concluded that s^2/n is unsatisfactory as an estimate of (8.3.5).

Considering the situations of non-response, let r_h be the number of respondents in post-stratum $h, r_h \leq n_h (h = 1, \ldots, H)$. In this case, $\mathbf{n}$ may or may not be known, depending on whether Z is known for non-respondents. It is assumed that non-response is ignorable (Little, 1982; Rubin,1976), in the sense that respondents within post-stratum h can be taken as a random subsample of sampled cases in post-stratum h. Thus, r_h has a binomial distribution

$$r_h \mid \mathbf{n} \sim Bin(n_h, \psi_h)$$

where ψ_h is the response probability in stratum h. A stronger assumption is that the non-response rate is the same across post-strata, i.e. $\psi_h = \psi \; \forall \; h$ and respondents are a random subsample of sampled cases over all h. This is called *missing completely at random* (MACR) assumption by Rubin (1976). The post-stratified sampling estimator of $\bar{y}$ is

$$\bar{y}_{ps} = \sum_{h=1}^{H} P_h \bar{y}_h' \tag{8.3.8}$$

where

$$\bar{y}_h' = \sum_{i=1}^{r_h} y_{hi}/r_h, \tag{8.3.9}$$

the respondent sample mean. Also

$$\bar{y}_{ps} = \sum_{i=1}^{r} w_i y_i$$

where the weight $w_i = P_h/r_h$ if i belongs to stratum h and $r = \sum_h r_h$. The conditional variance of $\bar{y}_{ps}$, proposed by Holt and Smith (1979) as a measure of performance by $\bar{y}_{ps}$, is

$$Var\{\bar{y}_{ps} \mid \mathbf{r}\} = \sum_h P_h^2 (1 - f_h') s_h'^2/r_h \tag{8.3.10}$$

where $\mathbf{r} = (r_1, \ldots, r_H), f_h' = r_h/N_h$ and $s_h'^2$ is the sample variance of y based on r_h respondents.

We now consider Bayesian approach. Consider the basic normal post-stratification model (BNPM)

$$(y_i \mid Z_i = h, \mu_h, \sigma_h^2) \sim NID(\mu_h, \sigma_h^2); f(\mu_h, \log \sigma_h) = \text{Constant} \tag{8.3.11}$$

where Z_i is the value of Z on unit i, identifying the post-stratum and f is the joint density of $(.,.)$ and $(\mu_h, \log \sigma_h)$ follows Jeffrey's non-informative prior. The *iid* assumption within post-strata is to be modified for designs involving cluster sampling and differential selection rates within post strata. Under BNMP assumption

$$E(\bar{y} \mid Z, y_s) = \bar{y}_{ps}$$

$$V(\bar{y} \mid Z, y_s) = \sum_h P_h^2(1 - f_h')\delta_h s_h'^2/r_h \qquad (8.3.12)$$

where $Z = (Z_1, \ldots, Z_N)', y_s = (y_i, i \in s)$, and $\delta_h = (r_h - 1)/(r_h - 3)$ is a small sample correction for estimating the variance.

The Bayesian analysis, therefore, suggests that the randomisation conditional variance (8.3.10) differs from (8.3.12) only in the substitution of sample estimates for the (unknown) population variance and in the attendant small sample correction.

If we assume the same distribution of y_i across the post-strata

$$(y_i \mid Z_i = h, \mu, \sigma^2) \sim NID(\mu, \sigma^2) \qquad (8.3.13.1)$$

$$f(\mu, \log \sigma^2) = \text{Constant} \qquad (8.3.13.2)$$

Bayes estimate of $\bar{y}$ is

$$E(\bar{y} \mid Z, y_s) = \bar{y}_s = \sum_h \sum_i y_{hi}/r \qquad (8.3.14)$$

with posterior variance

$$Var(\bar{y} \mid Z, y_s) = v = \delta(1 - f)s'^2/r$$

where s'^2 is the sample variance of y based on r units and $\delta = (r-1)/(r-3)$ and $f' = r/N$. Alternatively, ignoring the data on Z, we might assume

$$(y_i \mid \mu, \sigma^2) \sim NID(\mu, \sigma^2)$$

together with (8.3.13.2). Here also the Bayesian estimate is $\bar{y}$ with posterior variance v.

Therefore, Bayesian analysis under (8.3.11) leads to the posterior mean $\bar{y}_{ps}$ and variance v_{ps}. Bayesian analysis under (8.3.13.1) [or (8.3.15)] and (8.3.13.2) yields the posterior mean $\bar{y}_s$ and variance v. The Bayesian approach leads to the two natural measures of precision (v_{ps} and v) and also provides small-sample correction (δ, δ_h) for estimating the variance.

Doss, Hartley and Somayajulu (1979) developed an exact small sample theory for post-stratification (Exercise 7). Some related references are Jagers, et al (1985), Jagers (1986), Casady and Valliant (1993), Valliant (1987, 1993).

8.3.1 ALGORITHM FOR COLLAPSING POST-STRATA

The small post-strata that contribute excessively to the variance are some-times collapsed in order to reduce v_{ps} (given in (8.3.12)). Frequentist strate-gies for collapsing post-strata were considered by Tremblay (1986) and by Kalton and Maligwag (1991).

If two post-strata i and j are collapsed, the model (8.3.11) is modified by replacing the means and variances in these post-strata by a single mean and variance for the combined post-strata . The post-stratified estimator under the collapsed model is

$$\bar{y}_{ps}^{(ij)} = \sum_{h(\neq i,j)} P_h \bar{y}_{hs} + (P_i + P_j)\bar{y}_{(i+j)s} \qquad (8.3.16)$$

where

$$\bar{y}_{(i+j)s} = (r_i \bar{y}_i + r_j \bar{y}_j)/(r_i + r_j) \qquad (8.3.17)$$

The posterior variance of $\bar{y}$ (corresponding to (8.3.12)) is then given by

$$v_{ps}^{(ij)} = v_{ps} - \Delta v_{ij}$$

where

$$\Delta v_{ij} = \frac{P_i^2 \delta_i (1 - f_i)s_i^2}{r_i} + \frac{P_j^2 \delta_j (1 - f_j)s_j^2}{r_j}$$

If the correction terms δ_i and δ_j for estimating the variance are ignored, the expected reduction in Bayes risk of $\bar{y}_{ps}$ is

$$E(\Delta v_{ij}) = \frac{P_i^2 (1 - f_i)\sigma_i^2}{r_i} + \frac{P_j^2 (1 - f_j)\sigma_j^2}{r_j} - \frac{(P_i + P_j)^2 (1 - f_{ij})\sigma_{ij}^2}{r_i + r_j} \qquad (8.3.20)$$

where $E(s_i^2) = \sigma_i^2, E(s_{ij}^2) = \sigma_{ij}^2$, ($s_{ij}^2$ being defined similarly as s_i^2), expecta-tion being taken wrt the superpopulation distribution of y_i given in (8.3.11). If y and the stratifying variable Z are independent, then

$$\sigma_i^2 = \sigma_j^2 = \sigma_{ij}^2 = \sigma^2$$

and (8.3.20) reduces to

$$E(\Delta v_{ij}/\sigma^2) \simeq \frac{r_i r_j}{(r_i + r_j)}\{w_i - w_j\}^2 \qquad (8.3.21)$$

Hence, in this case the variance is always reduced by collapsing, provided the weights w_i and w_j in the collapsed strata are unequal. However, when y and Z are associated, this reduction is counteracted by an increase in variance σ_{ij}^2 from collapsing. The following algorithm has been suggested.

- Order the post-strata so that the neighbours are *a priori* relatively homogeneous.

- Collapse the post-stratum pair (i, j) that maximises (8.3.20), subject to the restriction that $j = i + 1$, i.e. only neighbouring pairs are considered.

- Proceed sequentially until a reasonable number of pooled post-strata remain or (8.3.20) becomes substantially positive.

Little (1993) gave an algorithm to implement the above rules and made a numerical study of the procedure by collpasing post-strata in the Los Angels Epiodemiological Catchment Area (LAECA) survey.

Some alternative strategies for collapsing are as follows.

If the stratum means can be regarded as exchangeable, one can assume the following prior under normal specification.

$$p(\mu_h \mid \sigma_h) \sim NID(\mu, \tau^2), \sigma_h^2 = \sigma^2 \; \forall \; h$$

$$f(\mu, \; \log \tau^2, \; \log \sigma^2) = \text{ Constant} \tag{8.3.22}$$

This along with (8.3.11) yields Bayes estimate of $\bar{y}$ that smooths $\bar{y}_{ps}$ towards $\bar{y}_s$.

If variances are not assumed constant, one may assume the model

$$p(\mu_h \mid \sigma_h) \sim NID(\mu, \tau^2); f(\mu, \; \log \tau^2, \; \log \sigma_h^2) = \text{ Constant} \tag{8.3.23}$$

the resulting posterior mean shrinks $\bar{y}_{ps}$ towards an estimator where contribution from post-stratum h are weighted by s_h^{-2}. One should find groups of post-stratification within which exchangeability is possible and find out Bayes estimate of $\bar{y}$ and its risk.

Another approach to smoothing is to model the response rates. Let θ be the overall probability of selection of a unit and $\pi_h = \theta \psi_h$ be the probability of selection and response in post-stratum h. Assume

$$r_h \mid N_h, \pi_h \sim Bin(N_h, \pi_h) \tag{8.3.24}$$

Assume that π_h has a beta distribution with mean π and variance $k\pi(1-\pi)$. Then posterior distribution of π_h is beta with mean

$$E(\pi_h \mid \text{ data }) = \hat{\pi}_h = \lambda_h \pi + (1 - \lambda_h)\frac{r_h}{N_h}$$

where

$$\lambda_h = \frac{1 - K}{1 - K + KN_h}, (K \text{ a suitable constant }) \tag{8.3.25}$$

which smooths the observed selection rate π_h towards π. Estimating π by r/N, smoothed weights from this model have the form

$$\tilde{w}_h \propto \frac{w_h}{1 - \tilde{\lambda}_h + \tilde{\lambda}_h w_h} \qquad (8.3.26)$$

obtained by replacing r_h by $N_h\pi_h$ in $\tilde{w}_h(= P_h/r_h)$ where $\tilde{\lambda}_h$ is derived from λ_h by replacing K by its estimate. With more than one post-stratifier, the likelihood of sparse or empty post-strata increases, so the need to modify the BNMP assumption becomes greater. Let Z_1, Z_2 be two categorical post-stratifying variables and let P_{hk} be the population proportion with $Z_1 = h$ and $Z_2 = k$. Then

$$\bar{y} = \sum_h \sum_k P_{hk}\bar{y}_{hk} \qquad (8.3.27)$$

where $\bar{y}_{hk}$ is the population mean of cell (h, k). Suppose that a random sample is taken, resulting in n_{hk} individuals in cell (h, k), r_{hk} of whom respond to y. Model-based inference about $\bar{y}$ involves two distinct components: inference about $\{P_{hk}\}$ based on a model for the joint distribution of Z_1 and Z_2 and inference about $\bar{y}$ based on a model for the distribution of y given Z_1 and Z_2. A direct extention of (8.3.11) for y given Z_1, Z_2 is the basic normal two-way post-stratification model

$$(y_i \mid Z_{1i} = h, \mu_{hk}, \sigma_{hk}^2) \sim NID(\mu_{hk}, \sigma_{hk}^2)$$

$$f(\mu_{hk}, \ \log \sigma_{hk}) = \text{ Constant} \qquad (8.3.28)$$

The BNPM model (8.3.28) with known P_{kh} yields

$$E(\bar{y} \mid Z, y_s) = \bar{y}_{ps} = \sum_h \sum_k P_{hk}\bar{y}_{hk} \qquad (8.3.29)$$

In some circumstances $\{P_{hk}\}$ is unknown but the marginal distributions P_{h+}, P_{+k} of Z_1, Z_2 are known; then a model is needed to predict the P_{hk} values. Model inference on $\bar{y}$ is then needed using the information on **n** (Little, 1991 b).

Suppose first that Z_1, Z_2 are observed for non-respondents so that n_{hk} is observed. Under random sampling, a natural model for **n** is

$$\{n_{hk}\} \mid \mathbf{n} \sim MNOM[\{P_{hk}\}, n] \qquad (8.3.30)$$

the multinomial distribution with index n and probabilities $\{P_{hk}\}$. With the Jeffrey's prior P_{hk}, the posterior distribution of $P_{hk} \mid n_{hk}$ is Dirichlet with parameter $\{n_{hk} + 1/2\}$.

The posterior distribution of P_{hk} given P_{h+}, P_{+k}, n_{hk} does not have a simple form. But its posterior mean can be approximated as

$$\{^{(1)}_{hk}\} = Rake[\{n_{hk}\}, \{P_{h+}\}, \{P_{+k}\}] \qquad (8.3.31)$$

which denotes the result of raking the sample counts $\{n_{hk}\}$ to the known margins. For further details the reader may refer to Little (1993).

8.4 DESIGN-BASED CONDITIONAL UN-BIASEDNESS

One way of taking into account the auxilary information at the estimation stage is to adjust the value of the estimator conditional on the observed value of an auxiliary statistic. Post-stratification, where the estimator is based on the observed sample size in the post-strata, is an example of using such a conditional approach. Mukhopadhyay (1985 b) obtained a post-sample estimator of Royall (1970)-type which remains almost (model-) unbiased with respect to a class of alternative polynomial regression models. Robinson (1987) compared the conditional bias of the ratio estimator $\hat{t} = \frac{\bar{y}_s}{\bar{x}_s}X$ for given values of $\bar{x}_s$ under assumption of normality and corrected the estimate by using an estimate of its conditional bias. The conditionally adjusted estimator proposed by him is

$$\hat{t}_a = \hat{t} + (\hat{R} - b)(\bar{x}_s - \bar{x})\bar{X}/\bar{x}_s$$

where $b = \sum_{i \in s}(y_i - \bar{y}_s)(x_i - \bar{x}_s)/\sum_{i \in s}(x_i - \bar{x}_s)^2$, $\hat{R} = \bar{y}_s/\bar{x}_s$ and a conditionally adjusted variance estimator for $\hat{t}$ is

$$v_a = v_c \bar{x}^2/\bar{x}_s^2$$

where

$$v_c = (1 - f) \sum_{i \in s}(y_i - \hat{R}x_i)^2/n(n - 1)$$

and $f = n/N$. Rao (1985, 1994) obtained a general method of estimation using auxiliary information and applied a conditional bias adjustment for mean estimator, the ratio estimator in simple random sampling and stratified random sampling. Conditional design-based approach is a hybrid between classical design-based approach and model-dependent approach. This approach restricts one to a relevant set of samples, instead of the whole sample space and leads to the conditionally valid inferences in the sense that the ratio of the conditional bias to conditional standard error (conditional bias ratio) goes to zero as $n \to \infty$. Also, approximately $100(1 - \alpha)\%$

of the realised confidence intervals in repeated sampling from the relevant subset of samples will contain the unknown total T (Rao, 1997, 1999). We shall, here, specifically consider the conditionally design-unbiased estimation when sampling is restriced to a relevant set of samples. Tille' (1995, 1998) proposed a general method based on conditional inclusion probabilities that allows one to construct directly an estimator with a small conditional bias with respect to a statistic.

Consider $\eta(s) = \eta(x_k, k \in s)$ a statistic (where x is an auxiliary variable). Since the population is finite, $\eta(s)$ takes only a finite number of values $\eta_1, \ldots, \eta_m$ (say). Let $\hat{\theta}$ be an unbiased estimator of θ. The conditional bias of $\hat{\theta}$ given $\eta(s)$ is

$$B(\hat{\theta} \mid \eta(s)) = E(\hat{\theta} \mid \eta(s)) - \theta \qquad (8.4.1)$$

Hence, a conditionally unbiased estimator of θ is

$$\hat{\theta}^* = \hat{\theta} - B(\hat{\theta} \mid \eta(s)) \qquad (8.4.2)$$

Also,

$$V(\theta^*) = V(\hat{\theta}) - V(B(\hat{\theta} \mid \eta(s))] \qquad (8.4.3)$$

because,

$$Cov(\hat{\theta}, B(\hat{\theta} \mid \eta(s))) = V[B(\hat{\theta} \mid \eta(s))]$$

Therefore, the unconditional variance of conditionally unbiased estimator $\hat{\theta}^*$ is not greater than that of the original (unbiased) estimator of $\hat{\theta}$. In this approach, however, the conditional bias $B(\hat{\theta} \mid \eta(s))$ has to be estimated.

NOTE 8.4.1

If $\eta(s)$ is a sufficient statistic then $E(\hat{\theta} \mid \eta(s))$ is the improved estimator over $\hat{\theta}$ obtained using Rao-Blackwellisation technique. Also,

$$V[B(\hat{\theta} \mid \eta(s))] = V[E(\hat{\theta} \mid \eta(s))] \le V(\hat{\theta})$$

NOTE 8.4.2

The conditional unbiasedness is a special type of calibration imposed on the estimation procedure.

Consider a linear estimator of $\bar{y}$,

$$\hat{\bar{y}} = \sum_{k \in s} w_k(s) y_k = \sum_{k=1}^{N} w_k(s) I_k(s) y_k \qquad (8.4.4)$$

where the weights $w_k(s)$ may depend on k, s and a function of $x = (x_1, \ldots, x_N)$ (but is independent of y) and $I_k(s)$ is an indicator function ($I_k(s) = 1(0)$) if $s \ni k$ (otherwise)). To make $\hat{\bar{y}}$ conditionally unbiased we must have

$$E(\hat{\bar{y}} \mid \eta(s)) = \sum_{k=1}^{N} y_k E\{w_k(s)I_k(s) \mid \eta(s)\} = \frac{1}{N} \sum_{k=1}^{N} y_k$$

or

$$E\{w_k(s)I_k(s) \mid \eta(s)\} = \frac{1}{N} \ \forall \ k = 1, \ldots, N \tag{8.4.5}$$

Note that a necessary and sufficient condition for a conditionally unbiased estimator (8.4.4) for $\bar{y}$ to exist is

$$E\{I_k(s) \mid \eta(s)\} > 0 \ \forall \ k = 1, \ldots, N \tag{8.4.6}$$

A solution of (8.4.5), provided (8.4.6) is satisfied, is

$$w_k(s) = \frac{1}{N E\{I_k(s) \mid \eta(s)\}} \ \forall k \ \in s \tag{8.4.7}$$

Generally, a conditionally unbiased estimator does not exist.

Now,

$$E(\sum_{k \in s} w_k(s)y_k \mid \eta = \eta_i) = \bar{y}$$

$$\Rightarrow \sum_{s \ni k; \eta(s) = \eta_i} w_k(s)p(s) = \frac{P(\eta = \eta_i)}{N} \ \forall \ k \text{ such that}$$

$$P[k \in s \mid \eta(s) = \eta_i] > 0 \tag{8.4.8}$$

Therefore, conditions of conditionally unbiasedness can be written in terms of linear constraints (8.4.8).

EXAMPLE 8.4.1

Consider a *srswor* of size $n(s)$ from a population of size N. Let $\eta(s) = n(s)$. Here, $E\{I_k(s) \mid \eta(s)\} = n(s)/N \ \forall \ k = 1, \ldots, N$. Therefore, a conditionally unbiased estimator of $\bar{y}$ exists. The estimator $\bar{y}_s = \frac{1}{n(s)} \sum_{k \in s} y_k$ is conditionally unbiased.

DEFINITION 8.4.1 An estimator $\hat{\bar{y}}$ in (8.4.4) is *almost* or *virtually conditionally unbiased* (*ACU* or *VCU*) if

$$E(\hat{\bar{y}} \mid \eta(s)) = \bar{y} + \sum_{k : E(I_k(s) \mid \eta(s)) = 0} \alpha_k(\eta)y_k \tag{8.4.9}$$

where the weights $\alpha_k(\eta)$ can depend on $\eta(s)$ and k.

We shall now try to obtain conditionally unbiased estimator of $\bar{y}$ by modifying the method of deriving Horvitz-Thompson estimator $\hat{\bar{y}}_\pi = \frac{1}{N} \sum_{i \in s} \frac{y_i}{\pi_i}$. Let us define the conditional inclusion probabilities

$$E(I_k(s) \mid \eta) = \pi_{k|\eta}, \; E(I_k(s)I_{k'}(s) \mid \eta) = \pi_{kk'|\eta} \; (k \neq k').$$

An estimator constructed with conditional inclusion probabilities will be called a *conditionally weighted (CW) estimator*. Consider the *simple CW (SCW) estimator*

$$\hat{\bar{y}}_{\pi|\eta} = \frac{1}{N} \sum_{k \in s} \frac{y_k}{\pi_{k|\eta}} \tag{8.4.10}$$

where we assume that $\pi_{k|\eta} > 0 \; \forall \; k = 1, \ldots, N$. The conditional bias of SCW estimator is

$$B\{\hat{\bar{y}}_{\pi|\eta}\} = -\sum_{k=1}^{N} \frac{y_k}{N} I[\pi_{k|\eta} = 0] \tag{8.4.11}$$

where

$$I[\pi_{k|\eta} = 0] = \left\{ \begin{array}{ll} 1, & \pi_{k|\eta} = 0 \\ 0 & \text{otherwise} \end{array} \right.$$

It readily follows from (8.4.11) that $\hat{\bar{y}}_{\pi|\eta}$ is almost conditionally unbiased. The estimator (8.4.10) is also obtained if HTE is corrected for its conditional bias (Exercise 13).

Consider some other CW estimators:

$$\tilde{\bar{y}}_{\pi|\eta} = \frac{1}{N} \sum_{k \in s} \frac{y}{k} h_k \pi_{k|\eta} \tag{8.4.12}$$

where

$$h_k = EI[\pi_{k|\eta} > 0] = P[\pi_{k|\eta} > 0]$$

The estimator (8.4.12) may be called corrected CW (CCW) estimator. Its conditional bias is

$$B(\tilde{\bar{y}} \mid \eta) = \frac{1}{N} \sum_k [\frac{I(\pi_{k|\eta} > 0)}{h_k} - 1] \tag{8.4.13}$$

The CCW estimator is not VCU but is unconditionally unbiased. Both SCW and CCW estimators are not translation invariant i.e. these estimators do not increase by the amount C if all the units y_k are increased by C. The following two ratio type estimators are translation invariant.

(i)The SCW ratio:

$$\hat{\bar{y}}'_{\pi|\eta} = \sum_{k \in s} \frac{y_k}{\pi_{k|\eta}} / \sum_{k \in s} \frac{1}{\pi_{k|\eta}} \tag{8.4.14}$$

(ii) The CCW ratio:

$$\tilde{\bar{y}}' = \sum_{k \in s} \frac{y_k}{h_k \pi_{k|\eta}} / \sum_{k \in s} \frac{1}{h_k \pi_{k|\eta}} \tag{8.4.15}$$

It may be noted that

$$E(I_k(s) \mid \eta = \eta_i) = \pi_k P(\eta = \eta_i \mid k \in s)/P(\eta = \eta_i) \tag{8.4.16}$$

Therefore, to know the conditional inclusion probabilities we must know the conditional as well as unconditional probability distribution of η.

EXAMPLE 8.4.2

Consider *srswor* of m units drawn from a population of size N. Let n_s be the number of distinct units. It is known (Des Raj and Khamis (1958), Pathak (1961)) that

$$E(n_s) = N(1 - \frac{(N-1)^m}{N^m}) \tag{8.4.17}$$

$$V(n_s) = \frac{(N-1)^m}{N^{m-1}} + (N-1)\frac{(N-2)^m}{N^{m-1}} - \frac{(N-1)^{2m}}{N^{2m-2}} \tag{8.4.18}$$

Suppose that n_s is an anciliary statistic. Given n_s, the sampling procedure is *srswor* and hence $\pi_{k|n_s} = n_s \ \forall \ k$. The unconditional inclusion probabilities are $\pi_k = E(\pi_{k|n_s}) = E(n_s)/N = 1 - \frac{(N-1)^m}{N^m}$. Now, Prob. $(\pi_k \mid n_s) > 0) = 1$ and hence all the four CW estimators are identical and given by $\bar{y}_s = \sum_{k \in s} y_k/n_s$.

The HTE is

$$\hat{\bar{y}}_s = \frac{n_s \bar{y}_s}{E(n_s)} = \{N(1 - \frac{(N-1)^m}{N^m})\}^{-1} \sum_{k \in s} y_k$$

Note that $\bar{y}_s$ is conditionally unbiased. Again, $\bar{y}_s$ is translation invariant while $\hat{\bar{y}}_s$ is not. Now

$$V(\bar{y}_s) = E(\frac{\sigma_y^2(N - n_s)}{n_s(N-1)}) = \frac{\sigma_y^2 \sum_{j=1}^{N-1} j^{m-1}}{N^{m-1}} \tag{8.4.19}$$

$$V(\hat{\bar{y}}) = \frac{\sigma_y^2}{N-1}[\frac{N}{E(n_s)} - \frac{E(n_s^2)}{\{E(n_s)\}^2}] + \frac{\bar{y}^2 V(n_s)}{(E(n_s))^2} \tag{8.4.20}$$

Hence, under certain conditions the CW-estimator $\bar{y}_s$ has a smaller variance than $\hat{\bar{y}}$.

One criterion of choice of η is to choose one for which the unconditional mean square of the conditionally unbiased estimator is as small as possible. Considering the HTE and its associated CW estimator, it is seen from (8.4.3) that

$$V[\hat{\bar{y}}_\pi - B(\hat{\bar{y}}_\pi \mid \eta)] = V(\hat{\bar{y}}_\pi) - V[E(\hat{\bar{y}} \mid \eta) - \bar{y}] \qquad (8.4.21)$$

The auxiliary statistic η must, therefore, be so chosen that $Var[E(\hat{\bar{y}}_\pi \mid \eta)]$ is as large as possible. The statistic η and $\hat{\bar{y}}_\pi$ must, therefore, be very dependent. Also, $\pi_{k\mid\eta}$ should be positive for all k, so that the conditional bias remains small.

Assuming the joint distribution of $\hat{\bar{y}}$, a design-based estimator of *bary* and $\hat{\bar{x}}$, a design-based estimator of a q-variate random vector $\bar{x} = (\bar{x}_1, \ldots, \bar{x}_q)'$ of mean of auxiliary variables $x_j (j = 1, \ldots, q)$ is mutivariate normal, Montanari (1999) considered the conditional distribution of $\hat{\bar{y}}$ given $\hat{\bar{x}}$ and hence found a bias-corrected conditionally unbiased estimator of $\bar{y}$ and studied its properties for several estimators. Mukhopadhyay (2000 b) considered the conditionally weighted estimator of a finite population variance (exercise 15).

8.5   EXERCISES

1. *Calibration estimator* Suppose that a measure of average distance between d_k and w_k is $D_p(w, d) = E_p(\sum_{k \in s}(w_k - d_k)^2/q_k d_k)$ where $d_k = 1/\pi_k$ and q_k are weights uncorrelated with w_k. The problem is to find calibration estimator $\hat{t}_{cal} = \sum_{k \in s} w_k y_k$ of T such that D_p is minimised subject to the calibration conditions

$$\sum_{k \in s} w_k x_k = t_x$$

Show that the minimisation leads to the equation

$$w_k = d_k(1 + q_k x_k' \lambda)$$

where λ is a vector of Lagrangian multipliers and is given by the equation

$$\lambda = T_s^{-1}(t_x - \hat{t}_{x\pi}),$$

$$T_s = \sum_{k \in s} d_k q_k x_k x_k'$$

and is assumed to be non-singular. The resulting calibration estimator of T is, therefore, $\hat{t}_{y,cal} = \hat{t}_{y\pi} + (t_x - \hat{t}_{x\pi})'\hat{B}_s$ where $\hat{B}_s = T_s^{-1}\sum_{k \in s} d_k q_k x_k y_k$ which is *greg*- predictor. In particular, when x_k is a scalar $(x_k > 0), q_k = 1/x_k$, show that $\lambda = t_x/\hat{t}_{x\pi}, w_k = d_k t_x/\hat{t}_{x\pi}$ and hence $\hat{t}_{y,cal} = t_x \hat{t}_{y\pi}/\hat{t}_{x\pi}$, ratio estimator of T.

(Deville and Sarndal, 1992)

2. *Calibration estimator* Consider the estimator $\hat{T} = \sum_{i \in s} b_{si} y_i$ of population total T where $b_{si} = 0$ if $i \notin s$. Define the distance function

$$\phi_s = \sum_{i \in s} \{b_{si} - b_{si}^*\}^2/q_{si}b_{si}$$

where q_{si} are known positive weights unrelated to b_{si}. Show that minimising ϕ_s subject to calibration conditions

$$\sum_{i \in s} b_{si}^* x_i = X$$

where $x_i = (x_{i1}, \ldots, x_{ip}), X = (X_1, \ldots, X_p), x_{ij}$ is the value of the auxiliary variable x_j on unit $i, X_j = \sum_{i=1}^N x_{ij}$, one gets the calibration estimator

$$\hat{T}_{gr} = \hat{T} + (X - \hat{X})'\hat{B}$$

with

$$\hat{B} = (\sum_{i \in s} b_{si}q_{si}x_i x_{i'})^{-1}(\sum_{i \in s} b_{si}q_{si}x_i y_i)$$

In particular, when $b_{si} = 1/\pi_i$ and $q_{si} = q_i, \hat{T}_{gr}$ reduces to $\hat{t}_{ygreg}$ given in (8.3.6).

(Rao, 1994)

3. *Calibration estimator of variance* Consider the variance of e_{HT}, HTE of population total $T(\text{y of } y, V\hat{T}_{HT})_{YG} = V_{YG}$ defined in (1.4.16) and its unbiased estimator v_{YG} defined in (1.4.17). Consider a calibration estimator of V_{YG},

$$v_1(e_{HT}) = \sum\sum_{i<j \in s} w_{ij}(y_i/\pi_i - y_j/\pi_j)^2 \qquad (i)$$

where w_{ij} are modified weights which are as close as possible in an average sense to $d_{ij} = (\pi_i\pi_j - \pi_{ij})/\pi_{ij}$ for a given distance measure and subject to the calibration constraint

$$\sum\sum_{i<j \in s} w_{ij}(x_i/\pi_i - x_j/\pi_j)^2 = V_{YG}(\hat{X}_{HT}) \qquad (ii)$$

where $\hat{X}_{HT} = \sum_{i \in s} x_i/\pi_i$, the HTE of population total X of an auxiliary variable x. Considerthe distance measure D between w_{ij} and d_{ij},

$$D = \sum \sum_{j<k=1}^{n} \frac{(w_{ij} - d_{ij})^2}{d_{ij}q_{ij}} \qquad (iii)$$

Show that minimisation of D subject to the condition (iii) leads to the optimal weights given by

$$w_{ij} = d_{ij} + \frac{d_{ij}q_{ij}(x_i/\pi_i - x_j/\pi_j)^2}{\sum \sum_{i<j=1}^{n}(x_i/pi_i - x_j/\pi_j)^4}$$

$$V_{YG}(\hat{X}_{HT} - \sum \sum_{i<j=1}^{n}(x_i/\pi_i - x_j/\pi_j)^2) \qquad (iv)$$

Substitution of (iv) in (i), therefore, leads to the following regression type estimator

$$v_1(\hat{T}_{HT}) = v_{YG}(\hat{T}_{HT}) + \hat{B}[V_{YG}(\hat{X}_{HT} - v_{YG}(\hat{X}_{HT})] \qquad (v)$$

where

$$\hat{B} = [\sum \sum_{i<j=1}^{n} d_{ij}q_{ij}(x_i/\pi_i - x_j/\pi_j)^2(y_i/\pi_i - y_j/\pi_j)^2]/[\sum \sum_{i<j=1}^{n} d_{ij}q_{ij}(x_i\pi_i - x_j/\pi_j)^4]$$

$$(vi)$$

Show that for $srswor$ and $q_{ij} = 1 \; \forall \; i \neq j$

$$\hat{v}_1(\hat{T}) = \frac{N^2(1-f)}{n}[s_y^2 + \hat{b}(S_x^2 - s_x^2)]$$

where

$$\hat{b} = \sum \sum_{i<j=1}^{n}(y_i - y_j)^2(x_i - x_j)^2 / \sum \sum_{i<j=1}^{n}(x_i - x_j)^4$$

and $f = n/N$. For $srswor$ and $q_{ij} = (x_i/\pi_i - x_j/\pi_j)^{-2}$,

$$v_1(\hat{T}) = v_{YG}(\hat{T}_{HT})[\frac{V_{YG}(\hat{X}_{HT})}{v_{YG}(\hat{X}_{HT})}]$$

(Singh, et al, 1999)

4. Consider the problem of estimating a finite population variance

$$S_y^2 = a_1 \sum_{i=1}^{N} y_i^2 - a_2 \sum \sum_{i \neq i'=1}^{N} y_i y_{i'}$$

where

$$a_1 = \frac{1}{N}(1 - \frac{1}{N}), \ a_2 = \frac{1}{N^2}.$$

It is known that the following Horvitz-Thompson type estimator due to Liu (1974) is unbiased for S_y^2,

$$\hat{S}_{y\pi}^2 = a_1 \sum_s d_k y_k^2 - a_2 \sum_s{}' d_{kk'} y_k y_{k'}$$

where $d_k = 1/\pi_k, d_{kk'} = 1/\pi_{kk'}$ and $\sum_s, \sum_s'$ denote $\sum_{k \in s}$ and $\sum_{k \neq k' \in s}$, respectively. It is desired to find calibration estimator of S_y^2,

$$\hat{S}_{y;cal}^2 = a_1 \sum_s w_k y_k^2 - a_2 \sum_s{}' w_{kk'} y_k y_{k'}$$

where the weights $\{w_k\}$ and $\{w_{kk'}\}$ are as close as possible to $\{d_k\}$ and $\{d_{kk'}\}$ (in some sense) and which satisfy the calibration constraints

$$\sum_s w_k(x_k^2) = \sum_{i=1}^{N} (x_i^2) \qquad (i)$$

$$\sum_s{}' w_{kk'}(x_k x_{k'}) = \sum_{i \neq i'=1}^{N} (x_i x_{i'}) \qquad (ii)$$

where

$$(x_k^2) = (x_{k1}^2, \ldots, x_{kp}^2)', \ (x_k x_{k'}) = (x_{k1} x_{k'1}, \ldots, x_{kp} x_{k'p})'$$

and it is assumed that the totals in the right hand side of (i) and (ii) are known. Here x_{kj} is the value of an auxiliary variable x_j on unit k in the population $(j = 1, \ldots, p; k = 1, \ldots, N)$.

Consider a distance function G with argument $z = w/d$ measuring the distance between w_k and d_k, $w_{kk'}$ and $d_{kk'}$ and consider

$$H_s = \sum_s d_k G(w_k/d_k) + \sum_s{}' d_{kk'} G(w_{kk'}/d_{kk'})$$

as a measure of distance between the weights w_k, d_k and $w_{kk'}, d_{kk'}$ for the whole sample. Considering the linear method of Deville and Sarndal (1992) where $G(z) = (z - 1)^2/2, z \in R^1$ show that the calibration estimator of S_y^2 is

$$\hat{S}_{y;cal}^2 = \hat{S}_{y\pi}^2 + a_1 \sum_{j=1}^{p} (\sum_{i=1}^{N} x_{ij}^2 - \sum_s d_k x_{kj}^2) \hat{B}_{sj} -$$

$$a_2 \sum_{j=1}^{p} (\sum_{i \neq i'=1}^{N} x_{ij}x_{i'j} - \sum_{s}^{'} d_{kk'}x_{kj}x_{k'j})\hat{T}_{sj}$$

where

$$\hat{B}_s = (\hat{B}_{s1}, \ldots, \hat{B}_{sp})', \ \hat{T}_s = (\hat{T}_{s1}, \ldots, \hat{T}_{sp})'$$

and $\hat{B}_s, \hat{T}_s$ are obtained by solving the sample-based equations

$$(\sum_{s} d_k(x_k^2)(x_k^2)')\hat{B}_s = \sum_{s} d_k(x_k^2)y_k^2$$

$$(\sum_{s}^{'} d_{kk'}(x_k x_{k'})(x_k x_{k'})')\hat{T}_s = \sum_{s}^{'} d_{kk'}(x_k x_{k'})y_k y_{k'}$$

(Mukhopadhyay, 2000 c)

5. Find a calibration estimator of a finite population variance

$$F_N(t) = \frac{1}{N} \sum_{i=1}^{N} \Delta(t - y_i)$$

where $\Delta(z) = 1(0)$ if $z \geq 0$ (elsewhere)

(Mukhopadhyay, 2000 b)

6. *Post-stratification* Show that under the superpopulation model

$$y_{hi} = \mu_h + \epsilon_{hi}, \ i = 1, \ldots, N_h; \ h = 1, \ldots, L$$

$$E(\epsilon_{hi}) = 0, E(\epsilon_{hi}^2) = \sigma_h^2, E(\epsilon_{hi}\epsilon_{hi'}) = 0, i \neq i'$$

the best linear unbiased prediction estimator of y_h is

$$\hat{y}_h^* = \sum_{i \in s_h} y_{hi} + (N_h - n_h)\bar{y}_{hs} = N_h \bar{y}_{hs}$$

and hence the BLUP-estimator of $\bar{y}$ is

$$\hat{\bar{y}}^* = \sum_{h} N_h \bar{y}_h / N = \bar{y}_{post}$$

The model bias of $\hat{\bar{y}}^*$ is $E[(\hat{\bar{y}}^* - \bar{y}) \mid \mathbf{n}] = \sum_{h} \mu_h(n_h/n - N_h/N)$. The prediction variance of $\hat{\bar{y}}^*$ is

$$E[(\hat{\bar{y}}^* - \bar{y})^2 \mid \mathbf{n}] = \sum_{h} (n_h/n)^2(1 - n_h/N_h)\sigma_h/n_h + \{\sum_{h} \mu_h(n_h/n - N_h/N)^2\}^2$$

$$+ \sum_h \sigma_h^2 (N_h/n^2)(n_h/N_h - n/N)^2$$

If, further, we assume that $\mu_h \sim NID(\nu, \tau^2)$, Bayes estimate of $\bar{y}$ is

$$\hat{\bar{y}}_B = \sum_h \{n_h \bar{y}_h + (N_h - n_h)\lambda_h \bar{y}_h + (N_h - n_h)(1 - \lambda_h)\bar{y}_h + (N_h - n_h)(1 - \lambda_h)\bar{y}\}/N$$

where

$$\bar{y}_s = \sum_h \lambda_h \bar{y}_{hs} / \sum_h \lambda_h, \quad \lambda_h = \frac{\tau^2}{\tau^2 + \sigma_h^2/n_h} \qquad (i)$$

Hence, show that when $\lambda_h \simeq 1$, i.e. when σ_h^2/n_h is very small compared to $\tau^2 \; \forall \; h$, (i) reduces to $\bar{y}_{post}$.

(Holt and Smith, 1979)

7. *Post-stratification* Assume that a population is divided into two strata. A *srs* of size n is drawn and the sampled units are identified as falling into either stratum 1 or 2. The following types of outcomes are also identified: (i) both the strata contain some sampled units (ii) one of the strata contains no sampled unit.

Consider the following estimator for the population mean $\bar{y}$.

For case (i), $\hat{\bar{y}} = \bar{y}_{post} = W_1 \bar{y}_1 + W_2 \bar{y}_2$ where $W_1 = N_1/N, W_2 = 1 - W_1, \bar{y}_i = $ sample mean for the ith stratum.

For case (ii),

$$\hat{\bar{y}} = \begin{cases} D_1 \bar{y}_{1s} & \text{if stratum 2 is empty} \\ D_2 \bar{y}_{2s} & \text{if stratum 1 is empty} \end{cases}$$

Show that $\hat{\bar{y}}$ is conditionally unbiased if $D_1 = W_1/P_1^*, D_2 = W_2/P_2^*$, where P_1^* is the probability that the stratum 2 will contain no sample element given that the case (ii) has happened and similarly for P_2^*.

Show that the conditional variance of $\hat{\bar{y}}$ for case (i) is

$$Var_I(\hat{\bar{y}}) = E_I[\sum_{i=1}^{2} W_i^2 \frac{N_i - n_i}{n_i N_i} S_i^2]$$

where E_I denotes expectation over all outcomes associated with case (i). Show also that the conditional variance of $\hat{\bar{y}}$ for case (ii) is

$$Var_{II}(\hat{\bar{y}}) = \frac{1}{n} \sum_{i=1}^{2} P_i^* D_i^2 (\frac{N_i - n_i}{N_i}) S_i^2 + (\sum_{i=1}^{2} \frac{W_i}{P_i^*} \hat{\bar{y}}_i^2 - \hat{\bar{y}}^2)$$

Hence, show that

$$Var(\hat{\bar{y}}) = P_I Var_I(\hat{\bar{y}}) + P_{II} Var_{II}(\hat{\bar{y}})$$

where P_I, P_{II}, respectively, are the probabilities that cases (i), (ii) occur. The common procedure in case (ii) is to combine or collapse the two strata. Denoting this estimator as $\hat{\bar{y}}_c$, show that the conditional bias of this estimator is

$$E_{II}(\hat{\bar{y}}_c - \bar{y}) = (P_1^* - W_1)\bar{y}_1 + (P_2^* - W_2)\bar{y}_2$$

$$MSE(\hat{\bar{y}}_c) = \sum_i \frac{N_i - n_i}{n_i N_i} S_i^2 P_i^* + \sum_i P_i^*(\bar{y}_i - \bar{y})^2$$

Therefore,

$$MSE_{II}(\hat{\bar{y}}_c) - Var_{II}(\hat{\bar{y}}) = \frac{1}{n} \sum_i P_i^*(1 - D_i^2)(\frac{N_i - n_i}{N_i}) S_i^2 +$$

$$+ \sum_i P_i^* \{(y_i - \bar{y})^2 - (D_i \bar{y}_i - \bar{y})^2\})$$

Hence, comment on the relative performance of $\hat{\bar{y}}_c$ and $\hat{\bar{y}}$ in case (ii).

(Fuller, 1965)

8. *Post-stratification* A simple random sample of size n is drawn from a population of size N divided into L post-strata of size $N_h(h = 1, \ldots, L)$. Let n_h be the sample size in the hth stratum. Show that the estimator

$$\bar{y}_s = \sum_h a_h P_h \bar{y}_{hs} / E(a_h)$$

where a_h is an indicator variable,

$$a_h = \begin{cases} 1 & \text{at least one sampled unit} \in \text{stratum } h \\ 0 & \text{elsewhere,} \end{cases}$$

$P_h = N_h/N$, $\bar{y}_{hs} = $ sample mean for the hth stratum if $n_h \neq 0$ ($\bar{y}_h$ if $n_h = 0$), is unbiased for $\bar{y}$.

Show that the estimator $\bar{y}_s$ is conditionally biased,

$$E(\bar{y}_s \mid \mathbf{n}) = \sum{}' P_h \bar{y}_h / E(a_h) \neq \bar{y},$$

$\sum'$ being over all post-strata with $n_h \neq 0$, with $\mathbf{n} = (n_1, \ldots, n_L)$.

Show also that if each y-value is changed to $y + c$(c, a constant), $V(\bar{y}'_s)$ (where $\bar{y}'_s$ denotes the estimator of $\bar{y}$ for the changed values) can be made as large as possible by the proper chioce of c. Show that the variance of the ratio estimator of $\bar{y}$,

$$\hat{R} = \frac{\sum_h a_h P_h \bar{y}_h / E(a_h)}{\sum_h a_h P_h / E(a_h)}$$

does not depend on the shift in origin of the y-values. Find an exact expression for the variance of $\hat{R}$ and its variance estimator.

Also, consider the estimator

$$\hat{R}' = \frac{\sum'_h W_h \bar{y}_{hs}}{\sum'_h W_h}$$

Find its variance and variance estimator. Show that $\hat{R}'$ is not unconditionally (over all **n**) consistent for $\bar{y}$.

(Doss, et al, 1979; Rao, 1985)

9. *Post-stratification* Suppose the population has been divided into L post-strata $U_1, \ldots, U_L$ and the relationship between y and x in the post-stratum (henceforth called stratum) is

$$y_{hj} = \alpha_h + \beta_h x_{hj} + \epsilon_{hj} \qquad (i)$$

$$E(\epsilon_{hj}) = 0, \ E(\epsilon_{hj}^2) = \sigma_h^2, \ E(\epsilon_{hj}\epsilon_{h'j'}) = 0, \ (h,j) \neq (h',j')$$

It is assumed that for every sampled unit its x-value and stratum-affiliation is known and for units not in the sample their x-values are known but not the stratum affiliation. Such a situation may arise if x is not the stratifying variable. Hence, the population size N and population mean $\bar{X}$ are known, but not the stratum means $\bar{X}_h$ and strata sizes N_h required for use of seperate regression estimates.

Now an optimal predictor of $\bar{y}$ is

$$\hat{\bar{y}}(N_h, \bar{X}_h) = \frac{1}{N}(\sum_{j \in s} y_j + \sum_{j \notin s} \hat{y}_j)$$

where $\hat{y}_j$ is the optimal model-dependent predictor of y_j. Again,

$$\hat{\bar{y}}(N_h, \bar{X}_h) = \frac{1}{N}(\sum_{j \in s} y_j + \sum_{j \in \bar{s}} \sum_{h=1}^{L} P_h(j \mid x_j)(\hat{\alpha}_h + \hat{\beta}_h x_j)\} \qquad (ii)$$

where $\hat{\alpha}_j = (\bar{y}_{hs} - \hat{\beta}_h \bar{x}_{hs}), \hat{\beta}_h$ is the ordinary least square estimator of $\beta_h, P_h(j \mid x_j)$ is the conditional probability that a unit j , selected at random out of $\bar{s}$ belongs to U_h, given that its x-value is x_j, i.e.

$$P_h(j \mid x_j) = \text{Prob. } \{j \in U_h \mid j \in \bar{s}, X_j = x_j\}$$

$$= \frac{\tilde{N}_h(x_j)}{\tilde{N}(x_j)}$$

where $\tilde{N}(x_j) = \sum_h \tilde{N}_h(x_j), \tilde{N}_h(x_j)$ is the number of units $j \in U_h, j \in \bar{s}$ with x-values equal to x_j. Hence, show that when $\bar{X}_h$ and N_h are unknown, an estimate of $\bar{y}$ is obtained by estimating $P_h(j \mid x_j)$ by $\frac{n_h^+(x_j)}{n^+(x_j)}$, where $n_h^+(x_j)$ is the number of units in the sample from stratum h falling in a pre-assigned interval of the x- values which include the value x_j and $n^+(x_j) = \sum_{h=1}^{L} n_h^+(x_j)$

(Pfeffermann and Krieger, 1991)

10. *Post-stratification* Consider the post-stratified estimator

$$\bar{y}'_{ps} = \sum_{h=1}^{L} W_{h\alpha}\bar{y}_{sh}$$

where $W_{h\alpha} = \alpha\frac{n_h}{n} + (1-\alpha)\frac{N_h}{N}$ and α is a suitably chosen constant (other symbols have usual meanings). Show that $\bar{y}'_{ps}$ is unbiased for $\bar{y}$ with variance, to terms of $0(n^{-2})$, given by

$$V(\bar{y}'_{ps}) = (\frac{1}{n} - \frac{1}{N}) \sum_{h=1}^{L} W_h S_h^2 + \alpha^2 (\frac{1}{n} - \frac{1}{N})$$

$$(S^2 - \sum_{h=1}^{L} W_h S_h^2) + (1-\alpha)^2 \frac{N-n}{n^2(N-1)} \sum_{h=1}^{L} (1-W_h) S_h^2$$

Also, the minimum variance is given by

$$V_{min}(\bar{y}'_{ps}) = (\frac{1}{n} - \frac{1}{N}) \sum_{h=1}^{L} W_h S_h^2 + (1-\alpha_{opt}) \frac{N-n}{n^2(N-1)} \sum_{h=1}^{l} (1-W_h) S_h^2$$

where α_{opt} denotes the optimal value of α and is given by

$$\alpha_{opt} = \frac{A_2}{A_1 + A_2}$$

$$A_1 = \frac{N-1}{N}(S^2 - \sum_{h=1}^{L} W_h s_h^2) \text{ and } A_2 = \frac{1}{n}\sum_{h=1}^{L}(1 - W_h)S_h^2$$

Hence, show that $\bar{y}'_{ps;opt}$ is more efficient than both $\bar{y}_{ps} = \sum_{h=1}^{L} W_h \bar{y}_{sh}$ and $\bar{y}_s (= \sum_{h=1}^{L} n_h \bar{y}_{sh}/n)$ to the above order of approximation.

(Agarwal and Panda, 1993)

11. Consider *srswr* of n draws from a population of size N, resulting in ν number of distinct units. Let $\bar{y}_\nu$ be the mean of the values on distinct units and $\bar{y}_{HT}$ the *HTE* of $\bar{y}$. Suppose we condition on the observed value of ν, i.e. the relevant reference set is $\mathcal{S}_\nu$ of $\binom{N}{\nu}$ samples of effective size ν. Show that

$$E_2(\bar{y}_\nu) = \bar{y}$$

$$E_2(\bar{y}_{HT}) = \frac{\nu}{E(\nu)}\bar{y} \neq \bar{y}$$

where E_2, V_2 denote, respectively, the conditional expectation and variance given ν. Show also that an estimate of $V_2(\bar{y}_\nu)$ is

$$v(\bar{y}_\nu) = (\frac{1}{\nu} - \frac{1}{N}s_{\nu y}^2)$$

where $s_{\nu y}^2 = \frac{1}{\nu-1}\sum_{i\in S}(y_i - \bar{y}_\nu)^2$, S being the set corresponding to the *wr* sample s. Another alternative estimator of $V_2(\bar{y}_\nu)$ is

$$v^*(\bar{y}_\nu) = [E(\frac{1}{\nu}) - \frac{1}{N}]s_{\nu y}^2$$

which is, however, conditionally biased.

(Rao, 1985)

12. *Outliers* Suppose the population contains a small (unknown) fraction W_2 of outliers (large observations), $W_1 >> W_2$ $(W_1 + W_2 = 1)$; $\bar{y}_2 >> \bar{y}_1, \bar{y}_i$ denoting the population mean of group $i, (i = 1, 2)$. If the observed sample contains no outliers (i.e. $w_2 = 0$), then $E_2(w_1\bar{y}_{1s} + w_2\bar{y}_{2s}) = \bar{y}_1 << \bar{y}$ where E_2 is the conditional expectation given (n_1, n_2), $w_i = n_i/(n_1 + n_2)$, n_i denoting the number of sampled units in group i and the other symbols have usual meanings. On the otherhand, if the observed sample contains a outlier $(w_2 > 0), E_2(\bar{y}_s) - \bar{y} = (w_1 - W_1)\bar{y}_1 + (w_2 - W_2)\bar{y}_2 >> 0$, since $w_2 > W_2$ and $\bar{y}_2$ large. For example, if $N_2 = 1, w_2 = 1/n >> W_2 = 1/N$. In this situation consider the modified estimator

$$\bar{y}_s^* = \frac{N - n_2}{N}\bar{y}_{1s} + \frac{n_2}{N}\bar{y}_{2s}$$

Show that the conditional relative bias of $\bar{y}_s^*$ is given by

$$\frac{B_2(\bar{y}_s^*)}{\bar{y}_2} = \delta(\frac{w_2 n}{N} - W_2)$$

where $\delta = (\bar{y}_1 - \bar{y}_2)/\bar{y}_2$. Again,

$$\frac{B_2(\bar{y}_s)}{\bar{y}_2} = (w_2 - W_2)\delta$$

Hence, show that the estimator $\bar{y}_s^*$ has less conditional bias than $\bar{y}_s$.

(Rao, 1985)

13.*Non-response* Suppose m responses are obtained in a simple random sample of size n and let $\bar{y}_1$ denote the population proportion in the response stratum (group of units who would have responded if selected in the sample), $\bar{y}_2$ the same in the non-response stratum, $\bar{y} = W_1\bar{y}_1 + W_2\bar{y}_2, W_2 = 1 - W_1$. Let p^* be the probability that a person when contacted responds. Show that under this situation, conditionally given m, the sample s_m of respondents is a simple random sample of size m from the whole population and hence the sample mean $\bar{y}_m$ is conditionally unbiased. On the otherhand, the Horvitz-Thompson estimator (p^* known)

$$\bar{y}_{HT} = \frac{m}{E(m)}\bar{y}_m = \sum_{i \in s_m} \frac{y_i}{n p^*}$$

is conditionally biased, although unbiased when averaged over the distribution of m.

(Oh and Scheuren, 1983; Rao, 1985)

14. *Domain Estimation* Under *srs*, the usual estimator of a sub-population mean $\bar{y}_i$ is

$$\bar{y}_{is} = \sum_{j \in s_i} y_j / n_i, \quad n_i > 0$$

where s_i is the sample falling in domain i and n_i is the sample size. The estimator $\bar{y}_{is}$ is conditionally unbiased (given n_i) if $n_i > 0$. The estimator is however, unstable for small domains with small n_i. Consider a modified estimator

$$\bar{y}'_{is} = \frac{a_i}{E(a_i)}\bar{y}_{is}, \quad n_i \geq 0$$

where $a_i = 1(0)$ if $n_i > 0$ (otherwise) and $\bar{y}_{is}$ is taken as $\bar{y}_i$ for $n_i = 0$. The estimator $\bar{y}'_{is}$ is, however, conditionally biased,

$$E_2(\bar{y}'_{is}) = \frac{a_i}{E(a_i)}\bar{y}_i$$

Sarndal (1984) proposed the following estimator in the context of small-area estimation:

$$\bar{y}_{iS} = \bar{y}_s + \frac{w_i}{W_i}(\bar{y}_{is} - \bar{y}_s), \ n_i \geq 0 \qquad (i)$$

where $\bar{y}_s = \sum_i n_i \bar{y}_{is}/n = \sum_i w_i \bar{y}_{is}$ is the overall sample mean and $W_i = N_i/N$. The conditional bias of $\bar{y}_{iS}$ is

$$E_2(\bar{y}_{iS} - \bar{y}_i) = (w_i/W_i - 1)(\bar{y}_i - \bar{y}')$$

where $\bar{y}' = \sum_i w_i \bar{y}_i$. If $n_i = 0$, the estimator $\bar{y}_{iS}$ reduces to $\bar{y}_s$. However, $\bar{y}_{iS}$ would have a larger absolute conditional bias (and a larger conditional MSE) than $\bar{y}_s$ if $w_i > 2W_i$.

Hidiroglou and Sarndal (1985) proposed:

$$\bar{y}_{iS}^* = \begin{cases} \bar{y}_{is} & \text{if } w_i \geq W_i \\ \bar{y}_s + (\frac{w_i}{W_i})^2(\bar{y}_{is} - \bar{y}_s) & \text{if } w_i < W_i \end{cases}$$

$\bar{y}_{iS}^*$ is conditionally unbiased if $w_i \geq W_i$, while its conditional absolute bias is smaller than that of $\bar{y}_s$ if $w_i < W_i$.

Drew et al (1982) proposed

$$\bar{y}_{iD} = \begin{cases} \bar{y}_{is} & \text{if } w_i \geq W_i \\ \bar{y}_{iS} & \text{if } w_i < W_i \end{cases}$$

If a concommitant variable x with known domain means $\bar{X}_i$ is available, then show that the ratio estimator

$$\hat{y}_{ir} = \frac{\bar{y}_{is}}{\bar{x}_{is}}\bar{X}_i$$

and a regression estimator

$$\bar{y}_{ilr} = \bar{y}_{is} + \frac{\bar{y}_s}{\bar{x}_s}(\bar{X}_i - \bar{x}_{is})$$

are both approximately conditionally unbiased.

(Rao, 1985)

15. *Conditionally Weighted Estimator* Find an conditionally unbiased estimator of bias of the HTE, $\hat{\bar{y}}_\pi$ and show that a conditionally weighted estimator of $\bar{y}$ is $\hat{\bar{y}}_{\pi|\eta}$.

16. *Conditionally Weighted Estimators* Suppose we want to use the HTE, conditioned on the value of $\hat{\bar{x}}_\pi = \frac{1}{N}\sum_{k \in s}\frac{x_k}{\pi_k}$ where x is an auxiliary

variable. Show that

$$E\{\hat{\bar{x}}_\pi \mid k \in s\} = \frac{1}{N} \sum_{l(\neq)k} \frac{x_l}{\pi_l} \pi_{l|k} + \frac{1}{N} \frac{x_k}{\pi_k} = \bar{x}_{|k} \text{ (say)}$$

$$N^2 V\{\hat{\bar{x}}_\pi \mid k \in s\} = \sum_{l(\neq)k} \frac{x_l^2}{\pi_l^2} \frac{\pi_{kl}}{\pi_k} (1 - \frac{\pi_{kl}}{\pi_k})$$

$$+ \sum \sum_{l \neq m(\neq k)} \frac{x_l x_m}{\pi_l \pi_m \pi_k} (\pi_{lmk} - \frac{\pi_{lk} \pi_{mk}}{\pi_k}) = V_{x|k} \text{ (say)}$$

Also, show that

$$E\{I_{ks} \mid \hat{\bar{x}}_\pi = z\} = \pi_k \frac{P\{\hat{\bar{x}}_\pi = z \mid k \in s\}}{P\{\hat{\bar{x}}_\pi = z\}} \qquad (i)$$

Assume that $\hat{\bar{x}}_\pi \sim N(\bar{x}, V(\hat{\bar{y}}_\pi) = V_x$ (say) $, \hat{\bar{x}}_\pi \mid k \in s \sim N(\bar{x}_{|k}, V_{x|k})$ and $\pi_k \simeq n/N \ \forall \ k$. Hence, using (i) show that

$$\pi_{k|\hat{\bar{x}}_\pi} \simeq \frac{N}{n} \frac{f(\hat{\bar{x}}_\pi)}{f_k(\hat{\bar{x}}_\pi)}$$

where $f(.)$ is the *pdf* of $\hat{\bar{x}}_\pi$ and $f_k(.)$ is the conditional *pdf* of $(\hat{\bar{x}}_\pi)$ given $k \in s$, as stated above. Hence, write an approximate expression for the conditional weighted estimator of $\bar{y}$.

(Tille, 1998)

17. *Conditionally Weighted Estimators* Find conditionally weighted estimators of a finite population variance $S_y^2 = \sum_{k=1}^{N}(y_k - \bar{y})^2/(N - 1)$ and study their properties with special emphasis on simple random sampling.

(Mukhopadhyay, 1999b)

References

Agarwal, M.C. and Panda, K.B. (1993) An efficient estimator in post-stratification. *Metron*, LI n. 3-4, 179 - 188.

Aitchison, J. and Dunsmore, I.R. (1975) *Statistical Prediction Analysis*, Cambridge University Press, London.

Arnold, S.F. (1981): *The Theory of Linear Models and Multivariate Analysis*, John Wiley & Sons, New York.

Arora, V., Lahiri, P. and Mukherjee, K. (1997) Empirical Bayes estimation of finite population means from complex surveys. *Journal of American Statistical Association*, **92**, 1535 - 1562.

Baranchick, A. J. (1970) A family of minimax estimators of the mean of a multivariate normal distribution. *The Annals of Mathematical Statistics*, **41**, 642 - 645.

Basu, D. (1958) On sampling with and without replacement. *Sankhya*, **20**, 287 - 294.

Basu, D.(1971): An essay on logical foundation of survey sampling. Part I. in *Foundations of Statistical Inference*, eds. Godambe, V.P. and Sprott, D.R.,Toronto, Holt, Rinehart and Winston, 203 - 242.

Basu, D. and Ghosh, J.K. (1967) Sufficient Statistics in sampling from a finite population. *Bulletin of the International Statistical Institute*, **42** (2), 85 - 89.

Berger, J.O. (1980) *Statistical Decision Theory: Foundations, Concepts and Methods*, Springer-Verlag, New York.

Berger, J.O. (1984) The robust Bayesian view point (with discussion). in *Robustness of Bayesian Analysis*, Ed. Kadane, J, North-Holland, Amsterdam, 63-124.

Berger, J.O. and Berliner, L.M. (1986) Robust Bayes and empirical Bayes analysis with ϵ-contaminated priors. *Annals of Statistics*, **14**, 461 - 486.

Bessel, G.W., Jr. and Saleh, Md. A.K.E. (1994) L_1-estimation of the median of a survey population. *Journal of Nonparametric Statistics*, **3**, 277 - 283.

Bethlehem, S.C. and Scheurhoff, M.H. (1984) Second-order inclusion probability in sequential sampling without replacement with unequal probabilities. *Biometrika*, **71**, 642 - 644.

Bhattacharyya, S. (1997) *Some studies on estimation of mean and variances in finite population sampling.* Unpublished Ph.D. Thesis submitted to Indian Statistical Institute, Calcutta.

Bickel, P.J. (1984) Parametric robustness or small biases can be worthwhile. *The Annals of Statistics*, **12**, 864 -879.

Blum, J.R. and Rosenblutt, J. (1967) On partial a priori information in statistical inference. *The Annals of Mathematical Statistics*, **38**, 1671 - 1678.

Bolfarine, H. (1987) Minimax prediction in finite population. *Communications in Statistics, Theory and Methods*, **16**(12), 3683 - 3700.

Bolfarine, H. (1989) A note on finite population under asymmetric loss functions. *Communications in Statistics, Theory and Methods*, **18**, 1863 - 1869.

Bolfarine, H.(1991) Finite population prediction under error-in-variables superpopulation models. *Canadian Journal of Statistics*, **19**(2), 191 - 207.

Bolfarine, H., Pereira, C.A.B. and Rodrigues, J. (1987) Robust linear prediction in finite population. - A Bayesin perspective. *Sankhya, Series B*, **49**, 23 - 35.

Bolfarine, H. and Sandoval, M. (1993) Prediction of finite population distribution function under Gaussian superpopulation model. *Australian Journal of Statistics*, **35** (2), 195 - 204.

Bolfarine, H. and Sandoval, M. (1994) On predicting the finite population distribution function. *Statistics and Probability Letters*, **19**, 339 - 347.

Bolfarine, H. and Zacks, S. (1991 a) Bayes and minimax prediction in finite population. *Journal of Statistical Planning and Inference*, **28**, 139 - 151.

Bolfarine, H. and Zacks, S. (1991 b) *Prediction theory of Finite Population*, Springer Verlag, New York.

Bolfarine, H., Zacks, S., Elian, S.N., and Rodrigues, J. (1994) Optimal prediction of finite population regression coefficients. *Sankhya, Series B*, **56**, 1-10.

Bolfarine, H., Zacks, S. and Sandoval, M.C. (1996) On predicting the population total under regression models with measurement errors. *Journal of Statistical Planning and Inference*, **35**, 63 - 76.

Breckling, J.U., Chambers, R.L., Dorfman, A.H., Tam, S.M., and Walsh, A.M. (1990) Maximum likelihood inference from sample survey data. *Australian National University Tech. Rep.No.*, *SMS-025-90*.

Brewer, K.R.W. (1963) A model of systematic sampling with unequal probabilities. *Australian Journal of Statistics*, **5**, 5 - 13.

Brewer, K.R.W. (1994) Survey sampling inference: Some past perspectives and present prospects. *Pakistan Journal of Statistics*, **10**, 213 -233.

Brewer, K.R.W., Early, L.J. and Joyce, S.F. (1972) Selecting several samples from a single population. *Australian Journal of Statistics*, **14**, 231 - 239.

Brewer, K.R.W. and Hanif, M. (1983): *Sampling with Unequal Probabilities* , Lecture Notes in Statistics Series, Springer-Verlag, New York.

Brewer, K.R.W., Hanif, M. and Tam, S. (1988) How nearly can model-based prediction and design-based estimation be reconciled? *Journal of the American Statistical Association*, **83**, 128 - 132.

Broemling, L.D. (1985) *Bayesian Analysis of Linear Models*, Marcel Drekker, New York.

Brunk, H.D. (1980) Bayesian least square estimation of univariate regression functions. *Communications in Statistics, Theory and Methods*, **A** 9(11), 1101 - 1136.

Butar, F. and Lahiri, P. (1999) Empirical Bayes estimation of finite population variances, *Sankhyā, Series B*, **61**, 305 - 314.

Butar, F. and Lahiri, P. (2000) Empirical Bayes estimation of several population means and variances under random sampling variances model. To appear in *Journal of Statistical Planning and Inference, (P.V.Sukhatme Memorial issue)*.

Carter, G.M. and Ralph, H. (1974) Empirical Bayes methods to estimating fire alarm probabilities. *Journal of the American Statistical Association*, **69**, 880 - 885.

Casady, R.J. and Valliant, R. (1993) Conditional properties of post-stratified estimators under normal theory. *Survey Methodology*, **19**, 183 - 192.

Cassel, C.M., Sarndal, C.E. and Wretman, J.H. (1976) Some results on generalised difference estimator and generalised regression estimator for finite population. *Biometrics*, **63**, 614 - 620.

Cassel, C.M., Sarndal, C.E. and Wretman, J.H. (1977) *Foundations of Inference in Survey Sampling*, Wiley, N.Y.

Chakravorty, M.C. (1963) On the use of incidence matrix for designs in sampling for finite universe. *Journal of Indian Statistical Association*, **1**, 78 -85.

Chambers, R.L., Dorfman, A. H. and Hall, P. (1992) Properties of the estimators of the finite population distribution function. *Biometrika*, **79** (3), 577 - 582.

Chambers, R.L., Dorfman, A.H., and Wehrly, T.E. (1993) Bias robust estimation in finite populations using nonparametric calibration. *Journal of the American Statistical Association*, **88**, 268 - 277.

Chambers, R.L. and Dunstan, R. (1986): Estimating distribution functions from survey data. *Biometrika* **73**, 597 - 604.

Chao, M.T. (1982) A general purpose unequal probability sampling plan. *Biometrika*, **69**, 653 - 656.

Chaudhuri, A. (1978) On estimating the variance of a finite population. *Metrika*, **23**, 201 - 205.

Chaudhuri, A. (1994) Small domain statistics: a review. *Statistica Netherlandica*, **48**(3), 215 - 236.

Chaudhuri, A. and Vos, J.W.E. (1988) *Unified Theory and Strategies of Survey Sampling*, North-Holland.

Chen, J. and Qin, J. (1993) Empirical likelihood estmation for finite populations and effective use of auxiliary information. *Biometrika*, **80** (1), 107 - 116.

Cocchi, D. and Mouchart, M. (1986) Linear Bayes estimation in finite population with a categorical auxiliary variable. Centre for Operations Research and Econometrics, Universite Catholique De Louvain, Belgium.

Cochran, W.G. (1946) Relative accuracy of systematic and stratified random samples for a certain class of populations. *The Annals of Mathematical Statistics*, **17**, 164 - 177.

Cochran, W.G. (1977) *Sampling Techniques*, Third Edition, Jhon Wiley & Sons, New York.

Cohen, M. and Kuo, L. (1985) Minimax sampling strategies for estimating a finite population distribution function. *Statistics and Decision,* **3**, 205 - 224.

Cox, D.R. and Hinkley, D.V. (1974) *Theoretical Statistics,* Chapman and Hall, London.

Das, A. and Tripathi, T.P.T. (1978) Use of auxiliary information in estimating the finite population variance. *Sankhya, Series C,* **40**, 139 - 148.

Datta, G.S. and Ghosh, M. (1991) Bayesian prediction in linear models. Application to small area estimation. *The Annals of Statistics,* **19**, 1748 - 1770.

David, H.A. (1981) *Order Statistics,* Second Edition, John Wiley & Sons, New York.

Deming, W. E. (1960) *Sampling Designs in Business Research,* John Wiley & Sons, New York.

Des Raj (1956) Some estimators in sampling with varying probabilities without replacement. *The Annals of Mathematical Statistics,* **29**, 350 - 357.

Des Raj and Khamis, H.S. (1958) Some remarks on sampling with replacement. *The Annals of Mathematical Statistics,* **29**, 350 - 357.

Deville, J. C. and Sarndal, C. E. (1992) Calibration estimators in survey sampling. *Journal of the American Statistical Association,* **87**, 376 - 392.

Deville, J.C., Sarndal, C.E., and Sautory, O. (1993): Generalised raking procedure in survey sampling. *Journal of the American Statistical Association,* **88**, 1013 - 1020.

Diaconis, P. and Yalvisaker, D. (1979) Conjugate priors for exponential families. *The Annals of Statistics,* **7**, 269 - 281.

Dorfman, A.H. (1993): A comparison of design-based and model-based estimators of the finite population distribution function. *Australian Journal of Statistics,* **35** (1), 29 - 41.

Doss, D. C., Hartley, H. O. and Somayajulu, G.R. (1979) An exact small sample theory for post-stratification. *Journal of Statistical Planning and Inference,* **3**, 235 - 248.

Drew, J. D., Singh, M. P. and Choudhry, G. H. (1982) Evaluation of small area estimation techniques for the Canadian Labour Force Survey, *Survey Methodology,* **8**, 17 - 47.

Dunstan, R. and Chambers, R. L. (1989) Estimating distribution functions from survey data with limited benchmark information. *Australian Journal of Statistics*, **31** (1), 1 - 11.

Dupont, F. (1995) Alternative adjustments where there are several levels of auxiliary information. *Survey Methodology*, **21**(2), 125 - 135.

Durbin, J. (1969) Inferential aspects of the randomness of sample size .in *New Developments in Survey Sampling*, eds. Jhonson, N.L. and Smith, H.,Jr., Jhon Wiley & Sons, New York.

Effron, B. and Morris, C. (1971) Limiting the risk of Bayes and Empirical Bayes estimators - Part I: The Bayes case. *Journal of the American Statistical Association*, **66**, 807 - 815.

Effron, B. and Morris, C. (1972) Limiting the risk of Bayes and Empirical Bayes estimators - Part I: The Empirical Bayes case. *Journal of the American Statistical Association*, **67**, 130 - 139.

Effron, B. and Morris, C. (1973) Stein's estimation rule and its competitors - an empirical Bayes approach. *Journal of the American Statistical Association*, **68**, 117 - 130.

Ericksen, E.P. (1973) A model for combining sample data and symptomatic indicators to obtain population estimates for local areas. *Demography*, **10**, 137 - 160.

Ericksen, E.P.(1974) A regression method for estimating populations of local areas. *Journal of the American Statistical Association*, **69**, 867 - 875.

Ericson, W.A. (1969 a): Subjective Bayesian models in sampling finite populations (with discussion). *Journal of the Royal Statistical Society, Series B*, **31**, 195 - 233.

Ericson, W.A. (1969 b): A note on posterior mean of a population mean. *Journal of the Royal Statistical Society, Series B*, **31**, 332 - 334.

Fahrmeir, L. and Tutz, G. (1994) *Multivariate Statistical Modelling based on Generalised Linear Models*, Springer Verlag, New York.

Fay, R.E. and Herriot, R.A. (1979) Estimates of income for small places: an application of James-Stein procedures to census data. *Journal of the American Statistical Association*, **74**, 269 - 277.

Francisco, C.A. and Fuller, W.A. (1991) Quantile estimation with a complex survey design. *The Annals of Statistics*, **19**(1), 454 - 469.

Fuller, W.A. (1966) Estimation employing post-strata. *Journal of the American Statistical Association*, **61**, 1172 - 1183.

Fuller, W.A. (1975) Regression analysis for sample surveys. *Sankhya, Series C*, **37**, 117 - 132.

Fuller, W.A. (1976) *Introduction to Statistical Time Series*. John Wiley & Sons, New York.

Fuller, W.A.(1987) *Measurement Error Models*, Jhon Wiley & Sons, New York.

Gabler, S. (1984) On unequal probability sampling: sufficient conditions for the superiority of sampling without replacement. *Biomoetrika*, **71**, 171 - 175.

Gasco, L., Bolfarine, H. and Sandoval, M.C. (1997) Regression estimators under multiplicative measurement error superpopulation models. *Sankhya, Series B*, **59**(1), 84 - 95.

Ghosh, M. (1992) Constrained Bayes estimation with applications. *Journal of the American Statistical Association*, **87**, 533 - 540.

Ghosh, M. and and Kim, D.H. (1993) Robust Bayes estimation of the finite population mean. *Sankhya, Series B*, **55** (3), 322 - 342.

Ghosh, M. and Kim, D.H. (1997) Robust Bayes competitors of the ratio estimator. *Statistics and Decision*, **15**, 17 - 36.

Ghosh, M. and Lahiri, P. (1987 a) Robust empirical Bayes estimation of means from stratified samples. *Journal of the American Statistical Association*, **82**, 1153 - 1162.

Ghosh, M. and Lahiri, P. (1987 b) Robust empirical Bayes estimation of variances from stratified samples. *Sankhya, Series B* **49**, 78 - 89.

Ghosh, M. and Maiti, T. (1999) Adjusted Bayes estimators with applications to small area estimation. *Sankhya, Series B*, **61**, 71 - 90.

Ghosh, M. and Meeden, G. (1986) Empirical Bayes estimation in finite population sampling. *Journal of the American Statistical Association*, **81**, 1058 - 1062.

Ghosh ,M. and Meeden, G. (1997) *Bayesian Methods for Finite Population Sampling*, Chapman and Hall, London.

Ghosh, M., Natarajan, K., Stroud, T. W. F. and Carlin, B. P. (1998) Generalised linear models for small area estimation. *Journal of the American Statistical Association*, **93**, 273 - 282.

Ghosh, M. and Rao, J. N. K. (1994) Small area estimation: an appraisal. *Statistical Science*, **6**(1), 55 - 93.

Godambe, V. P. (1955) A unified theory of sampling from finite populations. *Journal of the Royal Statistical Society, Series B*, **17**, 269- 275.

Godambe, V. P. (1966) A new approach to sampling from finite population, I, *Journal of the Royal Statistical Association, Series B*, **28**,310 - 328.

Godambe, V.P. (1989) Estimation of cumulative distribution function of a survey population. *Technical Report of the University of Waterloo, Canada.*

Godambe, V.P. (1999) Linear Bayes and optimal estimation. *The Annals of Mathematical Statistics*, **51** (2), 201 - 216.

Godambe, V.P. and Joshi, V.M. (1965) Admissibility and Bayes estimation in sampling from finite population, I. *The Annals of Mathematical Statistics*, **36**, 1707 - 1722.

Godambe, V.P. and Thompson, M.E. (1977) Robust near optimal estimation in survey practice. *Bulletin of the International Statistical Institute*, **47**(3), 129 - 146.

Goel, P.K. and DeGroot, M.H. (1980) Only normal distributions have linear posterior expectations in linear regression. *Journal of the American Statistical Association*, **75**, 895 - 900.

Goldstein, M. (1975) A note on some Bayesian nonparametric estimates. *The Annals of Statistics*, **3**, 736 - 740.

Good, I.J. (1963) *The Estimation of Probabilities*, MIT Press, Cambridge.

Gross, S.T. (1980) Median estimation in sample surveys. in *Proceedings of the Survey Research Methods Section, American Statistical Association*, 181 - 184.

Ha'jek, J. (1959) Optimum strategies and other problems in probability sampling. Casopis Pest. Mat, **84**, 387 - 423.

Hansen, M.H., Madow, W.G. and Tepping, B.J. (1983) An evaluation of model-dependent and probability sampling inference in sample surveys. *Journal of the American Statistical Association*, **78**, 776 - 793.

Hanurav, T.V. (1962 a) On Horvitz-Thompson estimator. *Sankhya, Series A,* **24**, 429 - 436.

Hanurav, T.V. (1962 b) An existance theorem in sample surveys. *Sankhya, Series A,* **24**, 327 -330.

Hanurav, T.V. (1966) Some aspects of unified sampling theory. *Sankhya, Series A,* **28**, 175 - 203.

Hartigan, J.A. (1969) Linear Bayes methods. *Journal of the Royal Statistical Society, Series B,* **31**(3), 454 - 464.

Hartley, H.O. and Rao, J.N.K. (1968) A new estimation theory for sample surveys. *Biometrika,* **55**, 547 - 557.

Hartley, H.O. and Rao, J.N.K. (1969) A new estimation theory for sample surveys II. in *New Developments in Survey Sampling,* eds. Johnson, N.L. and Smith, H., Jr., 147 - 169. Wiley Interscience, New York.

Hartley, H.O., Rao, J.N.K. and Silken, R.L., Jr (1975) A " superpopulation view point" for finite population sampling. *Biometrics,* **31**, 411 - 422.

Herzel, A. (1986) Sampling without replacement with unequal probabilities sample designs with pre-assigned joint inclusion-probabilities of any order. *Metron,* **XLIV**(1), 49 - 68.

Hidiroglou, M.A. and Sarndal, C.E. (1985) An empirical study of some regression estimators for small domains. *Survey Methodology,* **11**, 65 - 67.

Hill, P.D. (1985) Kernel estimation of a distribution function. *Communications in Statististics, Ttheory & Methods,* **14**, 605 - 620.

Holt, D. and Smith, T.M.F. (1979) Post stratification. *Journal of the Royal Statistical Society, Series A,* **142**, 33 - 46.

Horvitz, D.G. and Thompson, D.J. (1952) A generalisation of sampling without replacement from a finite universe. *Journal of the American Statistical Association,* **64**, 175 - 195.

Huber, P.J. (1973) The use of Choquet capacities in statistics. *Bulletin of the International Statistical Institute,* **45**, 181 - 191.

Hwang, J.T. (1986) Multiplicative errors-in-variables models with applications to recent data released by the US Department of energy. *Journal of the American Statistical Association,* **81**, 680 - 688.

Isaki, C.T. (1983) Variance estimation using auxiliary information. *Journal of the American Statistical Association*, **78**, 117 - 123.

Isaki, C.T. and Fuller, W.A. (1982) Survey designs under the regression superpopulation models. *Journal of the American Statistical Association*, **77**,89 - 96.

Jagers, P. (1986) Post-stratification against bias in sampling. *International Statistical Review*, 159 - 167.

Jagers, P., Oden. A. and Trulsson, L. (1985) Post-stratification and ratio estimation: usage of auxiliary information in survey sampling and opinion polls. *International Statistical Review*, 221 - 238.

Jewell, W.S. (1974) Credible means are exact Bayesian for exponential families. *The Astin Bulletin*, **8**, 77 - 90.

Joshi, V.M. (1965 a) Admissibility and Bayes estimation in sampling finite populations II. *The Annals of Mathematical Statistics*, **36**, 1723 - 1729.

Joshi, V.M. (1965 b) Admissibility and Bayes estimation in sampling finite populations III. *The Annals of Mathematical Statistics*, **36**, 1730 - 1742.

Joshi, V.M. (1966) Admissibility and Bayes estimation in sampling finite populations IV. *The Annals of Mathematical Statistics*, **37**, 1658 - 1670.

Joshi, V.M. (1969) Admissibility of the estimates of the mean of a finite population in *New Developments in Survey Sampling*, eds., Johnson, N.L. and Smith, H., Jr., Wiley Interscience, New York, 188 - 212.

Kalton, G. and Maligwag, D.S. (1991) A comparison of methods of weighting adjustments for non-response.in *Proceedings of the 1991 Annual Research Conference, US Bureau of Census*, 409 - 428.

Kass, R.E. and Staffey, D. (1989) Approximate Bayesian inference in conditionally independent hierarchical models (parametric empirical Bayes models). *Journal of the American Statistical Association*, **84**, 717 - 726.

Kleffe, J. and Rao, J.N.K. (1992) Estimation of mean square error of empirical best linear unbiased predictors under a random error variance linear model. *Journal of the Multivariate Analysis*, **43**, 1 - 15.

Kish, L. (1965) *Survey Sampling*. John Wiley & Sons, New York.

Konijn, H.S. (1962) Regression analysis in sample surveys. *Journal of the American Statistical Association*, **68**, 880- 889.

Kott, P.S. (1990) Estimating the conditional variance of a design consistent estimator. *Journal of the Statistical Planning and Inference*, **24**, 287 - 296.

Krieger, A.M. and Pfeffermann, D. (1992) Maximum likelihood estimation from complex surveys. *Technical Report of the Department of Statistics, University of Pennsylvania.*

Kuk, A.Y.C. (1988) Estimation of distribution functions and medians under sampling with unequal probabilities. *Biometrika*, **75**(1), 97 - 103.

Kuk, A.Y.C. (1993) A kernel method for estimating finite population distribution functions using auxiliary information. *Biometrika*, **80**(2), 385 - 392.

Kuk, A.Y.C. and Mak, T.K. (1989) Median estimation in the presence of auxiliary information. *Journal of the Royal Statistical Society, Series B*, **51**, 261 - 269.

Lahiri, D.B. (1951) A method of sample selection providing unbiased ratio estimates. *Bulletin of the International Statistical Institute*, **3**(2), 133 - 140.

Lahiri, P. (1990): Adjusted Bayes and Empirical Bayes estimators in finite population sampling. *Sankhya, Series B*, **52**, 50 -66.

Lahiri, P. and Peddada, S.D. (1992) Bayes and empirical Bayes estimation of finite population mean using auxiliary information. *Statistics and Decision*, **10**, 67 - 80.

Laird, N. and Louis, T.A. (1987) Empirical Bayes confidence intervals based on bootstrap samples. *Journal of the American Statistical Association*, **82**, 739 - 750.

La Motte, L. R. (1978) Bayes linear estimators. *Technometrics*, **20**, 281 - 290.

Lanke, J. (1975) *Some Contributions to the Theory of Survey Sampling*. AV Centralin i Lund.

Lehmann, E.L. (1977) *Testing Statistical Hypothesis*, John Wiley & Sons, New York (1959), Wiley Eastern, New Delhi, India (1977).

Liao, H. and Sedransk, J. (1975) Sequential sampling for comparison of domain means. *Biometrika*, **62**, 691 - 693.

Lindley, D.V. (1962) Comments on Stein's paper. *Journal of the Royal Statistical Society, Series B*, **24**, 285 - 287.

Lindleay, D.V. and Smith, A.F.M. (1972) Bayes estimates for the linear model (with discussion). *Journal of the Royal Statistical Society, Series B*, **34**, 1 - 41.

Little, R.J.A. (1982) Models for nonresponse in sample surveys. *Journal of the American Statistical Association*, **77**, 237 - 250.

Little, R.J.A. (1991 a) Inference with survey weights. *Journal of the Official Statistics*, **7**, 405 - 424.

Little, R.J.A. (1991 b) Discussion of session, "Estimation Techniques with Survey Data" in *Proceedings of the 1991 Annual Research Conference*, US Bureau of Census, 441 - 446.

Little, R.J.A. (1993) Post-stratification: a modeller's perspective. *Journal of the American Statistical Association*, **88**, 1001 - 1012.

Liu, T.P. (1974 a) A general unbiased estimator for the variance of a finite population, *Sankhya, Series C*, **36**, 23 - 32.

Liu, T.P. (1974 b) Bayes estimation for the variance of a finite population. *Metrika*, **21**, 127 - 132.

Liu, T.P. and Thompson, M. E. (1983) Properties of estimators of quadratic finite population functions: the batch approach. *The Annals of Statistics*, **11**, 275 - 285.

Louis, T. (1984) Estimating a population of parameter values using Bayes and Empirical Bayes methods. *Journal of the American Statistical Association*, **79**, 393 - 398.

Loynes, R.M. (1966) Some aspects of estimation of quantiles. *Journal of the Royal Statistical Society, Series B*, **28**, 497 - 512.

Lui, K.J. and Cumberland, W.G. (1989) A Bayesian approach to small domain estimation. *Journal of the Official Statistics*, **5**, 143 - 156.

Mak, T.K. and Kuk, A.Y.C. (1992) Estimators of distribution function and quantiles in the presence of auxiliary information. in *Nonparametric Statistics and Applied Topics*, ed. Saleh, A. K. Md. E., Elsevier Science, 385 - 398.

Malec, D. and Sedransk, J. (1985) Bayesian inference for finite population parameters in multistage cluster sampling. *Journal of the American Statistical Association*, **80**, 897 - 902.

McCarthy, P.J. (1965) Stratified sampling and distribution-free confidence intervals for a median. *Journal of the American Statistical Association*, **60**, 772 - 783.

Meeden, G. (1995) Median estimation using auxiliary information. *Survey Methodology*, **21**, 71 -77.

Merazzi, A. (1985) On controlled minimisation of the Bayes risk for the linear model. *Statistics and Decision*, **3**, 277 - 296.

Meyer, J.S. (1972) *Confidence intervals for quantities in stratified random sampling.* Unpublished Ph.D.Thesis, Iowa State University Library, Ames, Iowa.

Mickey, M.R. (1959) Some finite population unbiased ratio and regression estimators. *Journal of American Statistical Association*, **54**, 594 - 612.

Midzuno, H. (1950) An outline of the theory of sampling systems. *Annals of Institute of Statistical Mathematics*, **1**, 149 - 151.

Midzuno, H. (1952) On the sampling system with probability proportional to sum of sizes. *Annals of Institute of Statistical Mathematics*, **3**, 99 - 107.

Montanari, G.E. (1999) A study on the conditional properties of finite population mean estimators. *Metron*, LVII, n. 1-2, 21 - 36.

Morris, C. (1983) Parametric empirical Bayes inference: Theory and applications (with discussions). *Journal of the American Statistical Association*, **78**, 47 - 55.

Muirhead, R.J. (1982) *Aspects of Multivariate Statistical Theory*, John Wiley & Sons, New York.

Mukhopadhyay, P. (1972) A sampling scheme to realise a pre-assigned set of inclusion probabilities of first two orders. *Calcutta Statistical Association Bulletin*, **21**, 87 - 122.

Mukhopadhyay, P. (1975) An optimum sampling design to base HT-method of estimating a finite population total, *Metrika*, **22**, 119 - 127.

Mukhopadhyay, P. (1977 a) Robust estimation of finite population total under certain linear regression models. *Sankhya, Series C*, **39**, 71 - 87.

Mukhopadhyay, P. (1977 b) *Further Studies in Sampling Theory*, Unpublished Ph.D. Thesis submitted to the University of Calcutta, India.

Mukhopadhyay, P. (1978) Estimating the variance of a finite population under a superpopulation model. *Metrika*, **25**, 115 - 122.

Mukhopadhyay, P. (1982) Optimum strategies for estimating the variance of a finite population under a superpopulation model. *Metrika*, **29**, 143 - 158.

Mukhopadhyay, P. (1984) Optimum estimation of a finite population variance under generalised random permutation models. *Calcutta Statistical Association Bulletin*, **33**, 93 - 104.

Mukhopadhyay, P. (1985) Estimation under linear regression models. *Met rika*, **32**, 339 - 349.

Mukhopadhyay, P. (1986) Asymptotic properties of a generalised predictor of a finite population variance under probability sampling. Indian Statistical Institute Technical Report No. ASC/86/19.

Mukhopadhyay, P.(1990) On asymptotic properties of a generalised predictor of a finite population variance. *Sankhya, Series B*, **52**, 343 - 346.

Mukhopadhyay, P. (1991) Varying probability without replacement sampling designs: a review. Indian Statistical Institute Technical Report No. ASC/91/3.

Mukhopadhyay, P. (1992) On prediction in finite population under error-in-variables superpopulation models. Indian Statistical Institute Technical Report No. ASC/92/11.

Mukhopadhyay, P. (1994 a) Prediction in finite population under error-in-variables superpopulation models. *Journal of Statistical Planning and Inference*, **41**, 151 - 161.

Mukhopadhyay, P. (1994 b) Bayes and minimax procedures for finite population samlpling under measurement error models. *Communications and Statistics, Theory & Methods*, **23**(7), 1953 - 1961.

Mukhopadhyay, P. (1995 a) Bayes and minimax estimator for two-stage sampling from a finite population under measurement error models. *Communications and Statistics, Theory & Methods*, **24**(3), 663 - 674.

Mukhopadhyay, P. (1995 b) Prediction of finite population total using multi-auxiliary information under measurement error models. in *Probability and Statistics, essays in honour of Prof. A. K. Bhattacharyya*, eds.. Mukherjee, S.P., Chaudhuri, A., and Basu, S.K., Calcutta, India.

Mukhopadhyay, P. (1996) *Inferential Problems in Survey Sampling*, New Age International Publishers, New Delhi, India and London, United Kingdom.

Mukhopadhyay, P. (1997 a) Bayes estimation of small area totals under measurement error models. *Journal of Applied Statistical Sciences*, **5**, 105 - 111.

Mukhopadhyay, P. (1997 b) On estimating a finite population total under measurement error models. in *Proceedings of the International Conference on Quality Improvement through Statistical Methods*, Cochin, India.

Mukhopadhyay, P. (1998 a) Hierarchical and empirical Bayes estimation of a finite population total under measurement error models. *Journal of Applied Statistical Sciences*, **6**, 59 - 66.

Mukhopadhyay, P. (1998 b) Linear Bayes estimation of a finite population total under measurement error models. *Journal of Statistical Research*, **32**(1), 43 - 48.

Mukhopadhyay, P. (1998 c) Estimation of a finite population total under measurement error models. *Journal of Statistical Research*, **32**(2), 1 - 14.

Mukhopadhyay, P. (1998 d) Predicting a finite population distribution function under measurement error models. *International Journal of Mathematical and Statistical Sciences*, **7**, 1-15.

Mukhopadhyay, P. (1998 e) *Small Area Estimation in Survey Sampling*, Narosa Publishers, New Delhi, India and London, United Kingdom.

Mukhopahyay, P. (1998 f) *Theory and Methods of Survey Sampling*, Prentice Hall of India, New Delhi.

Mukhopadhyay, P. (1998 g) *Small Area Estimation of Population in Hugli, W.B., India*, a Survey Report, Indian Statistical Institute, Calcutta, India.

Mukhopadhyay, P. (1999 a) On prediction in finite population under error-in-variables superpopulation models. *International Journal of Mathematical and Statistical Sciences*, **8**(1), 89 - 101.

Mukhopadhyay, P. (1999 b) On conditionally weighted estimators of a finite population variance. *Journal of Statistical Research*, **33**(1).

Mukhopadhyay, P. (1999 c) Small area estimation of population for the district of Hugli, W.B., India. in *Small Area Estimation*, Proceedings of a Satellite Conference, Riga, Latvia, August, 1999, pp 263 - 268.

Mukhopadhyay, P. (2000 a) Bayesian estimation in finite population using asymmetric loss functions. *International Journal of Mathematical and Statistical Sciences*, **9**(1) (to appear)

Mukhopadhyay, P. (2000 b) Calibration estimators of a finite population variance. *Parisankhyan Samikkha* (to appear)

Mukhopadhyay, P. (2000 c) On estimating a finite population distribution function. *Parisankhyan Samikkha* (to appear)

Mukhopadhyay, P. and Bhattacharyya, S. (1989) On estimating the variance of a finite population under a superpopulation model. *Journal of Indian Statistical Association*, **27**, 37 - 46.

Mukhopadhyay, P. and Bhattacharyya, S. (1991) Estimating a finite population variance under general linear model with exchangeable errors. *Calcutta Statistical Association Bulletin*, **40**, 138 - 148.

Mukhopadhyay, P. and Bhattacharyya, S. (1994) Prediction under balanced samples. *Journal of Statistical Planning and Inference*, **39**, 85 - 93.

Mukhopadhyay, P. and Vijayan, K. (1996) On controlled sampling designs. *Journal of Statistical Planning and Inference*, **52**, 375 - 378.

Murthy, M.N. (1957) Ordered and unordered estimators in sampling without replacement. *Sankhya*, **18**, 379 - 390.

Murthy, M. N. (1963) Generalised unbiased estimation in sampling from finite population. *Sankhya, Series B*, **25**, 245 - 262.

Murthy, M.N. (1977): *Sampling Theory and Methods*, Second edition, Statistical Publishing Society, Calcutta, India.

Ogus, J.K. and Clark, D.F. (1971) The annual survey of manufacturers: A report on methodology. Technical Report No. 24, US Bureau of Census, Washington D C.

O'Hagan, A. (1986) On posterior joint and marginal modes. *Biometrika*, **63**, 329 - 333.

Oh, H.L. and Scheuren, F.J. (1983) Weighting adjustments for unit nonresponse. In *Incomplete Data in Sample Surveys, Vol. 2 - Theory and Bibliographies*, eds. Madow, W.G., Olkin, I. and Rubin, D.B., Academic Press, New York, 435 - 483.

Olkin, I. (1958) Multivariate ratio estimation for finite populations, *Biom etrika*, **45**, 154 - 165.

Olkin, I. and Ghurye, S.G. (1969) Unbiased estimates of some multivariate densities and related functions. *The Annals of Mathematical Statistics*, **40**, 1261 - 1271.

Pascual, J.N. (1961) Unbiased ratio estimators in stratified sampling. *Journal of the American Statistical Association*, **56**, 70 - 82.

Pathak, P.K. (1961) On the evaluation of moments of distinct units in a sample. *Sankhya, Series A*, **23**, 409 - 414.

Pereira, C.A.D.B. and Rodrigues, J. (1983) Robust linear prediction in finite populations. *International Statistical Review*, **51**, 293 - 300.

Pfeffermann, D. and Krieger, A.M. (1991) Post-stratification using regression estimates when information on strata means and sizes are missing. *Biometrika*, **78**, 409 - 419.

Quenouille, M.H. (1956) Notes on bias in estimation. *Biometrika*, **43**, 353 - 360.

Quin, J. and Chen, J. (1991) Empirical likelihood method in finite population and the effective usage of auxiliary information. Technical Report of the University of Waterloo, Canada.

Randles, R.H. (1982): On the asymptotic normality of statistics with estimated parameters. *The Annals of Statistics*, **10**, 462 - 474.

Rai, A. and Srivastava, A.K. (1998) Estimation of regression coefficients from survey data based on tests of significance. *Communications in Statistics, Theory & Methods*, **27**, (3), 761 - 773.

Rao, C.R. (1973) *Linear Statistical Inference*, Second edition, John Wiley and Sons, New York.

Rao, J.N.K. (1965) On two simple properties of unequal probability sampling without replacement. *Journal of Indian Society of Agricultural Statistics*, **3**, 173 - 180.

Rao, J.N.K. (1969) Ratio and regression estimators, in *New Developments in Survey Sampling*, eds. Johnson, N.L. and Smith, H. Jr., Wiley Interscience, New York.

Rao, J.N.K. (1985) Conditional inference in survey sampling. *Survey Methodology*, **11**, 15 - 31.

Rao, J.N.K. (1994) Estimating totals and distribution functions using auxiliary information at the estimation stage. *Journal of Official Statistics*, **10**, 153 - 165.

Rao, J.N.K. (1997) Developments in sample survey theory: an appraisal. *Canadian Journal of Statistics*, **25**, 1 - 21.

Rao, J.N.K. (1999) Some current trends in sample survey theory and methods. (with discussion), *Sankhya, Series B*, **61**, 1 - 57.

Rao, J.N.K., Kover, J.G. and Mantel, H.J. (1990) On estimating distribution functions and quantiles from survey data using auxiliary information. *Biometrika*, **77**, 365 - 375.

Rao, J.N.K. and Liu, J. (1992) On Estimating distribution functions from sample survey data using supplementary information at the estimation stage. in *Nonparametric statistics and Related Topics*, ed. A. K. Md. Saleh, 399 - 407, Elsevier Science Publishers, Amsterdam.

Rao, J.N.K. and Nigam, A.K. (1989) Controlled sampling with probability proportional to aggregate size. Technical Report No. 133, Laboratory to Research in Statistics and Probability, Carleton University, Ottawa, Canada.

Rao, J.N.K. and Nigam, A.K. (1990) Optimum controlled sampling designs. *Biometrika*, **77**, 807- 814.

Rao, J.N.K. and Singh, A.C. (1997) A ridge-shrinkage method for range-restricted weight calibration in survey sampling. in *Proceedings of Section on Survey Research Methods of American Statistical Association*, 57 - 65.

Robbins, H. (1955) An empirical Bayes approach to Statistics. in *Proceedings of the 3rd Berkley Symposium on Mathematical Statistics and Probability*, vol. 6, Berkley, University of California Press, 157 - 163.

Robinson, J (1987) Conditioning ratio estimates under simple random sampling. *Journal of the American Statistical Association*, **82**, 826 - 831.

Rodrigues, J. (1989) Some results on restricted Bayes least squares predictors for finite populations. *Sankhya, Series B*, **51**, 196 - 204.

Rodrigues, J., Bolfarine, H., and Rogatko, A. (1985) A general theory of prediction in finite population. *International Statistical Review*, **53**, 239 - 254.

Royall, R.M. (1970) On finite population sampling theory under certain linear regression models. *Biometrika*, **57**, 377 - 387.

Royall, R.M. (1971) Linear regression models in finite population sampling theory, in *Foundation of Statistical Inference*, eds. Godambe, V.P. and Sprott, D.R., Toronto, Holt, Rinehart and Winston, 259-277.

Royall, R.M. (1976) The linear least squares prediction approach to two-stage sampling. *Journal of the American Statistical Association*, **68**, 890 - 893.

Royall, R.M. and Eberherdt, J. (1975) Variance estimation for the ratio estimator. *Sankhya, Series C*, **37**, 43 - 52.

Royall, R.M. and Herson, J. (1973) Robust estimation in finite population I, *Journal of the American Statistical Association*, **68**, 880 - 889.

Royall, R.M. and Pfeffermann, D. (1982) Balanced samples and robust Bayesian inference in finite population sampling. *Biometrika*, **69**, 401 - 409.

Rubin, D.B. (1976) Inference and missing data. *Biometrika*, **63**, 581 - 592.

Sarndal, C.E. (1980 a) Two model-based inference argument in survey sampling. *Australian Journal of Statistics*, **22**, 314 - 318.

Sarndal, C.E. (1980 b) On π-inverse weighting versus best linear weighting in probability sampling. *Biometrika*, **67**(3), 639 - 650.

Sarndal, C.E. (1982) Implications of survey designs for estimation of linear functions. *Journal of Statistical Planning and Inference*, **7**, 155 - 170.

Sarndal, C.E. (1984) Design-consistent versus model-dependent estimators for small domains. *Journal of the American Statistical Association*, 79, 624 - 631.

Sarndal, C.E., Swensson, B. and Wretman, J.H. (1989) The weighted regression technique for estimating the variance of the generalised regression estimator. *Biometrika*, **76**, 527 - 537.

Sarndal, C.E., Swensson, B. and Wretman, J.H. (1992) *Model assisted Survey Sampling*, Springer-Verlag, New York.

Scott, A.J. (1975) On admissibility and uniform admissibility in finite population sampling. *The Annals of Statistics*, **3**, 489 - 491.

Scott, A.J., Brewer, K.R.W. and Ho, E.W.H. (1978) Finite population sampling and robust estimation. *Journal of the American Statistical Association*, **73**, 359‑ 361.

Scott, A. J. and Smith, T.M.F. (1969) Estimation in multistage surveys. *Journal of the American Statistical Association*, **64**, 830 - 840.

Sedransk, J. and Meyer, J.S. (1978) Confidence intervals for the quantiles of a finite population: simple random and stratified random sampling. *Journal of the Royal Statistical Society, Series B*, **40**, 239 - 252.

Sedransk, J.and Smith, P. (1983) Lower bounds for confidence coefficient for confidence intervals for finite population quantiles. *Communications in Statistics, Theory & Methods*, **11**, 1329 - 1344.

Sekkappan, R.M. and Thompson, M.E. (1975) On a class of uniformly admissible estimators in finite populations. *The Annals of Statistics*, **3**, 492 - 499.

Sen, A.R. (1952) Present status of probability sampling and its use in the estimation of a characteristic. (abstract), *Econometrika*, **20**, 103.

Sen, A.R. (1953) On the estimate of variance in sampling with varying probabilities. *Journal of the Indian Society of Agricultural Statistics*, **5**, 119-127.

Sengupta, S. (1988) Optimality of a design-unbiased strategy for estimating a finite population variance. *Sankhya, Series B*, **50**, 149 - 152.

Serfling, R.J. (1980) *Approximation Theory of Mathematical Statistics*, John Wiley and Sons, New York.

Shah, B.V., Holt, M.M., and Folsom, R.E. (1977) Inference about regression models from sample survey data. *Bulletin of the International Statistical Institute,***47** (3), 43 - 57.

Shah, D. N. and Patel, P. A. (1995) Uniform admissible estimators for the finite population variance. *Journal of the Indian Statistical Association*, **33**, 31 - 36.

Shah, D.N. and Patel, P.A. (1996): Asymptotic properties of a generalised regression-type predictor of a finite population variance in probability sampling. *Canadian Journal of Statistics*, **24**, 373 - 384.

Silva, P.L.D.N. and Skinner, C.J. (1995) Estimating distribution functions with auxiliary information under post-stratification. *Journal of Official Statistics*, **11** (3), 277 - 294.

Singh, P. and Srivastava, A.K. (1980) Sampling schemes providing unbiased regression estimators. *Biometrika*, **67**, 205 - 209.

Singh, S., Horn, S., Chowdhuri, S. and Yu, F. (1999) Calibration of the estimators of variance. *Australian and New Zealand Journal of Statistics*, **41**(2), 199 - 212.

Sinha, B.K. (1973) On sampling schemes to realise pre-assigned sets of inclusion-probabilities of first two orders. *Calcutta Statistical Association Bulletin*, **22**, 69 - 110.

Sivaganeshan, S. (1989) Range of posterior measures for priors with arbitrary contaminations. *Communications in Statistics, Theory & Methods*, **17** (5), 1581 - 1612.

Sivaganeshan, S. and Berger, J.O. (1989): Range of posterior measures for priors with unimodal contaminations. *The Annals of Statistics*, **17**, 868 - 889.

Smith, P. and Sedransk, J. (1983) Lower bounds for confidence coefficient for confidence intervals for finite population quantiles. *Communications in Statistics, Theory & Methods*, **11**, 1329 - 1344.

Smith, T.M.F. (1976) The foundations of survey sampling: a review. *Journal of the Royal Statistical Society, Series A* , **139**, 183 - 204.

Smouse, E.P. (1984) A note on Bayesian least squares inference for finite population models. *Journal of the American Statistical Association*, **79**, 390 - 392.

Srivastava, J.N. and Saleh, F. (1985) Need of *t*-design in sampling theory. *Utilitas Mathematica*, **25**, 5 - 7.

Stein, C. (1955) Inadmissibilty of usual estimates for the mean of a multivariate normal distribution. in *Proceedings of the third Berkley Symposium on Mathematical Statistics and Probability, vol.1*, Berkley: University of California Press, 197 - 202.

Stephan, F.F. (1945) The expected value and variance of the reciprocal and other negative powers of a positive Bernoullian variate. *Annals of Mathematical Statistics*, **16**, 50 - 61.

Strauss, I. (1982): On the admissibility of estimators for the finite population variance. *Metrika*, **29**, 195 - 202.

Strenger, H. (1977) Sequential sampling for finite populations. *Sankhya, Series C*, **39**, 10 - 26.

Sunter, A.B. (1977) List sequential sampling with equal or unequal probabilities without replacement. *Applied Statistics*, **26**, 261 - 268.

Tam, S. (1984) Optimal estimation in survey sampling under a regression superpopulation model. *Biometrika*, **71**, 645 - 647.

Tam, S. (1986) Characterisation of best model-based predictors in survey sampling. *Biometrika*, **73**, 232 - 235.

The'berge, A. (1999) Extension of calibration estimators in survey sampling. *Journal of American Statistical Association*, **94**, 635 - 644.

Thompson, J. (Jr.) and Thoday, J.M. (1979) *Quantitative Genetic Variation*, Academic Press, New York.

Thomsen, I. (1978) Design and estimation problems when estimating a regression coefficient from survey data. *Metrika*, **25**, 27 - 35.

Tin, M. (1965) Comparison of some ratio estimators. *Journal of the American Statistical Association*, **60**, 294 - 307.

Tiwari, R.C. and Lahiri, P. (1989) On robust empirical Bayes analysis of means and variances from stratified samples. *Communications in Statistics, Theory & Methods*, **18**(3), 921 - 926.

Tremblay, V. (1986) Practical criteria for definition of weighting classes. *Survey Methodology* ,**12**, 85 -97.

Tille', Y. (1995) Auxiliary information and conditional inference. in *Bulletin of the International Statistical Institute, Proceedings of the 50th session* ,**1**, 303 - 319.

Tille', Y. (1998) Estimation in surveys using conditional inclusion probabilities: simple random sampling. *International Statistical Review*, **66**, 303 - 322.

Valliant, R. (1987) Conditional properties of some estimators in stratified samples. *Journal of the American Statistical Association*, **82**, 509 - 519.

Valliant, R. (1993) Post-stratification and conditional variance estimators. *Journal of the American Statistical Association*, **88**, 89 - 96.

Varian, H.R. (1975) A Bayesian approach to real estate assessment. in *Studies in Bayesian Econometrics and Statistics in honour of Leonard J Savage*, eds. Fienberg, S.E. and Zellner, A., North-Holland, Amsterdam, pp. 195 - 208.

Vijayan, K. (1975) On estimating the variance in unequal probability sampling. *Journal of the American Statistical Association*, **70**, 713 - 716.

Wang, S. and Dorfman, A.H. (1996) A new estimator for the finite population distribution function. *Biometrika*, **83**(3), 639 - 652.

Warner, S.L. (1965) Randomized response: A survey technique for eliminating evasive answer bias. *Journal of the American Statistical Association*, **60**, 63 - 69.

Wright, R.L. (1983) Sampling designs with multivariate auxiliary information. *Journal of the American Statistical Association*, **78**, 879 - 884.

Yates, F. (1943) *Sampling Methods for Censuses and Surveys*: Charles Griffin, London.

Yates, F. and Grundy, P.M. (1953) Selection without replacement within strata with probability proportional to sizes. *Journal of Royal Statistical Society, Series B*, **15**, 253 - 261.

Zacks, S. (1971) *Theory of Statistical Inference*, John Wiley & Sons, New York.

Zacks, S. and Solomon, H. (1981) Bayes equivariant estimators of the variance of a finite population, Part I, simple random sampling. *Communications in Statistics, Theory & Methods*, **10**, 407 - 426.

Zellner, A. (1986) Bayesian estimation and prediction using asymmetric loss functions. *Journal of the American Statistical Association*, **81**, 446 - 451.

Author Index

Subject Index

Lecture Notes in Statistics

For information about Volumes 1 to 81,
please contact Springer-Verlag

Vol. 82: A. Korostelev and A. Tsybakov, Minimax Theory of Image Reconstruction. xii, 268 pages, 1993.

Vol. 83: C. Gatsonis, J. Hodges, R. Kass, N. Singpurwalla (Editors), Case Studies in Bayesian Statistics. xii, 437 pages. 1993.

Vol. 84: S. Yamada, Pivotal Measures in Statistical Experiments and Sufficiency. vii, 129 pages, 1994.

Vol. 85: P. Doukhan, Mixing: Properties and Examples. xi, 142 pages, 1994.

Vol. 86: W. Vach, Logistic Regression with Missing Values in the Covariates. xi, 139 pages, 1994.

Vol. 87: J. Müller, Lectures on Random Voronoi Tessellations.vii, 134 pages, 1994.

Vol. 88: J. E. Kolassa, Series Approximation Methods in Statistics. Second Edition, ix, 183 pages, 1997.

Vol. 89: P. Cheeseman, R.W. Oldford (Editors), Selecting Models From Data: AI and Statistics IV. xii, 487 pages. 1994.

Vol. 90: A. Csenki, Dependability for Systems with a Partitioned State Space: Markov and Semi-Markov Theory and Computational Implementation. x, 241 pages, 1994.

Vol. 91: J.D. Malley, Statistical Applications of Jordan Algebras. viii, 101 pages, 1994.

Vol. 92: M. Eerola, Probabilistic Causality in Longitudinal Studies. vii, 133 pages, 1994.

Vol. 93: Bernard Van Cutsem (Editor), Classification and Dissimilarity Analysis. xiv, 238 pages, 1994.

Vol. 94: Jane F. Gentleman and G.A. Whitmore (Editors), Case Studies in Data Analysis. viii, 262 pages, 1994.

Vol. 95: Shelemyahu Zacks, Stochastic Visibility in Random Fields. x, 175 pages, 1994.

Vol. 96: Ibrahim Rahimov, Random Sums and Branching Stochastic Processes. viii, 195 pages, 1995.

Vol. 97: R. Szekli, Stochastic Ordering and Dependence in Applied Probability. viii, 194 pages, 1995.

Vol. 98: Philippe Barbe and Patrice Bertail, The Weighted Bootstrap. viii, 230 pages, 1995.

Vol. 99: C.C. Heyde (Editor), Branching Processes: Proceedings of the First World Congress. viii, 185 pages, 1995.

Vol. 100: Wlodzimierz Bryc, The Normal Distribution: Characterizations with Applications. viii, 139 pages, 1995.

Vol. 101: H.H. Andersen, M.Højbjerre, D. Sørensen, P.S.Eriksen, Linear and Graphical Models: for the Multivariate Complex Normal Distribution. x, 184 pages, 1995.

Vol. 102: A.M. Mathai, Serge B. Provost, Takesi Hayakawa, Bilinear Forms and Zonal Polynomials. x, 378 pages, 1995.

Vol. 103: Anestis Antoniadis and Georges Oppenheim (Editors), Wavelets and Statistics. vi, 411 pages, 1995.

Vol. 104: Gilg U.H. Seeber, Brian J. Francis, Reinhold Hatzinger, Gabriele Steckel-Berger (Editors), Statistical Modelling: 10th International Workshop, Innsbruck, July 10-14th 1995. x, 327 pages, 1995.

Vol. 105: Constantine Gatsonis, James S. Hodges, Robert E. Kass, Nozer D. Singpurwalla(Editors), Case Studies in Bayesian Statistics, Volume II. x, 354 pages, 1995.

Vol. 106: Harald Niederreiter, Peter Jau-Shyong Shiue (Editors), Monte Carlo and Quasi-Monte Carlo Methods in Scientific Computing. xiv, 372 pages, 1995.

Vol. 107: Masafumi Akahira, Kei Takeuchi, Non-Regular Statistical Estimation. vii, 183 pages, 1995.

Vol. 108: Wesley L. Schaible (Editor), Indirect Estimators in U.S. Federal Programs. viii, 195 pages, 1995.

Vol. 109: Helmut Rieder (Editor), Robust Statistics, Data Analysis, and Computer Intensive Methods. xiv, 427 pages, 1996.

Vol. 110: D. Bosq, Nonparametric Statistics for Stochastic Processes. xii, 169 pages, 1996.

Vol. 111: Leon Willenborg, Ton de Waal, Statistical Disclosure Control in Practice. xiv, 152 pages, 1996.

Vol. 112: Doug Fischer, Hans-J. Lenz (Editors), Learning from Data. xii, 450 pages, 1996.

Vol. 113: Rainer Schwabe, Optimum Designs for Multi-Factor Models. viii, 124 pages, 1996.

Vol. 114: C.C. Heyde, Yu. V. Prohorov, R. Pyke, and S. T. Rachev (Editors), Athens Conference on Applied Probability and Time Series Analysis Volume I: Applied Probability In Honor of J.M. Gani. viii, 424 pages, 1996.

Vol. 115: P.M. Robinson, M. Rosenblatt (Editors), Athens Conference on Applied Probability and Time Series Analysis Volume II: Time Series Analysis In Memory of E.J. Hannan. viii, 448 pages, 1996.

Vol. 116: Genshiro Kitagawa and Will Gersch, Smoothness Priors Analysis of Time Series. x, 261 pages, 1996.

Vol. 117: Paul Glasserman, Karl Sigman, David D. Yao (Editors), Stochastic Networks. xii, 298, 1996.

Vol. 118: Radford M. Neal, Bayesian Learning for Neural Networks. xv, 183, 1996.